Membrane Dynamics and Signaling

FUNDAMENTALS OF MEDICAL CELL BIOLOGY
A Multi-Volume Work, Volume 5A

Editor: **E. EDWARD BITTAR,** *Department of Physiology, University of Wisconsin, Madison*

Fundamentals of Medical Cell Biology
A Multi-Volume Work

Edited by
E. Edward Bittar
Department of Physiology
University of Wisconsin, Madison, Wisconsin

Membrane Dynamics
and Signaling

Edited by   **E. EDWARD BITTAR**
Department of Physiology
University of Wisconsin
Madison, Wisconsin

 JAI PRESS, INC.

Greenwich, Connecticut   *London, England*

574.875 11

571.6 F

Library of Congress Cataloging-in-Publication Data

Membrane dynamics and signaling / edited by E. Edward Bittar.
    p. cm. -- (Fundamentals of medical cell biology ; v. 5A)
    Includes bibliographical references and index.
    ISBN 1-55938-309-7 : $78.50
    1. Cell membranes. 2. Cell receptors. 3. Cellular signal
transduction. 4. Ion channels. I. Bittar, E. Edward. II. Series.
QH581.2.F86 1991 vol. 5A
[QH601]
574.87 s--dc20
[574.87'5]   92-20214
                                      CIP

Copyright © 1992 by JAI PRESS INC.
55 Old Post Road, No. 2
Greenwich, Connecticut 06836

JAI PRESS LTD.
118 Pentonville Road
London N1 9JN
England

ISBN: 1-55938-309-7
ISBN: 1-55938-302-X (Treatise)
Library of Congress Catalog Card No.: 92-20214

Manufactured in the United States of America

CONTENTS   (Volume 5A)

CONTENTS   (Volume 5B)

III. Intracellular Communication

IV. Cell Motility and Contractility

LIST OF CONTRIBUTORS
(Volume 5A)

David J. Aidley

School of Biological Sciences
University of East Anglia
Norwich, England

Max M. Burger

Friedrich Miescher Institut
Basel, Switzerland

Carl E. Creutz

Department of Pharmacology
University of Virginia
School of Medicine
Charlottesville, Virginia

György Csaba

Department of Biology
Semmelweis University
Budapest, Hungary

Jörg Hagmann

Friedrich Miescher Institut
Basel, Switzerland

Lowell E. Hokin

Department of Pharmacology
University of Wisconsin
School of Medicine
Madison, Wisconsin

Morley D. Hollenberg

Department of Pharmacology
University of Calgary
Calgary, Alberta, Canada

E. Keith Matthews

Department of Pharmacology
University of Cambridge
Cambridge, England

Ole H. Petersen   The Physiological Laboratory
                                       University of Liverpool
                                       Liverpool, England

Lawrence Salkoff   Department of Anatomy and
                                       Neurobiology
                                       Washington University
                                       School of Medicine
                                       St. Louis, Missouri

David L. Severson   Department of Pharmacology
                                       University of Calgary
                                       Calgary, Alberta, Canada

David J. Triggle   Department of Biochemical
                                       Pharmacology
                                       SUNY-Buffalo
                                       Buffalo, New York

Thomas Wileman   Department of Medicine
                                       Beth Israel Hospital
                                       Boston, Massachusetts

H. Steven Wiley   Department of Pathology
                                       University of Utah
                                       Salt Lake City, Utah

PREFACE

It has been the habit of far too many medical educators until recently to dismiss the patent fact that students are apathetic and bored with the lecture system but enthusiastic about small group learning. If this enthusiasm is there to begin with, then the tutor is in a position to direct and strengthen this impulse in the student. Moreover, if education is the preparation of the student for the community, it is then not hard to see that an integrated curriculum, the essentials of which are molecular and cell biology and clinical medicine, and involving interdisciplinary teaching in small groups is in the public interest. I should not discuss this further here but there is one model of education, that of "problem-based, self-directed independent learning," that is worth mentioning. Such a model is found in the rather young medical schools, McMaster in Canada and Limburg in Holland. Students in small interacting groups are challenged to master basic medical principles, acquire skills in clear thinking and seek solutions to clinical problems. No precious time is wasted on irrelevant and peripheral material and trivia. Under this system there are no exams; tutors run discussions and evaluate student progress twice a week. Though it is but natural that the system requires improvement and definite strengthening, especially in the area of molecular and cell biology, it is essentially sound because it is not wholly utilitarian, and because it promotes self-reliance and self-education for life. This being so, it would be folly to belittle such a system of education.

Volumes 5A an 5B are a treatment of several key subjects, including cellular signaling. One profitable way to tackle both volumes is to read them with care but without endeavoring to master most of the details. Then should follow a more plodding study, chapter by chapter. Such a strategy is also urged in respect of the preceding volumes. Although each chapter is only as detailed as space permits, it should in the case of the student serve as an incentive to the study of the original papers.

I should like to take this opportunity to express special thanks to the scholars whose chapters have made both volumes possible.

E. EDWARD BITTAR

Part I

Membrane Dynamics

Chapter 1

Cytoskeleton-Plasma Membrane Interactions

JÖRG HAGMANN and MAX M. BURGER

INTRODUCTION

Liposomes, artificial vesicles bounded by a phospholipid bilayer of varying composition, assume a spherical configuration. Living cells, on the other hand, occur in many different shapes, although the plasma membranes surrounding them have the same basic bilayer as that of liposomal membranes. For example, epithelial cells may be flat and angular like cobblestones, whereas nerve cells have long processes. Muscle cells are fusiform and share with macrophages and polymorphonuclear leukocytes the ability to produce rapid and extensive changes in shape.

Fundamentals of Medical Cell Biology, Volume 5A
Membrane Dynamics and Signaling, pages 3–19
Copyright © 1992 JAI Press Inc.
All rights of reproduction in any form reserved.
ISBN: 1-55938-309-7

Non-spherical shapes are maintained by a rigid peripheral domain of the cytoplasm rich in actin filaments; this is termed the ectoplasm or cortex. To be effective, the microfilament (actin) network of the ectoplasm needs to be attached to the plasma membrane.

Besides the actin cytoskeleton, two other cytoskeletal systems dictate the shape of a cell: namely, the microtubules and the intermediate filaments (see Volume 2, Chapters 2 and 3). Both interact with cellular membranes; that is, the microtubules interact with the endoplasmic reticulum, as well as a number of small vesicles, whereas the intermediate filaments interact with both the nuclear membrane and plasma membrane. The nature of these interactions will be reviewed following some discussion of the actin cytoskeleton.

Changes in the shape and cytoskeletal architecture of a given cell are often accompanied by functional alterations. Granulosa cells are considered a prime example: Following stimulation by gonadotropins, they acquire the machinery for secreting steroids. At the same time, they lose their membrane attachment sites for stress fibers, as well as the sites for intermediate filaments, and express gap junctions instead (Amsterdam et al., 1989).

Another but more complicated example is the neoplastic cell, because it is distinguished from its normal parent cell not only by uncontrolled proliferation, but also by cytoskeletal alterations. Again, the junctions between the actin cytoskeleton and the membrane are the target of these alterations, leading to malignant morphology. It is not yet known whether these changes are a prerequisite for the observed changes in cell behavior or a consequence thereof. Both questions are of first importance and hence they are being addressed in many laboratories.

Reviews containing more detailed information on many aspects of cytoskeleton–embrane interactions are available (e.g., Niggli and Burger, 1987; Carraway and Carraway, 1988).

THE ACTIN-BASED CYTOSKELETON

Long (>1 µm) filaments of polymerized actin (microfilaments) are found in the cytoplasm of most cells. Gathered into bundles they terminate at the cell membrane in specialized structures that attach the cell to the substrate (focal contacts) or to adjacent cells (adherens type structures). Seemingly less structured networks of long microfilaments fill the peripheral parts of the cytoplasm (cortex) and are linked to the plasma membrane at multiple points of attachment. In erythrocytes, which lack the long actin filaments, a regular network of short (50 nm) filaments is closely apposed to the cytoplasmic face of the cell membrane. Though most conspicuous in erythrocytes, a similar membrane skeleton is found in blood platelets, and variations thereof are probably common to all cells.

Table 1. Membrane Skeleton-Associated Proteins of Erythrocytes and Non-Erythroid Cells

	Erythrocytes	*Non-Erythroid Cells*
Integral membrane proteins	anion transporter (band 3)	voltage-dependent Na channel (brain)
	glycophorin	Na-K ATPase (kidney)
	Rhesus antigen associated	glycoprotein Ic-IIa (platelet)
	proteins	glycoprotein Ib-IX (platelet)
		GP 250 (platelet)
Peripheral membrane proteins	actin	actin
	spectrin	non-erythroid spectrin (fodrin)
	ankyrin	ankyrin
	protein 4.1	protein 4.1
	tropomyosin	actin binding protein (platelet)
	adducin	adducin (brain)

The Membrane Skeleton

The biconcave shape of normal erythrocytes is maintained by the membrane skeleton, because its shape is conserved in the skeleton after dissolution of the membrane by non-ionic detergents. Table 1 lists the proteins involved in the maintenance of the membrane skeleton (for a review see Agre et al., 1988). A hexagonal array composed of spectrin heterodimers forms the base of the network. The heterodimers are joined head to head into tetramers. Five to eight tetramers meet at a junction made up of several proteins: protein 4.1, tropomyosin, adducin, and a short actin filament (Figure 1A). Both the junction and a domain close to the head of the dimer provide the sites linking the network to the membrane. Protein 4.1 connects the junction with the cytoplasmic portion of glycophorin, an integral (membrane spanning) glycoprotein that carries blood group determinants. The affinity of protein 4.1 for glycophorin is increased by phosphatidylinositol 4,5-bis-phosphate, a phospholipid involved in transmembrane signaling (Anderson and Marchesi, 1985). Ankyrin attaches the head of the β chain to band 3 of the erythrocyte membrane, the anion transporter. Electron microscopic examination of negatively stained erythrocyte skeletons reveals a structure resembling geodesic domes (Figure 1B).

Several congenital anemias are characterized by shape changes of the erythrocytes. In many instances the underlying defect could be traced to the membrane skeleton. In some patients with hereditary *spherocytosis* there seems to be a partial lack of or a defect in the spectrin molecules (Agre et al., 1982). Alterations of the amino-terminal region of spectrin or pathological variants of the protein 4.1 may cause hereditary *elliptocytosis* (Tchernia et al., 1981).

The membrane skeleton is most prominent in erythrocytes, and less so in other cell types. Proteins identical with or related to the proteins found in the erythrocyte membrane skeleton have been identified in blood platelets, neurons, and many

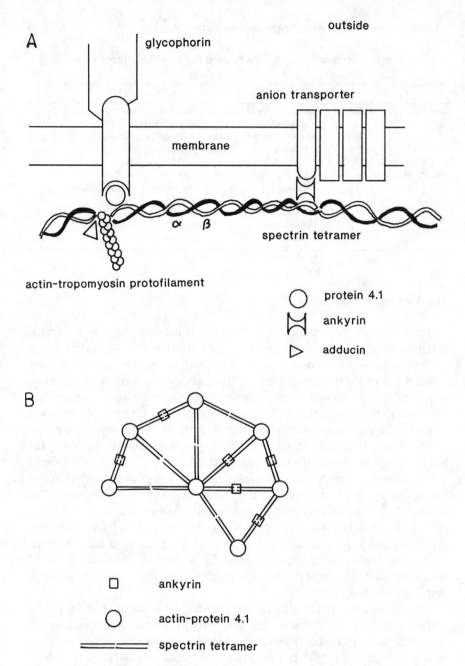

Figure 1. **A.** Model of the membrane skeleton in erythrocytes. The α,β heterodimers of spectrin associate to form tetramers and are linked to glycophorin and the anion transporter through protein 4.1 and ankyrin, respectively. Protofilaments consisting of actin and tropomyosin are anchored at sites where spectrin tetramers are joined. **B.** The erythrocyte membrane skeleton viewed from above.

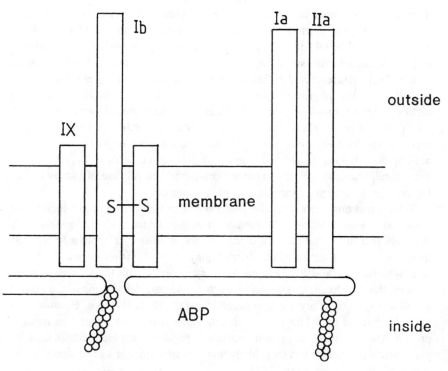

Figure 2. Model of the membrane skeleton in platelets. Short actin filaments are linked to the membrane by actin binding protein (ABP), which binds to the cytoplasmic domain of membrane spanning glycoproteins Ib-IX and Ia-IIa.

other cell types (see Table 1). It appears that in platelets an actin-binding protein (ABP) with a molecular mass of 250 kDa replaces spectrin (reviewed in Clemetson and Luescher, 1988; Fox and Boyles, 1988). Like spectrin, ABP links actin filaments to integral membrane proteins, which in this case are the glycoprotein complexes GPIb-IX and, to a lesser extent, GPIa-IIa and a glycoprotein with a molecular mass of 250 kDa (GP 250) (Figure 2). GPIb-IX is functionally very important because it is the receptor for von Willebrand factor and mediates the adhesion of platelets to the subendothelium of damaged blood vessel walls. GPIa-IIa seems to bind to collagen. After activation with thrombin, platelets change their shape and extend filopodia, which are small protrusions of the plasma membrane. Interestingly, this phenomenon coincides with the activation of a neutral, Ca^{2+}-dependent protease (calpain) and the degradation of ABP, a calpain substrate (McGowan et al., 1989). Destabilization of the membrane skeleton might therefore be a prerequisite for the shape changes observed in activated platelets.

Among the non-erythroid cell types containing proteins listed in Table 1, neurons and epithelial cells are particularly interesting. In both of them, ankyrin displays a characteristic subcellular distribution (Drenckhahn and Bennett, 1987). In neural cells it is restricted to the axon and the cell body but is absent in dendrites. In epithelial cells, ankyrin occupies the basolateral parts of the cell membrane. The picture is further complicated by the fact that brain and epithelial spectrin (=fodrin) binds both directly (as in thrombocytes) and indirectly via ankyrin (as in erythrocytes) to integral membrane proteins. The nature of the membrane proteins binding directly to epithelial and brain spectrin is not yet known. Ankyrin, on the other hand, links the skeleton to the sodium-potassium ATPase of kidney and to the voltage-dependent sodium channel of neuronal cells.

What functions other than the preservation of cell shape might membrane skeletons assume? Many cells contain stretch-activated cation channels (i.e., channels that allow the passage of cations when tension is applied to the cell membrane). Stretch-activated ion channels play a key role in the osmotic regulation of endothelial cells and in the mechanosensory system of cells that transmit signals through the movement of stereocilia, such as the hair cells of the auditory and vestibular systems. They are also found in many other cell types. Because it is unlikely that the lipid bilayer as such can transmit force to integral membrane proteins (the channels), a protein network is probably involved. Because of its geometrical properties, this most likely involves spectrin. For a review on stretch-activated channels see Sachs (1988).

The membrane skeleton may also play a role in maintaining the asymmetry of the lipid bilayer. Cell membranes are asymmetric with respect to their phospholipid composition: phosphatidylcholine is found in both the outer and the inner leaflet, whereas the anionic phospholipids, phosphatidylserine and phosphatidylinositol, are restricted to the cytoplasmic side of the bilayer. It has been shown that upon disruption of the membrane skeleton of erythrocytes, phosphatidylserine is redistributed into both leaflets, suggesting that it plays a role in restricting the distribution of this lipid to the cytoplasmic side of the membrane. Interestingly, phosphatidylserine is present on the surface of activated, but not of resting, thrombocytes. Because activation is accompanied by the breakdown of the membrane skeleton by calpain (see earlier), a causal relationship between the two phenomena has been suggested.

Bundled Microfilaments

Long parallel microfilaments of polymerized actin may be organized in bundles, which are anchored at the cell membrane (Geiger, 1985). Such bundles lie in the core of microvilli (e.g., the brush border microvilli of epithelial cells, the cortical microvilli of oocytes, and the stereocilia of cochlear hair cells) (Figure 3A). Less

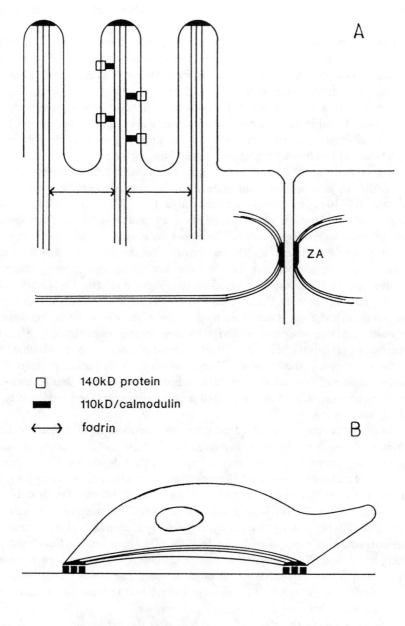

A

140kD protein
110kD/calmodulin
fodrin

B

ZA

fibronectin, vitronectin

Figure 3. Bundled microfilaments are anchored at the cell membrane through electron-dense plaques at the tip of microvilli and in zonulae adherents (ZA) of epithelial cells (**A**), and in focal contacts of tissue culture cells (**B**). The protrusion on the right end of the cell shown in **B** symbolizes the leading edge (front edge of a moving cell).

organized, non-polarized parallel assemblies of microfilaments occur as annular rings, reinforcing the zonulae adherentes where neighboring epithelial cells are attached to each other (Figure 3A). They are related to the stress fibers spanning the length of many tissue culture cells and terminating in focal contacts (Figure 3B). With the exception of endothelial cells, stress fibers and focal contacts are not commonly found in cells of normal tissues. However, they serve as models in studies of other structures, and much of our knowledge about the membrane attachment of bundled actin filaments comes from such studies.

The ventral surface of culture cells grown on plastic or extracellular matrix is not uniformly attached to the substrate. *Focal contacts* are small, elongated sites of contact, 2–10 μm long (reviewed in Burridge et al., 1988), which occupy a short distance of 10–15 nm in between the cell membrane and the substrate. They provide the anchoring site of stress fibers. *Close contacts* cover broader areas where the gap between the membrane and the surface of the culture dish is about 30 nm wide. Focal and close contacts can be observed directly by interference reflection microscopy, a technique furnishing information about the distance between the cell surface and the substrate. Focal contacts appear as black spots and close contacts as grey areas, and the unattached cell membrane surface in between as slightly grey or white, indicating a separation of >100 nm from the substrate (Figure 4). Electron microscopy reveals the presence of electron dense plaques at the cytoplasmic face of the membrane in focal contacts. They consist of closely packed proteins. Our understanding of how stress fibers are anchored at these sites and of how the turnover of these structures is regulated is still rudimentary, and models remain highly speculative.

Little controversy exists about the connection between focal contacts and the extracellular matrix. The nexus is formed by two members of a family of trans-membrane glycoproteins (integrins) that bind in a Ca^{2+}-dependent manner to two extracellular matrix proteins, viz. fibronectin and vitronectin. Like all integrins, the fibronectin receptor and the vitronectin receptor are heterodimers composed of an α and a β subunit. The α subunit consists of two disulfide-linked polypeptides (20 kDa and 120–140 kDa) generated by the proteolytic cleavage of a precursor. β subunits are single polypeptides of 90–140 kDa (reviewed in Ruoslahti and Pierschbacher, 1987). Integrins recognize the amino acid sequence arginine-glycine-aspartic acid (RGD) found on different extracellular matrix proteins. Glycoprotein Ib-IIa, which was mentioned in relation to the platelet membrane skeleton, is a member of the same family of receptors.

The structure of the electron-dense plaque at the cytoplasmic end of focal contacts remains poorly understood. To date, several plaque proteins have been purified and characterized (Table 2), mostly by employing monoclonal antibodies produced against the cytoskeleton. The picture that emerges is that talin binds to the cytoplasmic domain of the β chain of integrins (Horwitz et al., 1986), and vinculin binds strongly to talin and tensin (Burridge and Mangeat, 1984) but weakly to α-actinin, which cross-links actin filaments (Belkin and Kotelianski, 1987), and

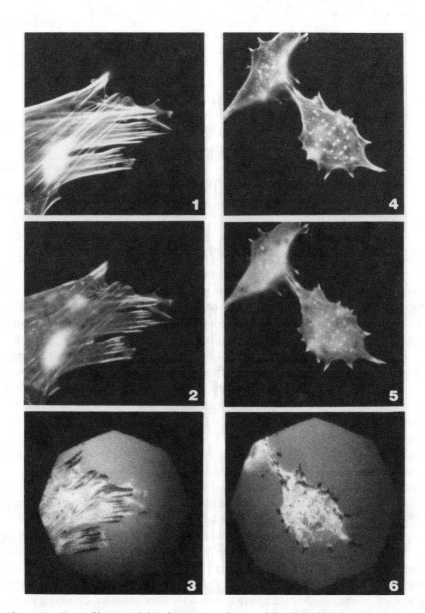

Figure 4. Stress fibers and focal contacts of normal (**1–3**) and Rous sarcoma virus transformed (**4–6**) chicken embryo fibroblasts. **1,4**: Actin filaments stained with phalloidin-rhodamin; **2,5**: focal contacts stained with fluorescein-labeled antivinculin antibodies; **3,6**: focal contacts appear as black streaks in interference reflection microscopy (from the work of the authors).

Table 2. Proteins of Focal Contacts and Adherens Type Structures

	Focal Contact	*Adherens Type Structure*
Integral membrane proteins	integrins: fibronectin receptor vitronectin receptor	cadherins: N-cadherin (A-CAM) E-cadherin (L-CAM, uvomorulin) P-cadherin
Peripheral membrane proteins	actin α-actinin fimbrin talin tensin vinculin	actin α-actinin plakoglobin vinculin
Associated enzymes	plasminogen activator (extracellular face) calpain 2 (intracellular face) pp60^{v-src} (intracellular face)	

possibly to actin filaments directly (Ruhnau and Wegner, 1988). Furthermore, electron microscopic preparations decorated with gold-labeled antibodies against vinculin, α-actinin, and talin show that vinculin is very closely apposed to the membrane bilayer, more closely than talin and α-actinin (Geiger et al., 1981; Volberg et al., 1986). Such findings support the view that there is close interaction between vinculin and anionic phospholipids (Niggli et al., 1986).

Based on these observations, models that explain the anchoring of stress fibers at focal contacts have been proposed (Figure 5). However, reconstitution experiments reveal that one or several elements are still missing. For example, vinculin is able to bind to focal contacts *in vitro* in the absence of talin. At best, this is a facultative link. However, vinculin is not bound prior to the formation of focal contacts with attached microfilaments *in vitro*, suggesting that vinculin plays a stabilizing rather than a central role in attaching stress fibers to the membrane. These observations favor the hypothesis that actin filaments bind directly to integrins (DePasquale and Izzard, 1987; Molony et al., 1987).

Adherens junctions are sites where neighboring cells are closely attached to each other, as evidenced by the zonulae adherentes of epithelial cells. Their basic structure resembles that of focal contacts: bundled microfilaments terminate in an electron-dense plaque apposed to the membrane bilayer containing integral membrane receptors. Although adherens junctions share a number of proteins with the focal contacts, there are some differences between them. Cadherins, not integrins, provide the link with the adjacent cells. Like integrins, they are a family of glycoprotein receptors that bind their ligands in a Ca^{2+}-dependent manner (Gumbiner, 1988). The main members, N-cadherin, E-cadherin, and P-cadherin, are expressed in different cell types. They seem to bind directly to their homologue on adjacent cells, but cells expressing different types do not adhere to each other. On the cytoplasmic side, α-actinin and vinculin are present whereas talin is

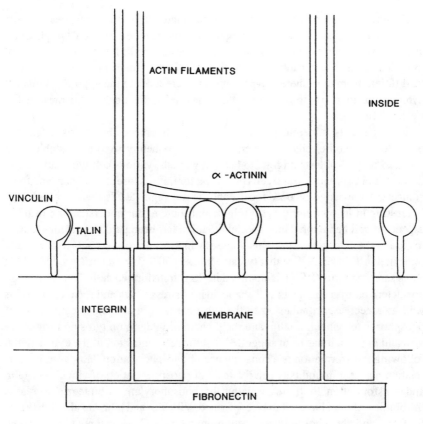

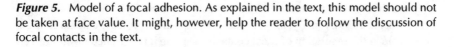

Figure 5. Model of a focal adhesion. As explained in the text, this model should not be taken at face value. It might, however, help the reader to follow the discussion of focal contacts in the text.

conspicuously absent. It is not known whether talin is replaced by a still uniden-
tified protein, or whether its presence in adherens junctions is not required. On the
other hand, plakoglobin, a peripheral membrane protein with a molecular mass of
83 kDa, is found in adherens type junctions but not in focal contacts (Cowin et al.,
1986). So far, plakoglobin is the only protein that is localized at both the adherens
type junction connecting microfilament bundles to the membrane, and desmosomes linking intermediate filaments to the membrane (see later). As is true of
focal contacts, detailed models of the molecular architecture of adherens type
junctions are not yet feasible. More recently, purified intact junctions have been
shown to contain seven uncharacterized proteins with molecular weights ranging
between 50 kDa and 400 kDa (Tsukita and Tsukita, 1989).

A rather striking finding is that talin accumulates at the adhesion sites between cytotoxic T lymphocytes and their target cells, and between helper T lymphocytes and B cells. Because the membrane receptors of these cells belong to the group of integrins rather than of cadherins, this type of cell–cell interaction is considered to be different from the adherens type. This lends credence to the hypothesis that the cytoplasmic domain of the receptor determines the localization of talin, presumably by direct binding.

Focal contacts, adherens junctions, and stress fibers are characteristic features of stable, non-motile cells. However, many cells frequently give up this stable state; for example, fibroblasts migrate to healing wounds, various cell types migrate in the developing organism, and cells preparing for mitosis lose stress fibers and focal contacts. Knowledge of how focal contacts are regulated is therefore not only interesting in its own right, but it might contribute to our understanding of basic aspects of cell behavior. Thus, special attention has been paid to enzymes known to modify the cytoskeletal proteins, some of which occur in focal contacts as already indicated in Table 2. Notice that tyrosine kinase $p60^{V-src}$, which is a product of the Rous sarcoma virus (RSV), is involved in RSV-transformed cells. These cells are characterized by a change in cell shape, loss of stress fibers and focal contacts, as well as a decrease in adhesion to the substrate (see Figure 4). Because $p60^{v-src}$ phosphorylates vinculin, talin, and the β chain of integrin on tyrosine residues *in vivo* and *in vitro*, it has been suggested that the reorganization of the cytoskeleton following transformation is a consequence of phosphorylation. However, no correlation has been found between the level of phosphorylation of vinculin or talin and transformation. Preliminary claims that phosphorylation of integrin decreases the binding of talin have yet to be confirmed. Vinculin and talin are also substrates for Ca^{2+}- and phospholipid-dependent protein kinase C, an enzyme that is known to be activated by phorbol esters with tumor promoting activity. Activation of this enzyme leads to the dissolution of focal contacts and stress fibers in many cell types (Turner et al., 1989). Finally, the Ca^{2+}-dependent, neutral protease calpain II might play a role in the regulation of focal contacts, because it is enriched at sites of cell adhesion (Beckerle et al., 1987). Though the range of substrates cleaved by calpain is limited, talin and vinculin are among its substrates. Calpain might therefore be involved in the loss of focal contacts observed after stimulation of certain tissue culture cells with growth factors that are known to increase intracellular Ca^{2+} concentration. (Herman and Pledger, 1985).

Actin filament bundles originating from the terminal web of microfilaments underlying the apical face of epithelial cells and extending into the core of microvilli are attached to the cell membrane via lateral bridges that are formed by a complex consisting of calmodulin (a Ca^{2+}-binding protein with a molecular mass of 17 kDa) and a 110 kDa protein (Glenney et al., 1982). How the 110 kDa protein is anchored is still unclear. Possibly a 140 kDa integral membrane protein acts as the link. How the electron-dense plaque lying at the tip of the microvillus into which filament bundles penetrate is attached to the membrane is also unknown.

Attachment of Long, Unbundled Microfilaments

The long actin microfilaments filling the peripheral regions of the cell (cortex or ectoplasm) are anchored to the cytoplasmic side of the plasma membrane at several points. This is indicated by evidence that particles bound to the cell surface move in synchrony with underlying waves of actin. It thus seems that a transmembrane linkage to the microfilament network does exist (Fisher et al., 1988). Although the molecular details of this linkage are mostly unknown, there are indications from studies with the slime mold *Dictyostelium discoideum* that a membrane protein with an apparent molecular mass of 24 kDa binds glutaraldehyde cross-linked microfilaments (Stratford et al., 1985). Whether a similar protein mediates the attachment of microfilaments in animal cells remains to be seen.

INTERMEDIATE FILAMENTS

The diameter of intermediate filaments falls in the range of 8–12 nm; that is, they lie between the smaller microfilaments and the larger microtubules, hence the term intermediate filaments (IF). IFs are present in the cytoplasm of most cells, where they radiate from the nucleus toward the cell membrane. Although the IFs of different cell types are alike in appearance, they contain different proteins: vimentin (53 kDa) is in mesenchymal cells; desmin (52 kDa) in smooth and striated muscle; glial fibrillary acidic protein (50 kDa) in astroglial cells; neurofilament protein (65, 105, and 135 kDa) in neurons, and keratins (40–70 kDa) in epithelial cells. The function of IFs is still obscure, but because they pervade the cytoplasm and are attached to many structures including desmosomes, nucleus, plasma membrane, dense bodies of smooth muscle cells, fat globules of adipocytes, and other cytoskeletal systems, they might well act as a framework stabilizing the cytomatrix.

In order to understand the interaction of IFs with nuclear and cell membranes, one needs to first examine the structure of IF proteins, which is shown in Figure 6. The central core domain of about 40 kDa is made up of α-helical subdomains that are linked by short, nonhelical sequences. It is flanked by the amino-terminal and carboxy-terminal tails, which are very different between the various types of IF proteins. Both the core and tails contribute to the interactions leading to the two-stranded coiled coils (i.e., the tetramer complexes and the protofilaments making up IFs). An interesting new observation is that the carboxy-terminal tail of vimentin binds to lamin B on nuclear membranes (Georgatos and Blobel, 1987a,b). Lamin B is complexed with lamin A and C to form the lamina on the inner surface of the nuclear membrane. It is thus likely that IFs interact with the lamina through the nuclear pores. At the other end, the amino-terminal tails of vimentin and desmin bind to the plasma membrane through ankyrin (Georgatos et al., 1987). This attachment of IFs to both the nuclear and plasma membranes raises the possibility of direct communication between the surroundings of the cell and the nucleus.

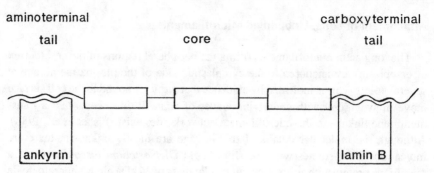

Figure 6. The basic structure of an IF subunit. The central core domain is flanked by amino-terminal and carboxy-terminal tails that are very different between the various types of IF proteins. The subunits assemble into filaments, and from 22 to 45 filaments are coiled round each other to form the mature intermediate filament. They thus resemble cables made up of large numbers of coiled wires.

MICROTUBULES

At first sight one might not consider microtubules as being associated with membranes. In most cells they are found to radiate from the perinuclear zone (where they are associated with microtubule organizing centers) into the periphery without reaching the plasma membrane. In fact, most studies show that they are excluded from the actin-rich cortex of the cell. In growth cones of *Aplysia* neurons, for example, there seems to be an inverse relationship between the degree of extension of microtubules and the width of the cortex. This is the case when internal cAMP levels are raised because the cortex is found to shrink and the microtubules to lengthen simultaneously (Forscher et al., 1987). The nature of the boundary lying between the actin and the microtubule domains is not yet known. Immunocytochemical studies reveal that a few microtubules are able to extend to stable focal contacts present in the cortex (Rinnerthaler et al., 1988), but the significance of this is far from clear.

Microtubules are known to be attached to internal membranes because they transport small vesicular bodies in a saltatory manner and play a decisive role in the formation and maintenance of the endoplasmic reticulum and lysosomes. The network of endoplasmic membrane cisternae and tubules present normally in cells collapses around the nucleus when microtubules are disrupted by drugs, notably colchicine. The process is reversible, and the endoplasmic reticulum reforms and is found to extend toward the periphery of the cell at a rate comparable to that of microtubules during growth (Terasaki et al., 1986). It is noteworthy that the endoplasmic reticulum can be reconstituted *in vitro*; Endoplasmic reticulum vesicles incubated on glass cover slips together with microtubules, ATP, and kinesin (which is a motility protein) are seen to form tubules. And at sites where microtubules cross each other, vesicles are seen to fuse and form a whole network (Vale and Hotani, 1988). By contrast, artificial vesicles fail to interact with microtubules, suggesting that an endoplasmic reticulum protein is required for the binding to microtubules. Similar mechanisms are probably involved in the formation of tubular extensions of lysosomes (Swanson et al., 1987).

CONCLUSIONS AND OUTLOOK

The study of membrane attachments of the cytoskeleton might at first strike the reader as rather arcane. However, it is clear that membrane attachments play a key role in the life of the cell, as evidenced by the fact that alterations in attachments go hand in hand with physiological and pathological changes, including differentiation (e.g., granulosa cells), migration (e.g., fibroblasts), and cell division. Although the role of unstable cytoskeleton–membrane attachments in malignant cells is not yet understood, there can be no question that the problem is of first importance. Thus, the present goal is to clarify the functions of cytoskeleton–membrane linkages not only in normal cells but also in malignant cells.

REFERENCES

Agre, P., Smith, B.L., Saboori, A.M., & Asimos, A. (1988). The red cell membrane skeleton: A model with general biological relevance but pathological significance for blood. In: Cell Physiology of Blood, pp. 91–100. The Rockefeller Univ. Press, New York.

Agre, P., Orringer, E.P., & Bennett, V. (1982). Deficient red cell spectrin in severe, recessively inherited spherocytosis. N. Engl. J. Med. 301, 1155–1161.

Amsterdam, A., Rotmensch, S., & Ben-Ze'ev, A. (1989). Coordinated regulation of morphological and biochemical differentiation in a steroidogenic cell: The granulosa cell model. TIBS 14, 377–382.

Anderson, R.A., & Marchesi, V.T. (1985). Regulation of the association of membrane skeletal 4.1 with glycophorin by a polyphosphoinositide. Nature 318, 295–297.

Beckerle, M.C., Burridge, K., De Martino, G.N., & Croall, D.E. (1987). Colocalization of calcium-dependent protease II and one of its substrates at sites of cell adhesion. Cell 51, 569–577.

Belkin, A.M., & Kotelianski, V.E. (1987). Interaction of iodinated vinculin, metavinculin and α-actinin with cytoskeletal proteins. FEBS Lett. 220, 291–294.

Burridge, K., & Mangeat, P. (1984). An interaction between vinculin and talin. Nature 308, 744–746.

Burridge, K., Fath, K., Kelly, T., Nuckolls, G., & Turner, C. (1988). Focal adhesions. Ann. Rev. Cell Biol. 4, 487–525.

Carraway, K.L. & Carraway, C.A.C. (1989). Membrane-cytoskeleton interactions in animal cells. Biochim. Biophys. Acta 988, 147–171.

Clemetson, K.J., & Luescher, E.F. (1988). Membrane glycoprotein abnormalities in pathological platelets. Biochim. Biophys. Acta 947, 53–73.

De Pasquale, J.A., & Izzard, C.S. (1987). Evidence for an actin-containing cytoplasmic precursor of the focal contact and the timing of incorporation of vinculin at the focal contact. J. Cell Biol. 105, 2803–2809.

Drenckhahn, D., & Bennett, V. (1987). Polarized distribution of Mr 210,000 and 190,000 analog of erythrocyte ankyrin along the plasma membrane of transporting epithelial, neurons and photoreceptors. Eur. J. Cell Biol. 43, 479–486.

Fisher, G.W., Conrad, P.A., De Biasio, R.L., & Taylor, D.L. (1988). Centripetal transport of cytoplasm, actin, and the cell surface in lamellipodia of fibroblasts. Cell Motility and the Cytoskeleton 11, 235–247.

Forscher, P., Kaczmarek, L.K., Buchanan, J., & Smith, S. (1987). Cyclic AMP induces changes in distribution and transport of organelles within growth cones of Aplysia bag cell neurons. J. Neuroscience 7, 3600–3611.

Fox, J.E.A., & Boyles, J.K. (1988). Structure and function of the platelet membrane skeleton. In: Cell Physiology of Blood, pp. 111–123. The Rockefeller Univ. Press, New York.

Geiger, B. (1985). Microfilament-membrane interaction. TIBS 10, 456–461.

Geiger, B., Dutton, A. H., Tokuyasu, K. T., & Singer, S. I. (1981). Immunoelectron microscope studies of membrane-microfilament interaction. J. Cell Biol. 91, 614–628.

Georgatos, S.D., & Blobel, G. (1987a). Two distinct attachment sites for vimentin along the plasma membrane and the nuclear envelope in avian erythrocytes: A basis for a vectorial assembly of intermediate filaments. J. Cell Biol. 105, 105–115.

Georgatos, S.D., & Blobel, G. (1987b). Lamin B constitutes an intermediate filament attachment site at the nuclear envelope. J. Cell Biol. 105, 117–125.

Georgatos, S.D., Weber, K., Geisler, N., & Blobel, G. (1987). Binding of two desmin derivatives to the plasma membrane and the nuclear envelope of avian erythrocytes: Evidence for a conserved site-specificity in intermediate filament-membrane interactions. Proc. Natl. Acad. Sci. USA 84, 6780–6784.

Glenney, J.R., Osborn, M., & Weber, K. (1982). The intracellular localization of the microvillus 110K protein, a component considered to be involved in side-on membrane attachment of F-actin. Exp. Cell Res. 138, 199–205.

Gumbiner, B. (1988). Cadherins, a family of Ca^{2+} dependent adhesion molecules. TIBS 13, 75–76.

Herman, B., & Pledger, W.J. (1985). Platelet-derived growth factor-induced alterations in vinculin and actin distribution in BALB/c-3T3 cells. J. Cell Biol. 100, 1031–1040.

Horwitz, A., Duggan, K., Buck, C., Beckerle, M.C., & Burridge, K. (1986), Interaction of plasma membrane fibronectin receptor with talin—a transmembrane linkage. Nature 320, 531–533.

McGowan, E.B., Becker, E., & Detwiler, T.C. (1989). Inhibition of calpain in intact platelets by the thiol protease inhibitor E-64d. Biochem. Biophys. Res. Com. 158, 432–435.

Molony, L., Kelly, T., & Burridge, K. (1987). Does integrin bind actin directly? J. Cell. Biol. 105, 177a.

Niggli, V., Dimitrov, D.P., Brunner, J., & Burger, M.M. (1986). Interaction of the cytoskeletal component vinculin with bilayer structures analyzed with a photoactivatable phospholipid. J. Biol. Chem. 261, 6912–6918.

Niggli, V., & Burger, M.M. (1987). Interaction of the cytoskeleton with the plasma membrane. J. Membrane Biol. 100, 97–121.

Rinnerthaler, G., Geiger, B., & Small, J.V. (1988). Contact formation during fibroblast locomotion: Involvement of membrane ruffles and microtubules. J. Cell Biol. 106, 747–760.

Ruhnau, K., & Wegner, A. (1988). Evidence for direct binding of vinculin to actin filament. FEBS Letters 228, 105–108.

Ruoslahti, E., & Pierschbacher, M.D. (1987). New perspectives in cell adhesion: RGD and integrins. Science 238, 491–497.

Sachs, F. (188). Mechanical transduction in biological systems. CRC Critical Rev. in Biomedical Engineering 16, 141–169.

Stratford, C.A., & Brown, S.S. (1985). Isolation of actin-binding protein from membranes of Dictyostelium discoideum. J. Cell Biol. 100, 727–735.

Swanson, J., Bushnell, A., & Silverstein, S.C. (1987). Tubular lysosome morphology and distribution within macrophages depend on the integrity of cytoplasmic microtubules. Proc. Natl. Acad. Sci. USA 84, 1921–1925.

Tchernia, G., Mohandas, N., & Shohet, S.B. (1981). Deficiency of skeletal membrane protein band 4.1 in homozygous hereditary spherocytosis. J. Clin. Invest. 81, 133–141.

Terasaki, M., Chen, L.B., & Fujiwara, K. (1986). Microtubules and the endoplasmic reticulum are highly interdependent structures. J. Cell. Biol. 103, 1557–1568.

Tsukita, S., & Tsukita, S. (1989). J. Cell Biol. 108, 31–41.

Turner, C.E., Pavalko, F.M., & Burridge, K. (1989). The role of phosphorylation and limited proteolytic cleavage of talin and vinculin in the disruption of focal adhesion integrity. J. Biol. Chem. 264, 11938–11944.

Vale, R.D., & Hotani, H. (1988). Formation of membrane networks *in vitro* by kinesin-driven microtubule movement. J. Cell Biol. 107, 2233–2241.

Volberg, T., Sabanay, H., & Geiger, B. (1986). Spatial and temporal relationship between vinculin and talin in the developing chicken gizzard smooth muscle. Differentiation 32, 34–43.

Wilkens, J.A., Risinger, M.A., Coffey, E., & Lin, S. (1987). Purification of a vinculin binding protein from smooth muscle. J. Cell Biol. 104, 130a.

Chapter 2

Membrane Fusion and Exocytosis

CARL E. CREUTZ

INTRODUCTION

Cells are separated from their environment and internally compartmentalized by membranes. These membranes have a lipid matrix that prohibits the free passage of molecules that are in aqueous solution either inside or outside of the cell.

Fundamentals of Medical Cell Biology, Volume 5A
Membrane Dynamics and Signaling, pages 21–39
Copyright © 1992 by JAI Press Inc.
All rights of reproduction in any form reserved.
ISBN: 1-55938-309-7

However, communication with the environment and between internal compartments is essential for many cell functions. Therefore, complex mechanisms have evolved to permit signal transduction across membranes and to catalyze the movement of solutes from compartment to compartment. In the case of small molecules such as inorganic ions, carbohydrates, amino acids, and so forth, protein molecules embedded in the membrane serve as molecular transport devices. However, even though some of these transport molecules are fairly complex multisubunit proteins, they generally are not capable of promoting the transbilayer movement of folded macromolecules. In eukaryotic cells more complex protein-based devices permit the transfer of mRNA through specialized pores in the nuclear membrane, and unique proteins are involved in catalyzing the transfer of newly synthesized, unfolded proteins into the lumina of the endoplasmic reticulum, mitochondria, or chloroplasts. However, an important general mechanism that eukaryotic cells rely on for transfer of material through compartments or, conversely, for the sequestration of macromolecules into compartments is the process of *membrane fusion*. This chapter will present an overview of membrane fusion events in cells and discuss general aspects of the mechanics of this phenomenon. Then current understanding of the molecular biology of protein secretion by exocytosis will be described in more detail. The membrane fusion events in the exocytotic pathway are just beginning to be revealed at the molecular level. Current research hypotheses will be presented. Although some of these hypotheses will be disproved in the near future, ultimately others will prove to be correct and will likely become the basis of novel pharmacological approaches to the treatment of disorders in intercellular communication.

MEMBRANE FUSION EVENTS BETWEEN AND WITHIN CELLS

Biological membrane fusion events can be grouped into two classes: those involving the fusion of the *extracytoplasmic* faces of membranes and those involving the *cytoplasmic* faces of membranes (Figures 1 and 2). The internal faces of the membranes limiting intracellular organelles, such as Golgi cisternae or secretory vesicles, are topologically equivalent in this context to extracytoplasmic, or "extracellular," faces. There are no obvious examples of fusion between a cytoplasmic face of one membrane with an extracytoplasmic face of another membrane. It is reasonable to consider these two classes of fusion separately because it is likely they are mediated by very distinct mechanisms. It has been demonstrated that both the plasma membrane and the membranes of intracellular organelles are asymmetric structures: The lipid compositions of each monolayer making up the bilayer structure are distinct, with phosphatidylcholine and glycolipids found in greater concentration in the extracytoplasmic face and acidic lipids such as phosphatidylserine and phosphatidylinositol in the cytoplasmic face. In addition, membrane

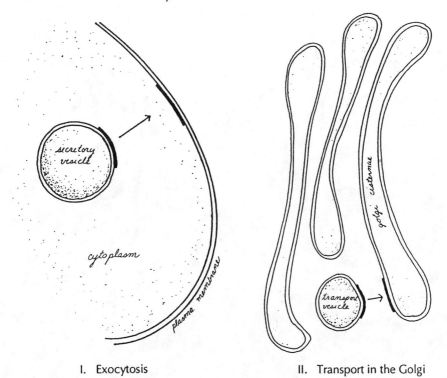

I. Exocytosis   II. Transport in the Golgi

Figure 1. Schematic illustration of two examples of membrane fusion events involving the *cytoplasmic* membrane faces. Drawing by Sandra L. Snyder.

proteins have unique orientations with regard to the cytoplasmic and extracytoplasmic faces. The environments of each of the different faces of a membrane are also dramatically different. The cytoplasmic face is bathed in a medium that has a very low concentration of Ca^{2+}, an ion that may serve an important signaling role, and is rich in soluble or cytoskeletal proteins that may be part of complex fusion control mechanisms. The extracytoplasmic face, on the other hand, is exposed to high calcium concentrations in a less controlled environment, probably devoid of regulatory proteins. Fusion events between extracytoplasmic faces may therefore be completely dependent upon macromolecules that are integral parts of the membranes themselves.

To emphasize the breadth of membrane fusion phenomena, it is useful to compile a list of some of the major occurrences of this process. Fusion between extracellular faces of membranes is less common because this event violates the integrity of the single cell. Such fusion events are nonetheless essential for the development of tissues that consist of syncytia. For example, during the development of muscle fibers, the membranes of individual myoblasts fuse to form

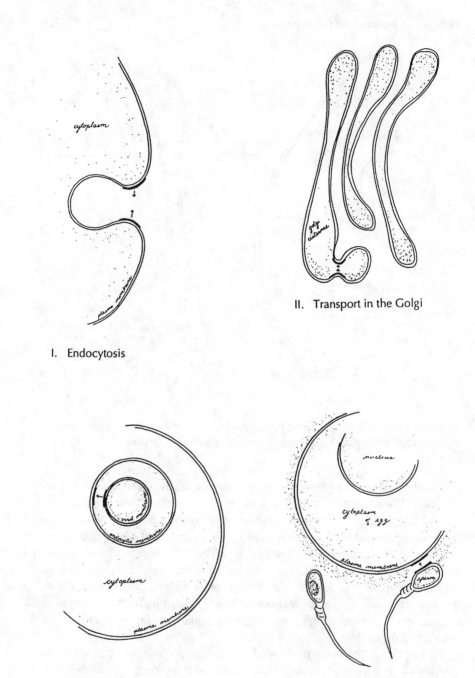

I. Endocytosis

II. Transport in the Golgi

III. Viral fusion in endosome

IV. Gamete fusion

Figure 2. Schematic illustration of four examples of membrane fusion events involving the *extracytoplasmic* membrane faces. Drawing by Sandra L. Snyder.

elongated, multinuclear, single muscle cells. The fusion of the membranes of gametes, such as sperm and egg, removes the critical barrier to the mixing of genetic information that is the basis of sexual reproduction. Extracellular membrane fusion is of broad medical importance as it provides the route of entry of enveloped viruses into target cells. The fusion may occur at the plasma membrane of the cell, although the more frequent pathway is fusion with the inner surface of an endocytic vesicle that has encapsulated the virus particle. The fusion event in the case of some viruses is understood at the molecular level (see the chapter on viral fusion in this volume). As anticipated for extracytoplasmic fusion, the viral fusion event does indeed depend on specific fusion proteins that are components of the viral membrane and specific target molecules that are part of the cell membrane.

As much of the organization of the eukaryotic cell is based on compartmentalization by membranes, it is not surprising that internal membrane fusion events are abundant and frequent between intracellular membranes. Secretory proteins and surface membrane proteins are transported from their site of origin in the rough endoplasmic reticulum (ER) to the Golgi by transport vesicles. These vesicles pinch off from the ER membrane by fusion of their inner (extracytoplasmic) membrane surfaces and then fuse with proximal Golgi cisternae. Transport through the varied manufacturing compartments of the Golgi is similarly accomplished by pinching off and fusing of transport vesicles. Finally, at the distal face of the Golgi, additional vesicles are formed by pinching off that go to the cell surface immediately or fuse with lysosomes or become storage vesicles destined for fusion with the plasma membrane at a later time as part of the regulated secretory pathway. Endocytic vesicles pinch off from the plasma membrane to bring solutes or membrane components into the cell. These vesicles fuse with one another to form acidic endosomes, which may in turn then fuse with lysosome or Golgi elements. In light of the abundance and complexity of this intracellular traffic it is clear that regulatory mechanisms and "chaperones" must have evolved to assure that only the appropriate partners undergo fusion. These regulatory mechanisms may obscure the features that are common to all fusion events. Because of the fundamental similarity in structure of biological membranes, it is likely that all fusion events in one class (i.e., involving intracellular or extracellular membrane surfaces) may have mediating proteins in common.

ELEMENTS OF MEMBRANE FUSION

The fusion of two biological membranes, in general terms, can be broken down into a series of events. These events are *translocation, recognition, attachment, disruption,* and *resolution.* Classifying the steps in such terms helps focus on events that are mechanistically distinct, may be subserved by different proteins, and probably are subject to different regulatory mechanisms.

Translocation

Membranes must approach one another to fuse. In some cases this translocation occurs as a simple result of diffusion, as when a virus particle approaches a cell. However, intracellular movement of membrane-bound organelles may be mediated by specific interactions with cytoskeletal components. Both actin filaments and microtubules can interact with some intracellular vesicles and these interactions can result in immobilization through gelation or guided movements. Secretory vesicles are frequently sequestered to a pole of a cell where their release will occur. After such localization simple diffusion across a small space may provide efficient translocation.

Recognition

A recognition step may be necessary to prevent nondiscriminant membrane fusion. Virus particles have receptor molecules that recognize specific glycoproteins on the surfaces of cells. No such specific targeting molecules have yet been identified that mediate intracellular fusion events although such proteins have been widely hypothesized to exist. Some proteins that promote membrane attachment (the annexins, see later) interact with specific lipids that are found on intracellular membrane surfaces. The sensitivity of these proteins to activation by calcium appears to depend upon the actual relative concentrations of specific lipids in the membrane. Thus some degree of targeting may result from the ability of these proteins to attach to membranes of specific lipid composition at a given intracellular calcium concentration. Rigid compartmentalization of organelles into relative positions, such as the stacking of the cisternae of the Golgi by cytoskeletal components, might in some circumstances alleviate the need for specific targeting molecules because transport vesicles would have little opportunity to go astray before fusing.

Attachment

When the appropriate fusion partners achieve recognition, true fusion begins with an intimate attachment of the membranes. The binding of targeting proteins to their targets may provide this attachment. Viral spike proteins may further secure attachment by inserting hydrophobic peptide extensions into the target cell membrane. The postulated intracellular attachment proteins, the annexins (see later), have multiple recognition sites for membrane lipids. Possibly by attaching different sites to lipids on different membranes, they create an effective bridge or attachment between membranes. Alternatively, annexin molecules bound to each membrane of a fusion pair might self-associate, thus resulting in membrane–membrane attachment.

Disruption

The internal structures of the attached membranes must be disrupted in order for fusion to occur. This is a consequence of the topological fact that an aqueous passageway must form through the region that once contained the hydrocarbon-like interior of the lipid bilayer. Again the viral spike protein is thought to be instrumental in this step in virus–cell fusion. Membranes attached by annexins seem to require the perturbing effects of free cis-unsaturated fatty acids for disruption and complete fusion to occur (Creutz, 1981; Drust and Creutz, 1988; see later discussion and Figure 5). These detergent-like agents may enhance the probability of the formation of "inverted micelles" in the bilayer. These hypothetical structures have the polar lipid head groups facing one another and the hydrophobic tails extended outward. They have been postulated to be intermediates in the fusion process. The need for free fatty acids can be met through the action of specific phospholipases that are activated during exocytosis. Such enzymes may be a critical control point in the fusion process, subject to regulation by calcium or G-proteins.

Resolution

After disruption of the apposed bilayers the membranes must follow a pathway of remodeling and reorganization that results in the appropriate final morphology. For the fusion of a secretory vesicle with the plasma membrane, this pathway leads to the incorporation of the vesicle membrane into the plasma membrane. Fusion starts at a single point of contact between the two membranes and results in the formation of a single continuous (although microscopically heterogeneous) bilayer. Topologically, this is very different from the pathway that the plasma membrane and endosome membranes follow after endocytosis. In this case, the extracellular face of a single membrane, the plasma membrane, invaginates and is drawn upon itself in an annulus: The membrane contact region results from the contraction of a circle to zero radius. After fusion two independent bilayers are formed by remodeling.

A classification of the steps in membrane fusion, as given here, serves to emphasize our ignorance of the molecular basis of the process: Most of these steps remain "black boxes." However, this logical scheme is helpful in organizing information about subcellular components that have been examined in various *in vitro* systems. The classification may also help to define future research goals in this field.

In order to discuss further molecular details of membrane fusion as we are beginning to understand them, it is helpful now to restrict discussion to a specific system that has been particularly fruitful and is of broad medical importance. This is the secretory pathway and exocytosis.

THE SECRETORY PATHWAY AND EXOCYTOSIS

The membrane fusion events in the secretory pathway have been examined in two very informative model systems that will be discussed here. The first approach involves the genetic analysis of the secretory pathway in the common baker's yeast, *Saccharomyces cerevisiae*. The second is the construction of *in vitro* model systems using the secretory vesicles of the bovine adrenal medullary chromaffin cell. These are obviously quite different systems, involving the biology of organisms separated by hundreds of millions of years of evolution, and a constitutive, unregulated secretory pathway in one case (yeast) compared to the highly regulated calcium-dependent release of hormone from a storage pool in the other (the chromaffin cell). However, common elements are coming to light in these systems, which may reflect some of the basic features of membrane fusion as described earlier. In addition, these systems are excellent examples of how research can be thrust forward by careful selection of models that possess unique features amenable to analysis using current technology.

The Yeast *sec* Mutants

Yeast maintain an active secretory pathway for the release of cell wall components and hydrolytic enzymes during growth as well as for the release of polypeptide mating pheromones. A series of temperature-sensitive mutants in the secretory pathway have been isolated based upon the ingenious concept that cells that cannot secrete proteins as they are synthesized will increase in density as their cytoplasm becomes increasingly clogged with dense protein molecules (Novick et al., 1980). After chemical mutagenesis of a yeast culture, secretion, or *sec*, mutants can be isolated on density gradients. Loss of the secretory pathway is lethal to yeast; therefore it has been necessary to isolate temperature-sensitive mutants in which secretion is blocked only at a certain, usually higher, temperature. Twenty-three non-complementing *sec* mutants were originally isolated, but additional essential complementation groups are continuing to be discovered. The morphology of these cells when shifted to the non-permissive temperature reveals that the *sec* mutations control various different steps in the overall secretory pathway, including movement from the ER to the Golgi (these mutants accumulate extensive ER) movement within the Golgi (these mutants elaborate extensive Golgi cisternae), and transport to and fusion with the plasma membrane (these mutants accumulate large numbers of secretory vesicles).

The *SEC4* gene product is an excellent example of the type of protein component of the secretion machinery this genetic approach has brought to light (Salminen and Novick, 1987). This protein is a peripheral component of the cytoplasmic face of both the secretory vesicle membrane and the plasma membrane. It is a small, 20 kDa protein that shows sequence similarity to the mammalian *RAS* gene product and to other GTP-binding proteins such as the coupling, or G, protein of the adenylate cyclase system. The amino acid residues involved in binding GTP are particularly conserved. The protein binds and hydrolyzes GTP. When one of the residues involved in hydrolysis is changed, the yeast cell cannot carry out secretion and fills up with secretory vesicles.

What could be the function of such a GTP-binding protein in secretion? By analogy with the role of the G protein in the adenylate cyclase system, the *SEC4* protein may couple a fusogenic protein or complex to a sensor that recognizes when appropriate membrane–membrane contact has occurred. Alternatively, by analogy with the role of elongation factor Tu in protein synthesis, the *SEC4* protein could act as a gate that must complete a cycle of hydrolysis of one molecule of GTP for every fusion event. Insight into which model may be correct comes from the experiments with the mutant *SEC4* protein that cannot hydrolyze GTP. In the cyclase system a G protein with non-hydrolyzable GTP bound is permanently turned "on" and irreversibly activates the cyclase. In protein synthesis, if elongation factor Tu cannot hydrolyze bound GTP, the regulating cycle is broken and protein synthesis cannot continue. Because mutation of the *SEC4* gene so that the *SEC4* protein cannot hydrolyze GTP results in the accumulation of secretory vesicles, it appears that *SEC4* acts as an essential one-way ratchet in a more complex process. Interestingly, small GTP-binding proteins of mass 20 kDa have now been identified in mammalian secretory tissues and even localized to the chromaffin granule membrane (Burgoyne and Morgan, 1985). It seems likely these proteins will be found to perform similar roles in the mammalian secretory pathway to that of *SEC4* in yeast secretion.

The yeast *SEC18* gene product is apparently a component of a fairly general intracellular fusion machine. This protein was independently investigated as an essential protein for the reconstitution of transport between components of the mammalian Golgi *in vitro* (Wilson et al., 1989). Its activity is sensitive to the sulfhydryl reagent *N*-ethylmaleimide (NEM) so it has been called NSF for NEM-Sensitive Factor. The native protein is a tetramer of 70 kDa subunits. It has been found to be essential for the *in vitro* reconstitution of ER to Golgi transport and for the fusion of endosomes with one another as well as intra-Golgi fusion events. Yeast cells with temperature-sensitive mutations in the *SEC18* gene exhibit greatly expanded ER in their terminal morphology at the non-permissive temperature. Although the protein may be essential at several steps in the secretory pathway, the terminal morphology may naturally reflect only the first step in the sequential pathway in which the gene product is essential. The actual function of the sec18 protein is completely obscure. Its structure suggests it is unlikely to insert into membranes and thus act to disrupt membrane structure. Each subunit has a consensus sequence common to several ATP-binding proteins, and indeed *in vitro* reconstitution systems dependent on this protein also require ATP. The protein does have some sequence similarity to an ATP-dependent bacterial protease and therefore it may be involved in activating another component of a complete "fusion machine" by proteolysis.

Regulated Secretion in the Adrenal Chromaffin Cell

The yeast cell does not have a large storage pool of secretory vesicles. Secretion occurs in a constitutive manner as secretory products are synthesized. However, endocrine and exocrine cells typically store large amounts of secretory product to be released upon specific stimulation of the cell by a hormone or neurotransmitter. The adrenal chromaffin cell has been a popular model for studies of the biochemistry of such "regulated" exocytosis because the secretory vesicles, called chromaffin granules, can readily be isolated in milligram quantities from bovine tissue. These are complex and interesting organelles (see Figure 3) that have on the order of 60 polypeptides in their membranes and another 20 (the *chromobindins*) that can reversibly associate with them in the presence of calcium (Creutz, et al., 1983). In addition, short-term primary cultures of chromaffin cells can be maintained and will respond to secretogogues with the release of catecholamines and secretory proteins. The process of exocytosis was in fact originally defined in this cell by morphological criteria (i.e., the fusion of secretory vesicle membrane with the plasma membrane [De Robertis and Vaz Ferreira, 1957]) and biochemical criteria (i.e., the release of vesicle contents but not membrane components [Kirshner et al., 1966]). It has been shown that the plasma membrane of the chromaffin cell can be permeabilized with detergents (e.g., digitonin [Dunn and Holz, 1983; Wilson and Kirshner, 1983] or saponin [Brooks and Treml, 1983]) or

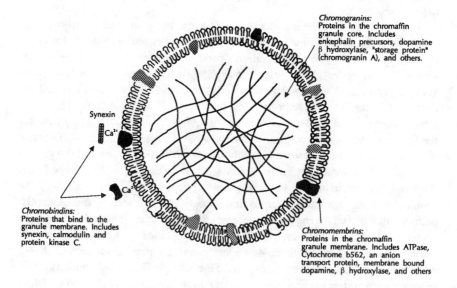

Chromogranins:
Proteins in the chromaffin granule core. Includes enkephalin precursors, dopamine β hydroxylase, "storage protein" (chromogranin A), and others.

Synexin

Ca²⁺

Ca²⁺

Chromobindins:
Proteins that bind to the granule membrane. Includes synexin, calmodulin and protein kinase C.

Chromomembrins:
Proteins in the chromaffin granule membrane. Includes ATPase, Cytochrome b562, an anion transport protein, membrane bound dopamine, β hydroxylase, and others

Figure 3. Schematic illustration of the chromaffin granule, the secretory vesicle of the adrenal medulla. The terms *chromogranins, chromomembrins,* and *chromobindins,* are defined. Drawing by Carol Ceriani. (From Creutz et al., 1990.).

by electrical discharge (Knight and Baker, 1982) yielding a model cell that still carries out exocytosis but in which the cytoplasmic milieu can by altered by changing the external medium. In this way the level of calcium necessary to activate the machinery of exocytosis has been determined to be on the order of 1 μM.

To enable a biochemical analysis of exocytosis one would ideally like to be able to reconstruct *in vitro* a model system involving minimal components in which secretory vesicles fuse with plasma membranes. Unfortunately a reliable model of this type has not yet been developed. Two major problems seem to be (1) obtaining properly oriented plasma membrane sheets or vesicles and (2) the development of a simple biochemical assay for events that are essentially morphological in nature. However, as detailed later, considerable progress has been made in developing models for the fusion of secretory vesicles with one another. These models may be physiologically relevant because in many activated secretory cells the vesicles cannot approach the plasma membrane because of the shear number of vesicles in the cytoplasm and therefore they fuse with one another. In this way they create channels through vesicles that have already fused with the plasma membrane and through which secretory products can escape.

Annexin-Mediated Chromaffin Granule Fusion

Isolated chromaffin granules have a negative surface charge and do not interact with one another unless exposed to unphysiologically high levels of divalent cations. Therefore it is apparent that during their isolation some factor is lost that permits their interaction with one another in the cell. A number of homologous soluble proteins, called annexins, have been isolated that are able to overcome this repulsive barrier and attach chromaffin granules to one another in the presence of calcium. The first member of this family to be identified was the 47 kDa protein synexin that was isolated from adrenal medullary cytosol using turbidity increases in chromaffin granule suspensions as an assay reflective of membrane aggregation (Creutz et al., 1978). The name synexin embodies the concept of membrane contact as it is based on the Greek word *synexis*, which means "a meeting." The other members of this protein family have been isolated in a variety of contexts. For example, lipocortin was isolated as a steroid-inducible inhibitor of phospholipase A2 (Wallner et al., 1986). Calpactin was isolated as a component of the sub-membranous cytoskeleton in the intestinal epithelial cell brush border (Gerke and Weber, 1984). Both lipocortin and calpactin were identified as major substrates for tyrosine-specific protein kinases: the transforming *src* kinase in the case of calpactin (Glenney and Tack, 1985) and the epidermal growth factor receptor associated kinase in the case of lipocortin (Schlaepfer and Haigler, 1987, 1988). Thus both proteins have been associated with membrane signaling events related to growth control. Endonexin I and 68 kDa calelectrin were identified as mammalian homologues of a electric ray protein, called calelectrin, that binds to membranes in the presence of calcium and might underlie the release of acetylcholine by exocytosis (Walker, 1982; Sudhof et al., 1984). Endonexin II was identified as a placental inhibitor of blood coagulation and another substrate for the epidermal growth factor receptor kinase (Kaplan, et al., 1988). It exhibits anticoagulant activity because of its ability to cover membrane surfaces essential for activation of clotting factors. The common element among this confusing array of activities is apparently the ability of all the annexins to bind to membranes in the presence of calcium. This property is shared by some unrelated proteins such as protein kinase C. However, the annexins are unusual (perhaps unique) in that their action is in a sense "bivalent": They can bind to two membranes simultaneously and bring them together.

That the different annexins are truly related is evident from their amino acid sequences, which show 40–60% identity. The proteins have a common structural theme: They have a core domain consisting of four homologous repeats approximately 70 amino acids long and each has a unique N-terminal region 20 to 50 amino acids long. However, there are some interesting variations on this theme. Synexin has an unusually long amino-terminal extension that is highly hydrophobic and repetitive in sequence (Creutz et al., 1988; Burns et al., 1989). The 68 kDa calelectrin was apparently formed by duplication of the gene for one of the other annexins as it has eight repeating domains rather than four. Calpactin associates with a small, 11 kDa subunit that is itself a dimer homologous in structure to the calcium-binding protein S-100 (Glenney and Tack, 1985). As a consequence, the native state of calpactin is apparently a tetramer composed of two 36 kDa heavy chains and two 11 kDa light chains. These structural features of the annexins are summarized schematically in Figure 4.

Although the annexins appear to be able to promote the fusion of pure lipid vesicles with one another (Hong et al., 1982), they are much less effective at promoting complete fusion of biological membranes such as that of the chromaffin granule. Thus, as discussed previously, they are essentially aggregating agents, capable of mediating the *attachment* step in fusion but dependent upon free fatty acids as cofactors to promote the *disruption* step (see Figure 5). Annexin- and free fatty acid-mediated fusion of chromaffin granules also apparently requires an increase in osmotic pressure within the granule (Zaks and Creutz, 1988). In model systems involving the fusion of pure lipid vesicles with planar bilayers, a requirement for intravesicular osmotic pressure increases has also been observed (Zimmerberg et al., 1980). It seems that the disruption of the attached bilayers requires additional forces to be applied to the lipids for proper reorganization to occur. Exocytosis from cells is also sensitive to suppression by hyperosmotic media (Pollard et al., 1984). Therefore, the swelling of secretory vesicles may be an essential part of the process of exocytosis. If so, then ion and water fluxes across the vesicle membrane are likely to be tightly controlled in the regulation of exocytosis. Mechanisms for accomplishing this have yet to be identified.

Although all of the annexins may be able to promote intermembrane contacts, they show a great variation in the levels of calcium needed to activate this process

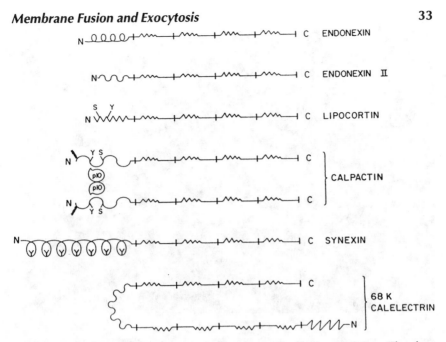

Figure 4. Schematic illustration of the structures of the annexins. The four homologous domains each contain the 17-amino acid endonexin fold sequence represented by a sawtooth line. The unique N-terminal structures are on the left (or right in the case of 68 kDa calelectrin). **Y** and **S** represent phosphorylation sites in the tails of calpactin and lipocortin. The calpactin tetramer is drawn showing the association of the N-termini of the heavy chains with the light chain dimer. The **Y**'s inside the loops in the tail of synexin represent characteristic repeating tyrosines. (From Creutz et al., 1990.)

in vitro. Calpactin is the most sensitive to calcium and promotes membrane contacts at levels of calcium (threshold at 0.7 μM) that are comparable to the levels of calcium that activate exocytosis from permeabilized chromaffin cell models (Drust and Creutz, 1988). Synexin and endonexin require 100 to 200 μM calcium for half maximal activation, which seems extraordinarily high for intracellular activation. Lipocortin and endonexin II are even less sensitive to calcium, which is one reason extracellular roles for these proteins in phospholipase or coagulation inhibition are still entertained. However, the precise sensitivity to calcium of an annexin depends upon the lipid surface with which the protein is interacting. Higher concentrations of phosphatidylserine or free fatty acids in the membrane result in higher affinities for calcium. This seems to indicate there is a tight coupling between the calcium binding site(s) and the lipid binding site(s). One model that could explain this behavior proposes that one of the phosphate groups of the phospholids in the bilayer provides additional direct coordination of the calcium in the calcium-binding pocket or "endonexin fold" (Taylor and Geisow, 1987). However, the three-dimen-

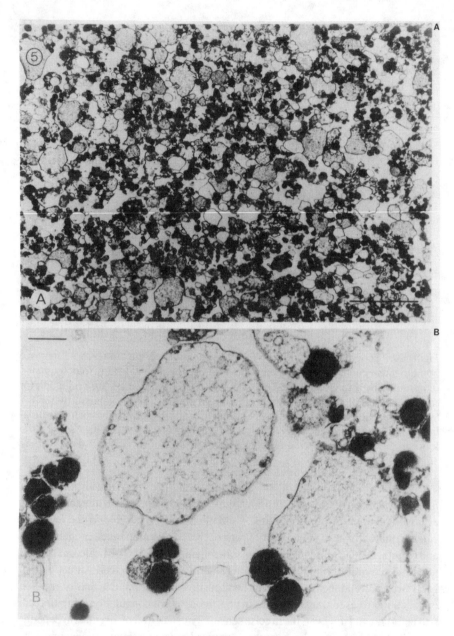

Figure 5. Low and high magnification electron micrographs of chromaffin granules that have undergone synexin-mediated attachment and arachidonic acid–dependent fusion. The fused vesicles swell to encompass the maximum volume. The flocculent material in the large vesicles represents the retained but diluted chromogranins (secretory product). This model system appears to recreate the membrane fusion events that occur between secretory vesicles undergoing compound exocytosis. Length of *bar* in **A**, 5 μm; in **B**, 0.3 μm. (From Creutz, 1981.)

sional structure of these proteins has not yet been determined, nor have the amino acid residues involved in chelating the bound calcium been identified. Because these proteins bear no sequence similarity to classical high-affinity calcium-binding proteins such as calmodulin or troponin C, the eventual elucidation of the mechanism by which they interact with calcium should prove to be quite interesting.

Although similar to one another in their basic structure, the annexins do seem to perform distinct functions in cells. This is suggested by the fact that although all members of the family seem to be present in all tissues, their relative concentrations vary widely. In addition, within single cells their subcellular distributions are unique. Analysis of the distribution of annexins in chromaffin tissue by sucrose gradient analysis of tissue homogenates has revealed that some members of the family exhibit a significant degree of association with membranes even in the absence of calcium (Drust and Creutz, 1991). Although endonexin and lipocortin are largely soluble, 68 kDa calelectrin and calpactin are predominantly membrane bound. Interestingly, significant amounts of calpactin (~20%) are found irreversibly associated with the secretory vesicle membrane. Such calcium-independent annexin–membrane interactions may be due to posttranslational modifications such as acylation or may be due to significant conformational changes in the proteins that permit them to enter the hydrophobic portion of the bilayer to some extent.

The differences between distribution and properties of the various annexins suggest they play non-overlapping roles in the cell. Perhaps all are involved in membrane fusion, but at different locations and involving different membranes. Alternatively, they may perform unrelated functions that have some common element, such as association with a lipid surface. The variations in sequence in the core domains could result in preferential interaction of each annexin with a membrane of different lipid composition. The unique N-terminal extensions could interact with unique proteins, just as the calpactin heavy chain interacts with the light chain. Of the group as a whole, perhaps the best candidate for a direct mediator of membrane fusion in exocytosis is the calpactin tetramer. This molecule has the most appropriate sensitivity to calcium for this role and has been localized by immunocytochemistry to the region of cytoplasm immediately underlying the plasma membrane (Burgoyne and Cheek, 1987).

If calpactin is indeed an important mediator of membrane contact in exocytosis there are some interesting ways in which its activity might be regulated in addition to its dependence on calcium. The N-terminal extension of the 36 kDa heavy chain contains nearly vicinal (2 residues apart) sites for phosphorylation on tyrosine by the *src* kinase and on serine by protein kinase C. Phosphorylation in this region alters the affinity of the heavy chain for the light chain (Johnsson et al., 1986), which is important because the isolated heavy chain (as opposed to the tetramer) is inactive at promoting membrane contacts at physiological levels of calcium. Furthermore, mapping of a monoclonal antibody epitope on the heavy chain indicates that these phosphorylation sites are physically adjacent to the first endonexin fold domain that may bind calcium or lipids (Johnsson et al., 1988). Phosphorylation on tyrosine has indeed been shown to alter the affinity of calpactin for lipid (Powell and Glenney, 1987), and phosphorylation of the corresponding region of lipocortin has been shown to alter the affinity of this protein for calcium (Schlaepfer and Haigler, 1987). Therefore, in the cell, calcium, free fatty acids, lipid concentrations, and protein kinases may all play roles in regulating the activity of calpactin or other annexins in exocytosis. It may be that the calcium requirement can be reduced to levels that would even permit members of this protein class to mediate constitutive membrane fusion events in resting cells.

Tests of the Annexin Hypothesis for Exocytosis

As outlined previously the annexins are potential candidates for essential mediators of membrane fusion in exocytosis. Direct tests of this suggestion need to be made in the future by biochemical or genetic means. Because the secretory apparatus can apparently be accessed in permeabilized cell models, it may be possible to directly alter annexin function with specific inhibitors such as antibodies or peptides that compete with annexins for membrane binding sites (Ali et al., 1989). Alternatively, the yeast cell may provide insights into the role of annexins in secretion or other membrane-associated events. A group of novel calcium-dependent membrane-binding proteins has been characterized in yeast (Creutz et al., 1991). Although is not yet known if they are true homologues of the vertebrate annexins, deletion or alteration of the genes for these proteins in yeast might result in phenotypic clues regarding the functions of calcium-dependent membrane-binding proteins in cells in general.

SUMMARY

The lipid bilayer membranes that separate cells and organelles undergo fusion to permit the exchange or compartmentalization of macromolecules. The fusion events involve contact between cytoplasmic or between extracytoplasmic membrane surfaces. These two types of fusion may depend on very different mechanisms because of the different environments of the membrane surfaces. The membrane fusion process can be broken down into the following steps: (1) translocation, the positioning of membranes adjacent to one another; (2) recognition, initial specific interaction between the appropriate fusion partners; (3) attachment, a close physical sealing between the two bilayers; (4) disruption, alteration of bilayer structure to create an aqueous passageway through the membranes; and (5) reorganization, the formation of new bilayer(s) from the disrupted membrane components. Any or all of these steps may be mediated by specific proteins and be subject to regulation.

The secretory pathway includes a number of steps that involve membrane fusion: pinching off and fusing of transport vesicles in the endoplasmic reticulum and the Golgi, pinching off of secretory vesicles from the Golgi, fusion of secretory vesicles with the plasma membrane during simple exocytosis or with one another during compound exocytosis, and pinching off of endocytic vesicles to recycle secretory vesicle membrane. The molecular basis of these events has been studied using two important model systems that can be subjected to biochemical analysis (exocytosis in the adrenal chromaffin cell) or genetic analysis (the secretory pathway in yeast). In the chromaffin cell the annexins are a group of calcium-dependent lipid-binding proteins that might mediate the fusion of the secretory vesicle with the plasma membrane during regulated secretion. A group of yeast genes (the *SEC* genes) have been identified that mediate several steps in the secretory pathway. These genes have mammalian counterparts that may play similar roles. The annexins and the *SEC* gene products are promising targets for the development of novel pharmacological agents to manipulate hormone or neurotransmitter release.

REFERENCES

Ali, S.M., Geisow, M.J., & Burgoyne, R.D. (1989). A role for calpactin in calcium-dependent exocytosis in adrenal chromaffin cells. Nature 340, 313–315.

Brooks, J.C., & Treml, S. (1983). Catecholamine secretion by chemically skinned cultured chromaffin cells. J. Neurochem. 40, 468–473.

Burgoyne, R.D., & Cheek, T. R. (1987). Reorganization of peripheral actin filaments as a prelude to exocytosis. Bioscience Rep. 7, 281–288.

Burgoyne, R.D., & Morgan, A. (1989). Low molecular mass GTP-binding proteins of adrenal chromaffin cells are present on the secretory granule. FEBS Lett. 245, 122-126.

Burns, A.L., Magenzo, K., Srivistana, M., Cheung, B., Seaton-Johnson, D., Shirvan, A., Alijani, R., Rojas, E., & Pollard, H.B. (1989). Purification of human synexin calcium channel protein and structure of the human synexin gene. Proc. Natl. Acad. Sci. USA 86, 3798–3802.

Creutz, C.E. (1981). Cis-unsaturated fatty acids induce the fusion of chromaffin granules aggregated by synexin. J. Cell Biol. 91, 247–256.

Creutz, C.E., Dowling, L.G., Sando, J.J., Villar-Palasi, C., Whipple, J.H., & Zaks, W.J. (1983). Characterization of the chromobindins: Soluble proteins that bind to the chromaffin granule membrane in the presence of calcium. J. Biol. Chem. 258, 14664–14674.

Creutz, C.E., Drust, D.S., Hamman, H.C., Junker, M., Kambouris, N.G., Klein, J.R., Nelson, M.R., & Snyder, S.L. (1990). Calcium-dependent membrane-binding proteins as potential mediators of stimulus-secretion coupling. In: Stimulus-Response Coupling: The Role of Intracellular Calcium (Dedman, J., & Smith, V., eds.) in press.

Creutz, C.E., Pazoles, C.J., & Pollard, H.B. (1978). Identification and purification of an adrenal medullary protein (synexin) that causes calcium-dependent aggregation of isolated chromaffin granules. J. Biol. Chem. 253, 2858–2866.

Creutz, C.E., Snyder, S.L., Husted, L.D., Beggerly, L.K., & Fox, J.W. (1988). Pattern of repeating aromatic residues in synexin. Similarity to the cytoplasmic domain of synaptophysin. Biochem. Biophys. Res. Commun. 152, 1298–1303.

Creutz, C.E., Snyder, S.L., & Kambouris, N.G. (1991). Characterization of calcium-dependent secretory vesicle binding proteins from yeast. Yeast 7, 229–241.

Creutz, C.E., Zaks, W.J., Hamman, H.C., Crane, S., Martin, W.H., Gould, K.L., Oddie, K., & Parsons, S.J. (1987). Identification of chromaffin granule binding proteins: Relationship of the chromobindins to calelectrin, synhibin, and the tyrosine kinase substrates p35 and p36. J. Biol. Chem. 262, 1860–1868.

De Robertis, E., & Vas Ferreira, A. (1957). Electron microscopic study of the excretion of catechol-containing droplets in the adrenal medulla. Exp. Cell. Res. 12, 568–574.

Drust, D.S., & Creutz, C.E. (1988). Aggregation of chromaffin granules by calpactin at micromolar levels of calcium. Nature 331, 88–91.

Drust, D.S., & Creutz, C.E. (1991). Differential subcellular localization of calpactin and other annexins in the adrenal medulla. J. Neurochem. 56, 469–478.

Dunn, L.A., & Holz, R.W. (1983). Catecholamine secretion from digitonin-treated adrenal medullary chromaffin cells. J. Biol. Chem., 258, 4989–4993.

Gerke, V., & Weber, K. (1984). Identity of p36k phosphorylated upon Rous sarcoma virus transformation with a protein purified from brush borders; calcium-dependent binding to non-erythroid spectrin and F-actin. EMBO J. 3, 227–233.

Glenney, J.R., & Tack, B.F. (1985). Amino terminal sequence of p36 and associated p10: Identification of the site of tyrosine phosphorylation and homology with S-100. Proc. Natl. Acad. Sci. USA 82, 7884–7888.

Hong, K., Duzgunes, N., Ekerdt, R., & Papahadjopoulos, D. (1982). Synexin facilitates fusion of specific phospholipid membranes at divalent cation concentrations found intracellularly. Proc. Natl. Acad. Sci. USA 79, 4642–4644.

Johnsson, N., Johnsson, K., & Weber, K. (1988). A discontinuous epitope on p36, the major substrate of src tyrosine-proteinkinase, brings the phosphorylation site into the neighborhood of a consensus sequence for calcium/lipid-binding proteins. FEBS Lett. 236, 201–204.

Johnsson, N., Van, P.N., Soling, H.D., & Weber, K. (1986). Functionally distinct serine phosphorylation sites of p36, the cellular substrate of retroviral protein kinase; differential inhibition of reassociation with pII. EMBO J. 5, 3455–3460.

Kaplan, R., Jaye, M., Burgess, W.H., Schlaepfer, D.D., & Haigler, H.T. (1988). Cloning and expression of CDNA for human endonexin II, A calcium and phospholipid binding protein. J. Biol. Chem. 263, 8037–8044.

Kirshner, N., Sage, W.J., Smith, W.J., & Kirshner, A.J. (1966). Release of catecholamines and specific protein from the adrenal gland. Science, 154, 529–531.

Knight, D.E., & Baker, P.F. (1982). Calcium dependence of catecholamine release from bovine adrenal medullary cells after exposure to intense electric fields. J. Membr. Biol. 68, 107–140.

Novick, P., Field, C., & Schekman, R. (1980). Identification of 23 complementation groups required for post-translational events in the yeast secretory pathway. Cell 21, 205–215.

Pollard, H.B., Pazoles, C.J., Creutz, C.E., Scott, J.H., Zinder, O., & Hotchkiss, A. (1984). An osmotic mechanism for exocytosis from dissociated chromaffin cells. J. Biol. Chem. 259, 1114–1121.

Powell, M.A., & Glenney, J.R. (1987). Regulation of calpactin I phospholipid binding by calpactin I light-chain binding and phosphorylation by p60v-src. Biochem. J. 247, 321–328.

Salminen, A., & Novick, P.J. (1987). A ras-like protein is required for a post-Golgi event in yeast secretion. Cell 49, 527–538.

Schlaepfer, D.D., & Haigler, H.T. (1987). Characterization of calcium-dependent phospholipid binding and phosphorylation of lipocortin I. J. Biol. Chem. 262, 6931–6937.

Schlaepfer, D.D., & Haigler, H.T. (1988). In vitro protein kinase c, phosphorylation sites of placental lipocortin. Biochemistry 27, 4253-4258.

Sudhof, T.C., Ebbecke, M., Walker, J.H., Fritsche, U., & Boustead, C. (1984). Isolation of mammalian calelectrins: A new class of ubiquitous Ca^{2+}-regulated proteins. Biochemistry 23, 1103–1109.

Taylor, W.R., & Geisow, M.J. (1987). Predicted structure for the calcium-dependent membrane-binding proteins p35, p36 and p32. Protein Engineering 1, 183–187.

Walker, J.H. (1982). Isolation from cholinergic synapses of a protein that binds to membranes in a calcium-dependent manner. J. Neurochem. 39, 815–823.

Wallner, B.P., Mattaliano, R.J., Hession, C., Cate, R.L., Tizard, R., Sinclair, L.K., Foeller, C., Chow, E.P., Browning, J.L., Ramachandran, K.L., & Pepinsky, R.B. (1986). Cloning and expression of human lipocortin, a phospholipase A_2 inhibitor with potential anti-inflammatory activity. Nature 320, 77–81.

Wilson, S.P., & Kirshner, N. (1983). Calcium-evoked secretion from digitonin-permeabilized adrenal medullary chromaffin cells. J. Biol. Chem. 258, 4994–5000.

Wilson, D.W., Wilcox, C.A., Flynn, G.C., Chen, E., Kuang, W.J., Henzel, W.J., Block, M.R., Ullrich, A., & Rothman, J.E. (1989). A fusion protein required for vesicle-mediated transport in both mammalian cells and yeast. Nature 339, 355–359.

Zaks, W.J., & Creutz, C.E. (1988). Membrane fusion in model systems for exocytosis: Characterization of chromaffin granule fusion mediated by synexin and calelectrin. In: Molecular Mechanisms of Membrane Fusion (Ohki, S., Doyle, D., Flanagan, T.D., Hui, S.W.; & Mayhew, E., eds.), pp. 325–340. Plenum Publishing Co., New York.

Zimmerberg, J., Cohen, F.S., & Finkelstein, A. (1980). Fusion of phospholipid vesicles with planar phospholipid bilayer membranes: I. Discharge of vesicle contents across the planar membrane. J. Gen. Physiol. 75, 241–250.

Chapter 3

Endocytosis

THOMAS WILEMAN

Fundamentals of Medical Cell Biology, Volume 5A
Membrane Dynamics and Signaling, pages 41–63
Copyright © 1992 by JAI Press Inc.
All rights of reproduction in any form reserved.
ISBN: 1-55938-309-7

INTRODUCTION

Cells are able to ingest extracellular fluid and solutes by a process called endocytosis. During endocytosis small droplets of fluid are taken up into vesicles that pinch off from the plasma membrane and enter the cell. Two types of endocytosis have been described. The passive ingestion of molecules dissolved in the extracellular medium is called *fluid-phase endocytosis*, whereas the uptake of molecules bound to specific receptors on the plasma membrane is called *receptor-mediated endocytosis*. Receptor-mediated endocytosis is very much more efficient than fluid-phase endocytosis. The receptors not only increase the local concentration of their respective ligands at the cell surface but also cause ligand molecules to accumulate in endocytically active areas of the plasma membrane called coated pits.

Membrane and Receptor Recycling

The endocytic activity of cells results in the uptake of surprisingly large amounts of membrane; for example, fibroblasts internalize 10% of their own fluid volume each hour and this requires the ingestion of 50% of their plasma membrane. For the most part, the contents of endocytic vesicles are transported to lysosomes where they are degraded. Interestingly, much of the ingested membrane is rescued from degradation and recycled back to the cell surface (Steinman et al., 1983). This capacity to recycle membrane means that cells can maintain constant dimensions during many hours of endocytic activity without having to synthesize new membrane on a massive scale. In many cases cells are able to rescue important molecules such as receptors from degradation and can return them to the plasma membrane. This process, called receptor recycling, is a good example of intracellular sorting. The endocytic pathway is able to sort membrane and membrane-associated material from the contents of endocytic vacuoles before they fuse with lysosomes.

Sorting of membranes and receptors takes place in an intermediate organelle called an endosome. Endosomes are able to lower the pH of their contents, and this change in pH can be used to regulate the intracellular sorting of endocytosed ligands. In some cases a fall in pH accelerates the dissociation of ligands from their receptors. Such ligands are no longer associated with the membrane of the endosome and cannot be recycled to the plasma membrane. They travel to lysosomes and are degraded. On the other hand, ligands that do not dissociate at low pH remain bound to their receptors and are returned to the plasma membrane.

Three General Categories of Receptor Recycling

Many different receptors and ligands enter cells by receptor-mediated endocytosis, and the way in which they are sorted in the endosome can vary markedly

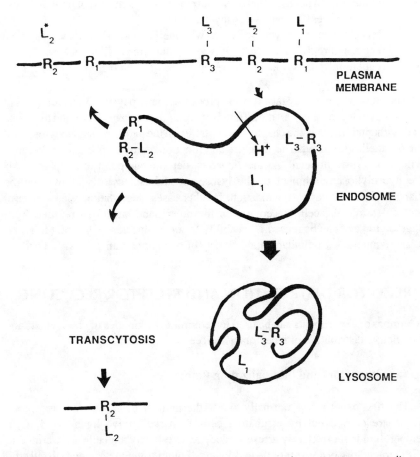

Figure 1. Three categories of receptor recycling. In each category receptor–ligand complexes are internalized through coated pits and coated vesicles and enter endosomes. (**1**) Ligands (L1) dissociate from receptors (R1); the receptors recycle to the plasma membrane while ligands travel to lysosomes where they are digested. (**2**) Ligands (L2) do not dissociate; receptor–ligand complexes (R2-L2) return to the plasma membrane. In polarized cells this can result in transcytosis. (**3**) Ligands (L3) remain bound to receptors (R3), and both are delivered to lysosomes where they are digested.

between receptor–ligand pairs and different cell types. Nevertheless, it is possible to place the pathways into three general categories (Figure 1).

1. Receptors recycle but ligands are targeted to lysosomes.
2. Receptors recycle but ligands are not transported to lysosomes. In this category ligands remain attached to the receptor and recycle to the plasma

membrane. In polarized cells this can result in the transport of ligands across cells; a process known as transcytosis.

3. Receptors do not recycle and travel with their ligand to lysosomes. Because receptors are degraded, this pathway is sometimes called receptor "down regulation" and is followed by receptors for growth factors.

The endocytic pathway takes up and processes many physiologically important molecules, but a review of all the different receptor–ligand pairs is beyond the scope of this chapter. Instead, the chapter concentrates on the pathway of endocytosis and the organelles involved. The structure and function of coated pits and endosomes will be described in detail. As mentioned earlier, the end point of the pathway is the hydrolytic environment of the lysosome, and endocytosis allows for the disposal of molecules by degradation. In other cases, degradation releases useful molecules into the cell. Receptors, on the other hand, are often rescued from degradation and can be reused. This ability to sort molecules to different intracellular destinations is a fundamental property of membrane transport within cells.

RECEPTOR DISTRIBUTION AND RECEPTOR RECYCLING

A number of experiments can be used to determine the kinetics of endocytosis and the relative distribution of receptors within cells.

Surface Receptors and Internalization Rates

The first experiments normally assess the number of cell surface receptors. These are carried out by incubating cells with radioactive ligand at 4 °C, a temperature that effectively arrests endocytic activity. When cells are warmed to 37 °C, receptors become mobile and ligands bound to them are internalized. If a low concentration of trypsin is added to the cells, it is possible to determine the number of ligand molecules that have moved into the cell during the warming period. Essentially, intracellular ligand molecules are protected from the action of the protease whereas those that have not been taken into the cell are released into the medium. By adding trypsin to cells at increasing time intervals, it is possible to determine the rate of internalization of ligands and receptors. Detailed studies have shown that most receptors enter the cell very rapidly at 37 °C, and calculated internalization half lives are a short as 1–2 minutes (Schwartz et al., 1982).

Endocytic Capacity and Receptor Recycling

The total endocytic capacity that a cell has for a particular ligand is a second important measure of endocytic activity. Cells are incubated with saturating concentrations of radiolabeled ligand at 37 °C and the quantity of ligand accumulated by the cell over a known period of time is determined. In most cases it is found that over a period of 15–30 minutes many more ligand molecules are internalized by cells than can be bound to the cell surface at 4 °C. It is apparent from such results that cell surface receptors somehow have to be reused. In fact, cells are able to take up ligands at a continuous rate for many hours, showing that receptors are reused many times. Observations like these were among the first to lead to the concept of receptor recycling (Goldstein et al., 1979). Most receptors recycle within 5–10 minutes of internalization. A well-characterized example is the asialoglycoprotein receptor of hepatocytes; the receptor returns to the cell surface within 7.5 minutes of its internalization and recycles a total of 250 times in its lifetime (Schwartz and Rup, 1983). If, in parallel experiments, the media containing the cells is analyzed for the appearance of degraded ligand (i.e., TCA-soluble radioactive peptides), then the time taken for the internalized ligands to reach lysosomes can be determined. In general, it takes 20–30 minutes for ligands to be transported from the plasma membrane to lysosomes.

There Are Internal Pools of Receptors

Many studies have demonstrated that it takes at least 5–10 minutes for an internalized receptor to get back to the plasma membrane. Interestingly, there is little loss of cell surface binding activity during the intervening period (Ciechanover et al., 1983a). This subtle point implies that internalized receptors can be replaced rapidly by receptors that move from a preexisting intracellular pool. The movement of internal receptors to the cell surface can be demonstrated directly by cooling cells to 4 °C and removing their surface receptors irreversibly with trypsin (Stahl et al., 1982). At this point, cells are unable to bind ligand unless they are warmed to 37 °C. Warming allows receptors to move from inside the cell to the plasma membrane and binding activity is recovered. In many cases, the internal pools of receptors have been found to be quite large, for example, hepatocytes have approximately 860,000 asialoglycoprotein receptors, of which only 80,000 are at the plasma membrane; similarly, 80% of mannose-6-phosphate receptors of fibroblasts are intracellular.

Morphology of Endocytic Pathway

The pathways taken by ligands and receptors have also been studied extensively using the electron microscope. In these experiments ligands are first adsorbed to electron-dense compounds such as colloidal gold or ferritin and are bound to the cell. At 4 °C, ligands bind to receptors located in specialized areas of the plasma membrane known as *coated pits*. When the cells are warmed, coated pits invaginate to form *coated vesicles*, which enter the cell. These vesicles transport ligands to vesicles and tubules near the plasma membrane called *endosomes*. This entry through coated pits and coated vesicles is a unique feature of receptor-mediated endocytosis. The details of this and subsequent steps in the pathway are given in the next section.

COATED PITS AND COATED VESICLES

Both fluid-phase and receptor-mediated endocytic pathways result in the uptake of macromolecules from the extracellular fluid. Receptor-mediated endocytosis is more efficient than fluid-phase endocytosis because the receptors involved become concentrated into coated pits (Goldstein et al., 1979). When the plasma membrane is viewed in cross section by an electron microscope, a protein coat, which will be described in detail later, can be seen attached to the cytoplasmic face of the pit. Coated pits have the ability to invaginate rapidly, and they pinch off from the plasma membrane to form coated vesicles. It has been calculated that between 1500 and 3000 pits can enter a single cell each minute, and consequently, coated pits and vesicles can move receptors and ligands quickly into the endocytic pathway. Although the coated pits can exclude some plasma membrane proteins, they do not appear to show any preference for individual receptors, and many different receptors can be found in a single pit (Carpentier et al., 1982). This selective incorporation of receptors at the expense of other plasma membrane proteins has led to the pits being thought of as molecular filters.

Coated Pits and Coated Vesicles Are Coated with Clathrin

It is possible to isolate pure coated vesicles from tissue sources such as placenta and brain, and this has allowed the vesicles and their protein components to be analyzed in detail (Pearse, 1987). When examined by the electron microscope, isolated coated vesicles are seen to be covered by a complex latticework of polygons. These same polygons can be seen attached to the cytoplasmic face of coated pits budding from the plasma membrane of cultured cells (Heuser and Evans 1980) (Figure 2). It is now known that coated pits and coated vesicles are coated with the same material, the major component of which is clathrin.

Much research has focused on the biochemistry and biophysics of this protein (Brodsky, 1988). Clathrin is composed of two chains, a heavy chain of 180 kDa, and a light chain of 30 kDa. When these proteins are dissociated from the membrane vesicles and allowed to reassociate in the absence of membrane, they assemble together into characteristic three-legged structures called triskelions. Soluble clathrin in the cytoplasm also assembles into triskelions. A triskelion is made from three heavy chains with their carboxy-termini bound together to form a hub (Figure 3A). The light chains bind in the center of the triskelion while the remainder of the heavy chains extend to form the legs. If triskelions are mixed under the right conditions, they will condense to form polygonal clathrin cages. Experiments such as these have shown that the triskelion is the building block of the polygonal coat

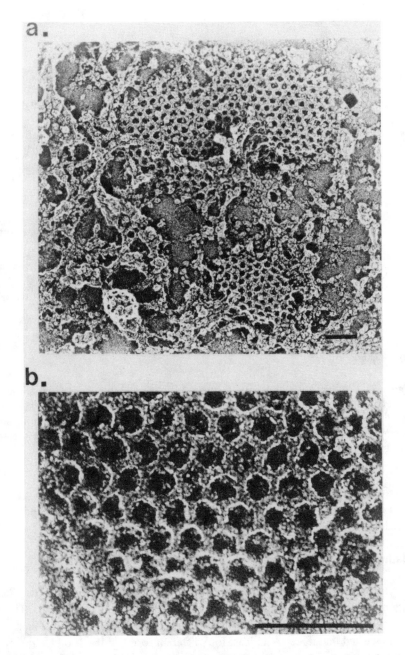

Figure 2. Coated pits. Electron micrographs showing the cytoplasmic face of coated pits. **a.** The view shows three coated pits attached to the plasma membrane of a fibroblast; the honeycomb of the coat is clearly visible. Bar, 0.1 μM. **b.** High magnification of the coat. The coat is seen arranged into regular polygons, each is approximately 30 nm across. Bar, 0.1 μM. (From Heuser and Evans, 1980.)

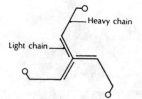

a).   **Triskelion**

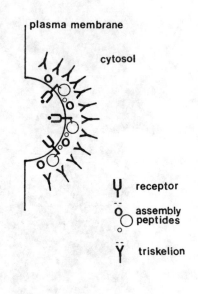

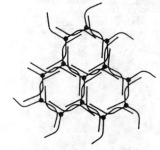

b).   **Clathrin lattice**

c).   **Coated pit**

Figure 3. Clathrin, triskelions, and coated pits. Diagrams showing **(a)** the arrangement of clathrin heavy and light chains in a triskelion, **(b)** the packing of triskelions
into the polygonal clathrin lattice, and **(c)** organization of receptors, clathrin, and
assembly/adaptor proteins in a coated pit. The cytoplasmic tails of receptors interact
with the assembly proteins. (Adapted from Pearse, 1987.)

and that by packing together correctly (Figure 3B) triskelions have the ability to
form the five- and six-sided clathrin polygons seen attached to coated pits and
vesicles.

Clathrin Coats Contain Assembly and Adaptor Proteins

A more detailed analysis of coated vesicles has revealed additional polypeptides
with molecular masses of approximately 100, 50, and 16 kDa, respectively. These
have been referred to as adaptor or assembly proteins. Immunochemical analysis
has suggested that there are at least four different forms of the 100 kDa protein,
and biochemical experiments point to a heterogeneity within the 50- and 16-kDa
peptides. It has been speculated that the presence of different forms of these proteins
could introduce some flexibility in coat construction. This is an attractive proposi-
tion because the construction of heterogeneous populations of coated pits could
allow segregation of pits to functionally different parts of the cell.

At the outset it was emphasized that in order to enter a cell by receptor-mediated endocytosis, a ligand has to bind to a receptor that is incorporated into a coated pit. A molecular basis for the association of receptors with coated pits has been described recently. In these experiments, an affinity matrix to which the cytoplasmic tail of the low density lipoprotein (LDL) receptor (see later) had been bound was shown to bind adaptor/assembly proteins directly. This binding could be inhibited by peptides corresponding to the cytoplasmic domains of other receptors known to enter coated pits along with LDL. Significantly, the cytoplasmic tail of the receptor bound only a subgroup of the adaptor proteins passed over the column, suggesting that the eluted adaptor proteins have a specificity for different receptors. It appears that receptors are incorporated into coated pits through an interaction between their cytoplasmic tails and the adaptor/assembly proteins and not by direct interaction with clathrin. A model showing such an arrangement is presented in Figure 3C.

Coated Pits Will Assemble on Purified Plasma Membranes

Although it has been established that the coated pits are the entrance to the endocytic pathway, the details of how they form have not yet been determined. For example, it is not known whether pits assemble around the tails of clustered receptors or form spontaneously at assembly sites on the plasma membrane. Assembly is undoubtedly a complicated process because the coats are made from at least six different proteins (the clathrin heavy and light chains and the assembly/adaptor proteins) that have to be recruited from the cytoplasm. Some insight into this process has come from studying the assembly of coated pits *in vitro*. These experiments are performed by adding cytosol to purified plasma membranes (Moore et al., 1987). These experiments show that plasma membranes contain a finite number of assembly sites. These have a high affinity for clathrin and will seed the rapid formation of coated pits from the components of the cytosol. Interestingly, the growth of the coat is an ordered process, and the first polygons to form are hexagonal whereas those added later form pentagonal structures. The coat soon stops growing, even if more cytosol is added, indicating that the coats will reach a finite size and cannot themselves seed assembly.

Removal of Clathrin Coats Requires an ATPase

It has been known for many years that coated vesicles are transient structures that lose their clathrin coats within minutes of pinching off from the plasma membrane. At first glance this would appear to be in conflict with the biochemical studies described earlier showing a spontaneous coating of the plasma membrane *in vitro*. The uncoating of vesicles probably requires energy. An enzyme that is able to remove clathrin coats from coated vesicles has been purified from cytosol. The enzyme uses ATP as an energy source and is called uncoating ATPase

(Schlossman et al., 1984). The enzyme binds to intact clathrin cages, hydrolyzes ATP, and releases itself in a stoichiometric complex with intact triskelions. The enzyme–triskelion complexes are unable to reassociate into cages and presumably stay in the cytoplasm until the clathrin is released for another round of assembly. The way in which clathrin is prepared for reuse is not known.

ENDOSOMES

Coated vesicles fuse with small vesicles and tubules located near the cell surface. These structures are called peripheral endosomes (Wall et al., 1980; Hopkins and Trowbridge, 1983), and they control the subsequent movement of receptors and ligands within the cell.

Sorting of Receptors and Ligands in Endosomes Requires Endosomal Acidification

Receptors and ligands are separated from one another in endosomes. This allows the receptor to be rescued from degradation and it can be reused. The first hint of a mechanism for this intracellular segregation came from pharmacological experiments demonstrating that receptor-mediated endocytosis could be inhibited by amines. The uncharged forms of weak bases such as chloroquine, NH_4Cl, and methylamine diffuse into cells. If they encounter an acidic intracellular compartment, they are protonated and consequently will no longer pass through membranes. Amines therefore accumulate in acidic intracellular vesicles and raise their internal pH. The observation that amines are able to inhibit receptor-mediated endocytosis provides a link between endosome function and an ability to acidify. The second clue comes from studies of the pH-dependence of receptor-ligand interactions. Studies with purified receptors and ligands show that those ligands that are sorted from their receptors during receptor–mediated endocytosis dissociate rapidly from their receptors at low pH. It appears from these experiments that by lowering its internal pH, the endosome can accelerate the dissociation of receptor–ligand complexes. Direct evidence for endosomal acidification was provided shortly afterwards. By monitoring fluorescence emitted from fluorescein-labeled ligands as they entered cells, it is possible to show that ligands encounter an acidic compartment within minutes of internalization (Tyco and Maxfield, 1982).

Recycling Receptors Are Concentrated into Specialized Membrane Extensions of Endosomes

The physical separation of receptors and ligands is further enhanced by the morphological changes that take place in endosomes. This process is demonstrated elegantly using immunogold electron microscopy (Geuze et al., 1983). Antibodies to the asialoglycoprotein receptor and its ligand asialoorosomucoid are linked to small and large colloidal gold beads, respectively. They are then incubated with tissue sections under conditions that would allow them to bind to their respective antigens. The different sizes of the gold beads allowed the localization of receptors and ligands to be distinguished from one another in electron micrographs.

Micrographs of cells that have been allowed to internalize ligand for a few minutes show separate distributions of large and small gold particles. The bulk of the ligand (large beads) are contained in the endosome lumen whereas high concentrations of receptors (small beads) are found in armlike extensions protruding from the endosome. This tubulovesicular structure is referred to as Compartment for Uncoupling of Receptor and Ligand (CURL). It is not known how receptors become concentrated in the arms of CURL; they may move there actively and become clustered in domains of the membrane in a manner analogous to coated pit formation. Alternatively, because 90% of the membrane of CURL is found in the arms, segregation could be passive and merely reflect the membrane distribution within the organelle. Regardless of the mechanism it is now thought highly likely that it is the arms of the endosome that concentrate receptors into the vesicles that return them to the plasma membrane. It is probable that, in addition to the receptors, any other components of the endosome, such as ion pumps and channels, that need to be saved from degradation will also be removed from the pathway at this point.

Endosome Biogenesis

Ligands that eventually travel to lysosomes move from the tubulovesicular peripheral endosomes to larger vesicles and multivesicular bodies. The transit time in this compartment can vary, but within 20–60 minutes most of its content will have been delivered to lysosomes for digestion. In some cells these intermediate vesicles accumulate near the nucleus and are called perinuclear endosomes. It is not clear how the ligand is transported between early and late endocytic vesicles. Endosomes can be stable organelles and function much like the Golgi apparatus and use a series of vesicular shuttles to move ligands from early to late endosomes and thence to lysosomes (Helenius et al., 1983). Or alternatively, even though peripheral endosomes might mature by recycling essential proteins to the plasma membrane, the bulk of the later endosomes are transient and eventually fuse with lysosomes. In practice experiments aimed at resolving these issues of endosome biogenesis are very hard to do. A vesicle shuttle would be an efficient means of transporting clustered receptors back to the plasma membrane from CURL, but because of their high surface to volume ratio, these vesicles would probably be too inefficient for the transport of the endosomal contents to lysosomes. The current model proposes that the bulk of the sorting and recycling takes place in the

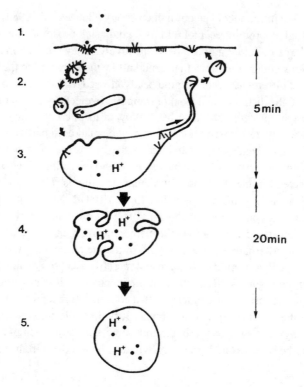

Figure 4. Receptor recycling and endosome biogenesis. (1) Ligands bind to receptors and enter the cell via coated pits and vesicles. (2) Coated vesicles lose their coats, and receptor–ligand complexes enter tubulovesicular structures called peripheral endosomes. (3) Endosome acidification leads to receptor–ligand dissociation. Receptors concentrate in arm-like extensions of the endosome and are recycled to the plasma membrane. (4) The remaining endosome vesiculates and eventually fuses with lysosomes. (5) Ligand molecules are degraded.

peripheral endosomes. The residual endosome membrane and content is then transferred to lysosomes (Figure 4).

A focus of many recent experiments has been to determine whether there are any permanent endosomal structure that organize the sorting of proteins or whether the endosomal apparatus operates continuously in conveyor-belt fashion. One means of doing this is to introduce ligands into the cell at closely spaced time intervals and see if they can catch up with one another (Saltzman and Maxfield, 1988). The rationale being that if they can, then the contents of their respective uptake vesicles are mixed in an organelle that has been receiving incoming vesicles for at least as long as the time interval between addition of ligands. Experiments such as these have shown that ligands added within 5 minutes of one another mix, but those added later do not. It appears that some early endosomes are able to accumulate ligands added to the cell at different times. There is, however, a very narrow time window after which they move from the pathway of incoming ligand and are presumably on their way to lysosomes.

One striking difference between early and late endosomes is a differential sensitivity to drugs that depolymerize microtubules. The drugs do not affect the recycling of receptors through the peripheral endosome, but they do seem to prevent the subsequent degradation of endocytosed ligands (Matteoni and Kreis, 1987; Gruenberg et al., 1989). The results suggest that depolymerization of microtubules causes endocytosed ligand to accumulate at the junction between the early and late endosomes and suggest that microtubules may organize the transport of ligands out of the recycling pathway toward lysosomes.

Coated Vesicles and Endosomes Contain ATP-Dependent Proton Pumps

As described earlier, the acidification of the interior of endosomes accelerates receptor–ligand dissociation and allows the physical separation of receptors and ligands. Experiments using fluorescent probes that emit strongly at acidic pH but weakly at neutral pH were the first to demonstrate the acidic nature of endosomes. Fluorescent ligands that are allowed to enter the cell for a short period of time label endosomes but not lysosomes (lysosomes are also acidic). When labeled cells are observed using a video intensification microscope, the fluorescence emission from endocytic vacuoles can be assessed. Endosomes have an average pH of approximately 5.5 (Tyco and Maxfield, 1982). More recent analysis at higher resolution has shown that not all endocytic compartments have the same pH (Yamashiro et al., 1984, Mellman et al., 1986). The peripheral endosomes are slightly less acidic (pH 6.0) than late endosomes and lysosomes (pH 5.5).

Acidification Requires a Proton-Translocating ATPase

Once it was established that fluorescent probes could be used to calculate intravesicular pH accurately, they were used in further experiments to determine the mechanism of acidification. The probes are internalized into endocytic vesicles and then either the vesicles are isolated from homogenized cells using density centrifugation (Galloway et al., 1983) or the plasma membrane of the cell is permeabilized specifically using mild detergent (Yamashiro et al., 1983). The object of both manipulations is to remove the endosome from the cell cytoplasm so that the environment immediately surrounding the endosome can be manipulated experimentally. When purified endosomes, or endosomes in permeabilized cells, are mixed with physiological buffers, they have a neutral pH, but significantly, when ATP is added they acidify. It is concluded from these experiments that endosomes actively lowered their internal pH by means of ATP-dependent proton pumps.

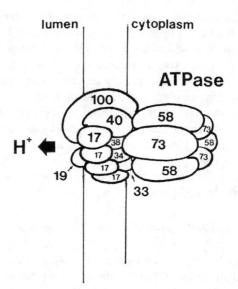

Figure 5. Vacuolar ATPase. Diagram showing the proposed arrangement of the many subunits of the vacuolar proton translocating ATPase. (From Arie et al., 1988.)

Acidification experiments are also carried out on coated vesicles. Practical limitations prevent fluorescent ligands from being introduced into coated vesicles in quantities that can be measured. Instead the ability of isolated vesicles to trap amines is used as an alternative measure of proton translocation (Forgac et al., 1983; Stone et al., 1983). Acidification of coated vesicles also requires addition of ATP, and extensive biochemical analysis of endosome and coated vesicle acidification suggests that they use similar proton pumps. This proton-translocating ATPase is likely to be the same as the proton pump that acidifies lysosomes, some Golgi-derived vesicles, and chromaffin granules (Al-Awqati, 1986). These "vacuolar" pumps are different from the proton ATPases found in mitochondria.

Structure of the Proton-Translocating ATPase

It is possible to prepare coated vesicles in large quantities from calf brain, and this tissue has provided a resource large enough for the isolation of the pump itself (Arai et al., 1987, 1988). The pump is a large macromolecular complex with a molecular mass of 750 kDa. It contains nine polypeptides with molecular weights between 17,000 and 100,000 (Figure 5). The smallest (17 kDa) subunit is very hydrophobic and there are six copies present in each pump. It is likely that these subunits form an ion channel in the vesicle membrane. The larger 58 and 73 kDa subunits are exposed to the cytoplasm and are likely to be ATP-binding proteins. The function of the other subunits has not been determined.

EXAMPLES OF ENDOCYTIC PATHWAYS

Endocytosis of LDL and Cholesterol Metabolism

Cells make a receptor that recognizes low density lipoprotein (LDL), this lipoprotein carries most of the cholesterol of human plasma. The receptor, which is localized in coated pits, binds to the protein component of LDL and takes the particle into the cell by receptor-mediated endocytosis (Brown and Goldstein, 1986). After uptake and transfer to endosomes, the LDL particle dissociates from its receptor and is transferred to lysosomes. Meanwhile, the receptor recycles to the plasma membrane. The lipoprotein particle is broken down in lysosomes and free cholesterol is released into the cytosol where it becomes available for the synthesis of new membrane. Interestingly, the gene for the LDL receptor is regulated by the level of cholesterol in the cell, and high levels of cholesterol in the cytosol suppress transcription from the gene. This means that the cell can regulate the synthesis of the LDL receptor in accordance with a metabolic require-ment for cholesterol. In short, if the cell needs to get cholesterol from the plasma it will increase the synthesis of the LDL receptor.

Individuals with familial hypercholesterolemia inherit a mutated gene encoding the LDL receptor and consequently are unable to take up LDL from the blood. High levels of cholesterol accumulate in the blood, and afflicted individuals become predisposed to atherosclerosis and heart attacks early in life. In many cases the molecular basis for the disease has been determined. The most common mutations result in very low levels of LDL receptor gene expression or in the synthesis of receptors that are unable to function. Most informative have been studies on receptors that are unable to cluster in coated pits. Recent developments in recom-binant DNA technology have allowed the genes for these receptors to be cloned, and mutations have been localized to their cytoplasmic tails. In one case, the J.D. mutation, a single amino acid substitution at position 807 is sufficient to exclude the receptor from the endocytic pathway. Other more extensive mutations that introduce frame shifts or premature stop codons into the cytoplasmic domain of the receptor result in a similar phenotype.

Endocytosis of Polypeptide Toxins and Enveloped Viruses

Several polypeptide toxins enter cells by endocytosis (Olsnes and Sandvig, 1985). These include the plant lectins ricin, abrin, and modeccin and several bacterial toxins of which the diphtheria, cholera, and shigella toxins are examples. All these toxins have two polypeptide chains. One chain, the A chain, is responsible for the toxic activity of the protein, whereas the other, the B chain, allows the toxin to bind to cells. In some cases the binding sites on cells have been defined; for example, the B chains of the plant lectins bind to galactose residues of glycoproteins exposed at the plasma membrane, whereas the B chain of cholera toxin binds to GM_1 gangliosides.

When these toxins are incubated with cells, they bind to their respective "receptors" on the plasma membrane, but surprisingly, the toxins are unable to affect the cell unless they are taken up by endocytosis. The toxins need to be exposed to a low pH before they can pass across the cell membrane. During endocytosis the polypeptide toxins are exposed to a low pH in endosomes and this causes their respective B chains to change conformation and expose hydrophobic domains. These domains are able to penetrate the endosome membrane and form a channel to facilitate the translocation of the attached toxic A chain into the cytoplasm.

The successful infection of cells by some enveloped viruses, for example, semliki forest virus, vesicular stomatitis virus, and influenza virus, also requires their transit through the endocytic pathway (Marsh, 1984). The reasons for this are similar to those described for the polypeptide toxins. The virus particles bind to the surface of cells and are taken up by endocytosis. When they reach the acidic environment of the endosome, their surface glycoproteins undergo conformational changes that again expose hydrophobic domains. These domains allow the membrane envelope of the virus to fuse with the membrane of the endosome and the viral genome passes into the cytosol.

Ligand Recycling—The Transferrin Cycle

When a ligand fails to dissociate from its receptor at low pH, it remains with the receptor throughout the recycling pathway and is returned to the cell surface. Transferrin is an example of a ligand that has evolved to exploit the recycling pathway to shuttle iron into the cell (Klausner et al., 1983; Ciechanover et al., 1983b; Hopkins, 1983). The affinity of transferrin for its receptor is dependent on both pH and the presence of bound iron. Apotransferrin (transferrin without iron) binds to its receptor at low pH but not at physiological pH. When transferrin complexes with iron in the blood, its binding affinity changes such that it can bind to the transferrin receptor at neutral pH. As a consequence, iron-loaded transferrin is taken out of the blood and into the cell by receptor-mediated endocytosis (Figure 6). Iron cannot remain bound to transferrin at low pH, and iron dissociates from transferrin as the endosome acidifies. It then passes across the endosome membrane into the cell cytoplasm where it is sequestered. This leaves apotransferrin bound to the receptor. It remains on the receptor as long as the pH is low and is recycled to the plasma membrane, here it encounters physiological pH and it dissociates from the cell into the blood. The cycle is able to repeat itself because the released

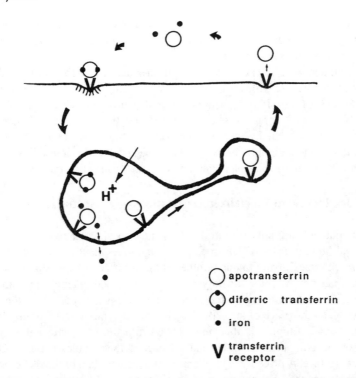

Figure 6. The transferrin cycle. Apotransferrin binds to iron at pH 7.4, is converted into diferric transferrin, and acquires high affinity for its receptor. Transferrin receptor–ligand complexes are transported to endosomes, which acidify. At low pH, iron dissociates from transferrin and enters the cell; the resulting apotransferrin remains bound to the receptor at low pH and recycles to the plasma membrane. At physiological pH, apotransferrin dissociates from the transferrin receptor and is released into the circulation where it can again bind to iron.

apotransferrin can now bind more iron and return to the cell-surface transferrin receptor.

Receptors That Do Not Recycle—Receptor "Down Regulation"

The epidermal growth factor receptor is unable to enter coated pits in the absence of its ligand (EGF). When EGF is added to cells, it binds to its receptor and receptor–ligand complexes cluster in coated pits; this allows the growth factor to be taken into the cell by receptor-mediated endocytosis. Instead of being recycled, EGF ligand–receptor complexes remain with the endosome and are transported to lysosomes where they are degraded (Carpenter, 1987). This mechanism results in the removal of EGF receptors from the cell surface and prevents cells from being stimulated constitutively by the mitogen. The process is sometimes referred to as receptor "down regulation," and it is thought that cell-surface insulin receptors may be regulated by a similar mechanism.

The immunoglobulin (Fc) receptor of macrophages can also be degraded in lysosomes, but its fate depends on the nature of the ligand bound to the receptor (Ukkonen et al., 1986). If macrophages are incubated with polyvalent ligands such as soluble immune complexes, the receptor and the ligand are delivered to lysosomes and degraded. On the other hand, if the cells are incubated with monovalent immunoglobulin or monovalent receptor-specific antibodies, then the receptor recycles to the plasma membrane with the ligand still attached.

Receptors Recycle to Specialized Destinations—Transcytosis

Epithelial cells are polarized in that they have specialized apical (upper) and basolateral (lower) surfaces. This specialization implies that they are able to direct some proteins specifically to one or other of these surfaces. This is true for some receptors, and the receptor for polymeric IgA is a good example (Mostov and Simister, 1985). After being synthesized the receptor is transported via the Golgi to the basolateral surface of the cell where it binds polymeric IgA (Figure 7). The receptor–ligand complex is then internalized into the endocytic pathway, but instead of recycling back to the basolateral surface, the complex is transported to the apical surface. A proteolytic clip removes the ectodomain of the receptor from its membrane anchor, and the receptor fragment, called the secretory component, and bound IgA are released from the cell. The net result is the transport of IgA across the cell. In whole tissues this corresponds to the uptake of IgA from the blood and its incorporation into a number of secretions such as milk, saliva, and bile.

Transcytosis can work in the opposite direction. Suckling rats can take up monomeric IgG from milk. The immunoglobulin binds to a receptor on the apical surface of the gut epithelia (Figure 7). The receptor has a high affinity for IgG at pH 6, the pH of the gut, but a low affinity at physiological pH. The receptor/IgG complexes formed at the apical surface are taken into the cell by endocytosis; they enter endosomes and then recycle to the plasma membrane. The receptors can only release immunoglobulin when they recycle to the basolateral membrane because here they encounter a pH of 7.4, which allows the ligand to dissociate. In this way a pH gradient across the epithelium can direct the one-way transport of a ligand.

Adsorptive Endocytosis via Non-Coated Pits

Not all invaginations of the plasma membrane are coated with clathrin, and it is interesting that these uncoated areas of membrane are also able to participate in endocytosis. The list of molecules that follow this non-coated pathway is relatively short; examples that have been studied in detail include HLA molecules cross-

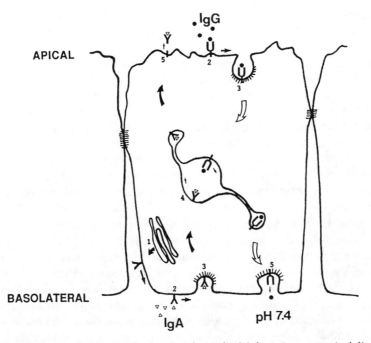

Figure 7. Transcytosis of **IgA**. After leaving the Golgi (1) the IgA receptor is delivered to the basolateral surface of the cell where it binds IgA (2). The receptor–ligand complex is transported via coated pits (3) to endosomes (4) and is sorted for transport to the apical surface. Here the receptor is clipped and is secreted from the cell attached to its ligand (5). Transcytosis of **IgG**. IgG binds to gut epithelia receptors at pH 6 (2) and enters the cell via coated pits (3). Because the pH is still low, the receptor–ligand complexes remain intact in the endosome (4) and can be delivered to apical or basolateral surfaces. At the basolateral surface, the pH is 7.4 and IgG is released from the receptor into the circulation (5).

linked by antibody (Huet et al., 1980) and the cholera and tetanus toxins (Montesano et al., 1982). Although clustering of adsorbed molecules in these uncoated patches of membrane does not appear to require clathrin, the endocytosed material is nevertheless delivered to the same endosomes as ligands entering through coated pits (Tran et al., 1987).

FUTURE DIRECTIONS

Recombinant DNA technology is making the isolation and characterization of cell surface receptors easier. It is now possible to use powerful expression vectors to prepare ligands of interest in large quantities and these can be used to screen cDNA expression libraries for genes encoding receptors. The isolation of increasing numbers of cell surface receptors will no doubt increase the list of those that follow the endocytic pathway.

The main challenges of the future will focus on the biochemistry and physiology of membrane recycling and protein sorting. Recombinant DNA technology will be applied routinely to determine the primary sequences of the proteins involved. The genes for many receptors have been cloned already and their primary amino acid sequences are known. Although it has been possible in some cases to determine the domains of the receptors that are responsible for ligand binding and incorporation into coated pits, it is not yet known which domains control the movement of receptors once they have entered the endocytic pathway. For example, some receptors recycle to the plasma membrane whereas others are degraded in lysosomes.

Recombinant DNA methods are allowing receptors and domains of receptors to be manufactured in large quantities (Greenfield et al., 1988; Domingo and Trowbridge, 1988). The tertiary and quaternary structure of receptors can now be probed by X-ray crystallography. In other experiments purified cytoplasmic domains can be used to identify proteins that interact with receptors as they travel through the cell. These same methods will be used to analyze the structural and functional domains of other proteins isolated from the compartments of the endocytic pathway. At present it is not known how the proton-translocating ATPase becomes incorporated selectively into the endocytic and lysosomal apparatus; this information may reside in one of the many subunits of the pump.

Another exciting area of research concerns the reconstitution of the membrane fusion events that take place during endocytosis. For example, coated vesicles fuse with endosomes, and endosomes fuse with both lysosomes and the plasma membrane. Some of these membrane interactions have already been reconstituted *in vitro*, and the assembly of coated pits onto isolated plasma membranes has been described. In addition, experiments have shown that fusion between endocytic vesicles requires ATP and cytosolic proteins. One of these proteins, *N*-ethyl-maleimide-sensitive factor (NSF), catalyzes fusion events throughout the secretory pathway (Diaz et al., 1989). The next challenge is to find the proteins that direct fusion of membrane vesicles to specific targets. These proteins may be the ones that regulate the specificity of membrane sorting.

REFERENCES

Al-Awqati, Q. (1986). Proton translocating ATPases. Ann. Rev. Cell Biol. 2, 179–200.

Arai, H., Terres, G., Pink, S., & Forgac, M. (1988). Topography and subunit stoichiometry of the coated vesicle proton pump. J. Biol. Chem. 263, 8796–8802.

Arai, H., Berne, M., Terres, G., Terres, H., Puopolo, K., & Forgac, M. (1987). Subunit composition and ATP site labeling of the coated vesicle proton-translocating Adenosinetriphosphotransferase. Biochemistry 26, 6632–6638.

Brodsky, M.F. (1988). Living with clathrin; Its role in intracellular membrane traffic. Science 242, 1396–1402.

Brown, M.S., & Goldstein, J.L. (1986). A receptor-mediated pathway for cholesterol homeostasis. Science 232, 34–47.

Carpenter, G. (1987). Growth factor receptors. Ann. Rev. Biochem. 56, 881–914.

Carpentier, J-L., Gorden, P., Anderson, R.G.W., Goldstein, J.L., & Brown, M.S. (1982). Co-localization of ^{125}I epidermal growth factor and ferritin-low density lipoprotein receptor in coated pits: A quantitative electron microscope study in normal and mutant fibroblasts. J. Cell. Biol. 95, 73–77.

Ciechanover, A., Schwartz, A. L., Dautry-Varsat, A., & Lodish, H.F. (1983a). Kinetics of internalization and recycling of transferrin and the transferrin receptor in a human hepatoma cell line. J. Biol. Chem. 258, 9681–9689.

Ciechanover, A., Schwartz, A., & Lodish, H.F. (1983b). The asialoglycoprotein receptor internalizes and recycles independently of the transferrin and insulin receptors. Cell 32, 267–275.

Diaz R., Mayorga L.S., Weidman, P.J., Rothman, J.E., & Stahl, P. D. (1989). Vesicle fusion following receptor-mediated endocytosis requires a protein active in Golgi transport. Nature (Lond.) 339, 398–400.

Domingo, D.L., & Trowbridge, I.S. (1988). Characterization of the human transferrin receptor produced in a baculovirus system. J. Biol. Chem. 263, 13386–13392.

Forgac, M., Cantley, L., Wiedenmann, B., Altsteil, L., & Branton, D. (1983). Clathrin-coated vesicles contain an ATP-dependent proton pump. Proc. Natl. Acad. Sci. USA 80, 1300–1303.

Galloway, C.J., Dean, G.G.E., Marsh, M., Rudnick, G., & Mellman, I. (1983). Acidification of macrophage and fibroblast endocytic vesicles in vitro. Proc. Natl. Acad. Sci. USA 80, 3334–3338.

Geuze, H.J., Slot, J.W., Strous, G.A.M., Lodish, H., & Schwartz, A.L. (1983). Intracellular site of asialoglycoprotein receptor-ligand uncoupling; double labeling immune-electron microscopy during receptor mediated endocytosis. Cell 32, 277–287.

Goldstein, J.L., Anderson, R.G.W., & Brown, M.S. (1979). Review Article: Coated pits, coated vesicles and receptor-mediated endocytosis. Nature (Lond.) 279, 679–684.

Greenfield, C., Patel, G., Clark S., Jones, N., & Waterfield, M.D. (1988). Expression of the human EGF receptor with ligand stimulatable kinase activity in insect cells using a baculovirus vector. EMBO J. 7, 139–146.

Gruenberg, J., Griffiths, G., & Howell, K.E. (1989). Characterization of the early endosome and putative endocytic carrier vesicles in vivo with an assay of vesicle fusion in vitro. J. Cell. Biol. 108, 1301–1316.

Helenius, A., Mellman, I., Wall, D., & Hubbard, A. (1983). Endosomes. Trends Biochem. Sci. 8, 245–250.

Heuser, J., & Evans, L. (1980). Three-dimensional visualization of coated pit formation. J. Cell. Biol. 84, 560–583.

Heuser, J.E. (1989). Mechanisms behind the organization of membraneous organelles in cells. Current Opinion in Cell Biology 1, 98–102.

Heut, C., Ash, J.F., & Singer, S.J. (1980). The antibody-induced clustering and endocytosis of HLA antigens on cultured human fibroblasts. Cell 21, 429–438.

Hopkins, C.R. (1983). Intracellular routing of transferrin and transferrin receptors in A431 cells. Cell 35, 321–330.

Hopkins, C.R., & Trowbridge, I.S. (1983). Internalization and processing of transferrin and transferrin receptors in human carcinoma A413 cells. J. Cell. Biol. 97, 508–521.

Klausner, R.D., Van Renswoude, J., Ashwell, G., Kempf, C., Schechter, A.N., & Bridges, K.R. (1983). Binding of apotransferrin to K562 cells: Explanation of the transferrin cycle. J. Biol. Chem. 258, 4715–4724.

Marsh, M. (1984). Review article; the entry of enveloped viruses into cells by endocytosis. Biochem. J. 218, 1–10.

Matteoni, R., & Kreis, T.E. (1987). Translocation and clustering of endosomes and lysosomes depends on microtubules. J. Cell. Biol. 105, 1253–1265.

Mellman, I., Fuchs, R., & Helenius, A. (1986). Acidification of endocytic and exocytic pathways. Annu. Rev. Biochem. 55, 663–700.

Montesano, R., Roth, A., Robert, A., & Orci, L. (1982). Non-coated membrane invaginations are involved in binding and internalization of cholera and tetanus toxins. Nature (Lond.) 296, 651–653.

Moore, M.S., Mahaffey, D.T., Brodsky, F.M., & Anderson, R.G.W. (1987). Assembly of clathrin coated pits onto purified plasma membranes. Science 236, 558–563.

Mostov, K., & Simister, N.E. (1985). Transcytosis. Cell 43, 389–390.

Olsnes, S., & Sandvig, K. (1985). Entry of polypeptide toxins into animal cells. In: Endocytosis (Pastan, I., & Willingham, M.C., eds.). Plenum Publishing Corp., New York.

Pearse, B.M.F. (1987). Clathrin and coated vesicles. EMBO J. 6, 2507–2511.

Saltzman, N.H., & Maxfield, F.R. (1988). Intracellular fusion of sequentially formed endocytic compartments. J. Cell. Biol. 106, 1083–1091.

Schlossman, D.M., Schmid, S.L., Braell, W.A., & Rothman, J.E. (1984). An enzyme that removes clathrin coats: Purification of an uncoating ATPase. J. Cell. Biol. 99, 723–733.

Schmid, S.L., Fuchs, R., Male, P., & Mellman, I. (1988). Two distinct subpopulations of endosomes involved in membrane recycling and transport to lysosomes. Cell 52, 73-83.

Schwartz A., & Rup, D. (1983). Biosynthesis of the human asialoglycoprotein receptor. J. Biol. Chem. 258, 11249–11255.

Schwartz, A., Fridovich, S.E., Knowles, B., & Lodish, H.F. (1982). Kinetics of internalization and recycling of asialoglycoprotein receptor in a hepatoma cell line. J. Biol. Chem. 257, 4230–4237.

Stahl, P.D., Schlesinger, P.H., Sigardson, E., Rodman, J.S. & Lee, Y.C. (1982). Receptor-mediated pinocytosis of mannose glycoconjugates by macrophages: Characterization and evidence for receptor recycling. Cell 19, 207–215.

Steinman, R. M., Mellman, I.S., Muller, W.A., & Cohn, Z. (1983). Endocytosis and recycling of plasma membrane. J. Cell. Biol. 96, 127.

Stone, D., Xie, X.-S., & Racker, E. (1983). An ATP-driven proton pump in clathrin-coated vesicles. J. Biol. Chem. 258, 4059–4062.

Tran, D., Carpentier, J-L., Sawano, F., Gorden, P., Orci, L. (1987). Ligands internalized through coated or non-coated invaginations follow a common intracellular pathway. Proc. Natl. Acad. Sci. USA 84, 7957–7961.

Tyco, B., & Maxfield, F.R. (1982). Rapid acidification of endocytic vesicles containing α2-macroglobulin. Cell 28, 643–651.

Ukkonen P., Lewis, V., Marsh, M., Helenius, A., & Mellman, I. (1986). Transport of macrophage Fc receptors and Fc receptor-bound ligands to lysosomes. J. Exp. Med. 163, 952–971.

Wall, D.A., Wilson, G., & Hubbard, A. (1980). The galactose-specific recognition system of mammalian liver; the route of ligand internalization in rat hepatocytes. Cell 21, 79–93.

Yamashiro, D., Fluss, J., & Maxfield, F.R. (1983). Acidification of endocytic vesicles by an ATP-dependent proton pump. J. Cell. Biol. 97, 929–934.

Yamishiro, D.J., Tyco, B., Fluss, J., & Maxfield, F.R. (1984). Segregation of transferrin to a mildly acidic para-Golgi compartment in the recycling pathway. Cell 37, 789–800.

RECOMMENDED READINGS

Al-Awqati, Q. (1986). Proton translocating ATPases. Ann. Rev. Cell Biol. 2, 179–200.

Brodsky, M.F. (1988). Living with clathrin; Its role in intracellular membrane traffic. Science 242, 1396–1402.

Brown, M.S., & Goldstein, J.L. (1986). A receptor-mediated pathway for cholesterol homeostasis. Science 232, 34–47.

Goldstein, J.L., Anderson, R.G.W., & Brown, M.S. (1979). Review article: Coated pits, coated vesicles and receptor-mediated endocytosis. Nature (Lond.) 279, 679–684.

Marsh, M. (1984). Review article; the entry of enveloped viruses into cells by endocytosis. Biochem. J. 218, 1–10.

Mellman, I., Fuchs, R., & Helenius, A. (1986). Acidification of endocytic and exocytic pathways. Annu. Rev. Biochem. 55, 663–700.

Mostov, K. & Simister, N.E. (1985). Transcytosis. Cell 43, 389–390.

Pearse, B.M.F. (1987). Clathrin and coated vesicles. EMBO J. 6, 2507–2511.

Steinman, R. M., Mellman, I.S., Muller, W.A., & Cohn, Z. (1983). Endocytosis and recycling of plasma membrane. J. Cell. Biol. 96, 127.

Wileman, T., Harding, C., & Stahl, P.D. (1985). Review article: Receptor-mediated endocytosis. Biochem. J. 232, 1–14.

Part II

Membrane Signaling

Chapter 4

Membrane Ionic Channels

DAVID J. TRIGGLE

INTRODUCTION

To optimize their cellular potential cells must receive and respond to a wide range of informational inputs. This broad receptivity is achieved through a variety of receptors that are sensitive to chemical and physical inputs and that are coupled to effectors that translate and amplify the information of the incoming signal. The control of ion movements represents one important mechanism mediating the control of cellular excitability (Hille, 1984).

Cells maintain a highly asymmetrical distribution of ions across membranes. This asymmetrical distribution of ions is maintained through the selectively permeable properties of the membranes and by selective ion transporters that serve to restore ionic gradients post-stimulus and to maintain the ionic gradients in the face

Fundamentals of Medical Cell Biology, Volume 5A
Membrane Dynamics and Signaling, pages 67–82
Copyright © 1992 by JAI Press Inc.
All rights of reproduction in any form reserved.
ISBN: 1-55938-309-7

of leakage. The maintenance of these ionic gradients serves at least two principal purposes: it permits the cell to use their discharge as a current-carrying device with which to alter membrane potential and thus control cellular excitability, and it permits the ions themselves to serve as second messenger species. The ion currents may serve as the end response as in action potential generation, or they may serve to couple ion flow and cellular events, as is the case with Ca^{2+} currents mediating stimulus–response coupling through the intermediacy of Ca^{2+} binding proteins.

Ion channels represent necessary and early events in cellular evolution. The lipid bilayer represents an extreme barrier to the free passage of ions, but the uncontrolled movement of ions is a lethal signal that will destroy cellular excitability. Ion channels must, therefore, be regulated species. The control of ion movements represents one of the ways in which the cell, as an excitable system, responds to informational inputs. Ion channels, as components of both plasma and intracellular membranes, serve to permeate the physiological ions Na^+, K^+, Ca^{2+}, and Cl^- in response to diverse cell stimuli including light, heat, pressure, and changes in chemical and electrical potential. Ion channels represent one major class of biological transducers, but they should not be viewed as totally independent from other systems because they both modulate and are modulated by other effector systems.

HOW ION CHANNELS ARE STUDIED

Contemporary studies of ion channels owe much to the voltage clamp technique introduced by Hodgkin and Huxley (1952) for squid giant axons. This technique permits, through the ability to hold membrane potentials constant at defined levels for defined times, the resolution of both the magnitudes and the time courses of current flow. This is illustrated in Figure 1 for the Na^+ and K^+ currents underlying action potential generation. This figure also reveals that ion channel kinetics may differ markedly according to permeant ion and the class of channel. The Na^+ channel activates rapidly but also inactivates rapidly despite the maintenance of constant membrane potential. In contrast, the K^+ channel ("delayed rectifier") activates less rapidly but inactivates very slowly. Differences in the time- and voltage-dependencies of ion channel activation and inactivation represent one important component of ion channel distinction (see p. 72). Subsequent developments have permitted the extension of this principle to membrane and cell patches in a variety of configurations and the detection and recording of single channel events (Sakman and Neher, 1983) (Figure 2).

COMMON CHARACTERISTICS OF ION CHANNELS

Most cells maintain a generally similar distribution of ions across the plasma membrane to maintain the cell in a polarized state with an approximate resting membrane potential of some –50 to –80 mV. Typically, the cell maintains a high

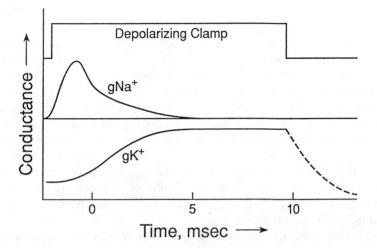

Figure 1. Representation of axonal Na$^+$ and K$^+$ currents following a depolarizing clamp. There is a rapid and transient rise in Na$^+$ conductance and a slower and sustained rise in K$^+$ conductance ("delayed rectifier"). Modified from A. L. Hodgkin and A. F. Huxley, J. Physiol. [Lond]. 117, 500, 1952.

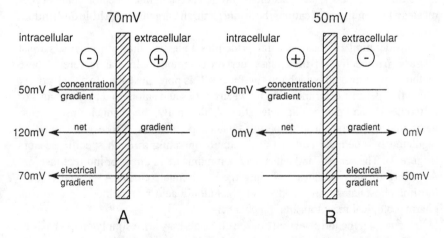

Figure 2. The electrochemical and concentration gradients for Na$^+$ distribution. **A.** At a membrane potential of –70 mV (resting) the driving force for the inward movement of Na$^+$ is made up of two components—an inward concentration gradient and an inward electrochemical gradient. **B.** At a membrane potential of +50 mV (depolarized) the inward concentration gradient and the outward electrochemical gradients balance and there is no net movement of ions.

intracellular K^+ concentration (approximately 150 mM) relative to the exterior, a low intracellular Na^+ concentration (approximately 10 mM) relative to the exterior, and an extremely low level of free intracellular ionized Ca^{2+} ($< 10^{-7}$ M) relative to the exterior. Each ion is subject to both electrical and chemical forces tending to dissipate the gradients. Accordingly each ion has a potential, the equilibrium potential, at which the concentration and electrical gradients are in equilibrium. Activation of an ion channel will thus shift the cellular potential to the equilibrium potential for the particular permeable ionic species (Figure 3). Thus, opening of Na^+ or Ca^{2+} channels will represent depolarizing, excitatory stimuli, whereas opening of K^+ or Cl^- channels will be inhibitory or hyperpolarizing stimuli. Conversely, the closing of a K^+ channel will represent a disinhibitory (excitatory) signal leading to cell excitation.

Ion channels represent one of several pathways through which ions cross cell membranes. Although ion channels are similar to other transporting systems including vectorial pumps and carriers in one important respect, namely ionic selectivity, they differ in other important characteristics. In particular, ion channels are extremely efficient molecular devices and permeate ions at rates close to diffusion controlled. Accordingly, channel densities may be comparatively low relative to those of other excitable proteins. Ion channels possess varying degrees of ionic selectivity. The Na^+ and K^+ channels involved in the action potential generation depicted in Figure 1 exhibit mutual discrimination ratios of approximately 10:1, but Ca^{2+} channels discriminate against Na^+ by a factor of approximately 1000:1, whereas the nicotinic acetylcholine receptor-channel permeates a large number of cations, both inorganic and organic, with little discrimination.

Consideration of these several principles indicates that ion channels must possess, irrespective of their class, certain common molecular features. These include a pore through which ions permeate. This pore must contain a "selectivity filter" that permits discrimination between cations and anions and between ions of different size and charge characteristics. Additionally, the channel must possess gates that open and close to permit channel permeation. These gates must be regulated by chemical and physical stimuli operating through specific sensors (Figure 4). The sensors and gates may communicate in direct or indirect fashion. A variety of intermediates including the cyclic nucleotides cAMP and cGMP, inositol phosphates, fatty acids, and arachidonic acid metabolites serve to link pharmacological receptors and ion channels.

In principle, the ion selectivity of ion channels may arise simply on the basis of size, whereby the channel serves as a molecular sieve, rejecting ions of greater than certain critical dimensions. Alternatively, and more plausibly, ionic selectivity arise from the existence of specific binding sites both within and without the channel that define the permeations. Thus, permeation through the voltage-dependent Ca^{2+} channel can be described by a model in which two (or more) binding sites for divalent cations exist within the channel: double occupancy of these sites

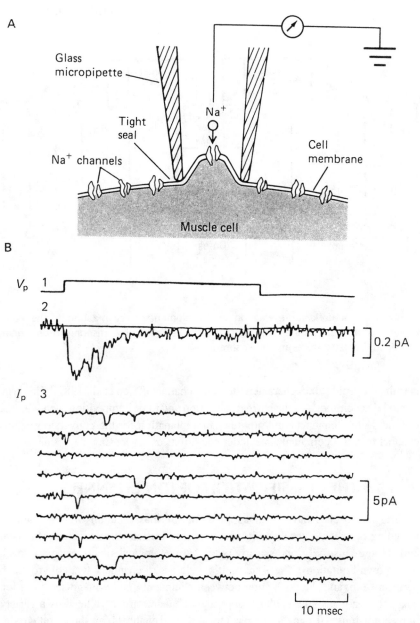

A. Glass micropipette

Tight seal

Na⁺ channels

Na⁺

Cell membrane

Muscle cell

B.

V_p 1

2

0.2 pA

I_p 3

5 pA

10 msec

Figure 3. **A.** The patch clamp technique for measuring current through single channels in which a small patch of membrane containing a single ion channel (in this example Na⁺) is isolated from the rest of the cell or membrane by the patch electrode. **B.** Recordings of single Na⁺ channels showing the time course of a 10-mV depolarizing clamp (1), the sum of the Na⁺ current during 300 trials (2), and individual trials showing the individual channel openings (3). Reproduced with permission from F. J. Sigworth and E. Neher, (1980). Nature 287, 447.

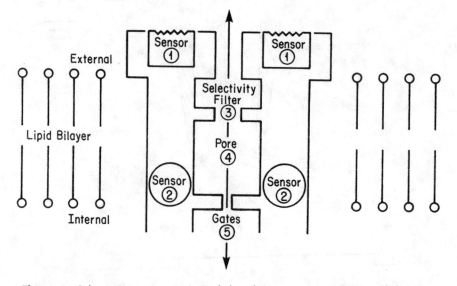

Figure 4. Schematic representation of the elementary organization of the components of an ion channel. The sensors indicated, **1** and **2**, respond to electrical, chemical, light, pressure, and other stimuli.

permits the rapid fluxes characteristic of the channel (Tsien et al., 1987). This type of model accommodates the current saturation and ionic competition seen with the Ca^{2+} and other channels and provides an explanation for the conversion of this channel to a Na^+ permeant state under conditions of severe Ca^{2+} depletion.

THE CLASSIFICATION OF ION CHANNELS

Several approaches have been adopted for the classification of ion channels including permeant ion selectivity, activation and inactivation kinetics, chemical and voltage sensitivity, and sensitivity to pharmacological agents. Each approach has its own limitations. There are multiple classes of channels available for each permeant ion, and the distinction between chemically and voltage-sensitive ion channels is not absolute. Voltage-dependent channels are modulated by a variety of receptor-initiated signals, and receptor-operated channels can show voltage-dependent behavior. Nonetheless, the distinction between voltage-dependent and receptor-operated channels is a useful one in the sense that they represent channel categories that are principally sensitive to changes in electrical or chemical potential, respectively.

The operation of voltage-dependent channels may be modulated through drug–reception interactions. The voltage-dependent cardiac Ca^{2+} channel is activated by

Receptor Channel

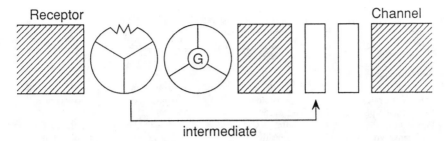

intermediate

Figure 5. The organization of receptor-operated ion channels. There are various ways in which the receptor and channel functions may communicate. These functions may be part of the same molecule or assembly of molecules, or, as indicated, they may be separate components that are linked through one or more biochemical signals or through coupling via G proteins.

depolarizing stimuli, and the probability of this channel activation event is modulated through the β-adrenoceptor–adenylate cyclase complex serving to phosphorylate the channel or a component thereof through the cAMP-dependent protein kinase A. Voltage-dependent channels may be viewed as possessing a common component, the voltage sensor; a conclusion strengthened from structural studies (see p. 80). However, receptor-operated channels may be constructed according to several models; the primary distinction between which is the possession of an intrinsic or extrinsic sensor. The receptor may be a component of the channel itself or may be linked to the channel by one or more cytosolic or membrane diffusible messengers (Figure 5).

A further classification of ion channels is provided by the drugs with which they interact. Ion channels are generally sensitive to a variety of chemical agents in

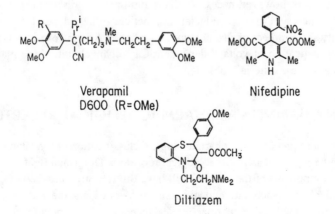

Figure 6. Structural formulas of Ca^{2+} channel antagonists.

Table 1. Properties of Plasmalemmal Ca^{2+} Channels

	Channel Class		
	L	*T*	*N*
Activation range, mV	−10	−70	−30
Inactivation range, mV	−60 to −10	−100 to −60	−120 to −30
Inactivation rate	very slow	rapid	moderate
Conductance, pS	25	8	13
Permeation	$Ba^{2+} > Ca^{2+}$	$Ba^{2+} = Ca^{2+}$	$Ba^{2+} > Ca^{2+}$
Cd^{2+} sensitivity	sensitive	insensitive	sensitive
1,4-DHP sensitivity	sensitive	insensitive	sensitive
w-Conotoxin	sensitive (neurons)? insensitive (muscle)	insensitive	sensitive

addition to those that act at the defined receptors for neurotransmitters and related agents. These specific drugs provide a further basis for channel classification. Thus, tetrodotoxin and tetraethylammonium represent agents that provide selective blockage of Na^+ and K^+ channels, respectively (Figure 1). Naturally occurring toxins of both plant and animal origin as well as synthetic agents have been useful in providing a pharmacological characterization of ion channels. A number of these agents have prominent therapeutic applications. Thus, the heterogeneous groups of agents known as the Ca^{2+} channel antagonists and including verapamil, nifedipine, and diltiazem (Figure 6) enjoy significant use in the treatment of a number of cardiovascular disorders. New toxins, derived particularly from the venoms of molluscs and spiders, are continuing to provide valuable agents for the further classification and characterization of both Na^+ and Ca^{2+} channels.

With the increasing realization that multiple classes of ion channels exist for single ions, it is, however, necessary to use multiple criteria for the classification of ion channels and to use conductance, kinetics, voltage dependence, and drug sensitivity as the indices of channel type and subtype. A useful example is provided by the classification of voltage-dependent Ca^{2+} channels into three major types according to these criteria (Table 1).

ION CHANNELS AS PHARMACOLOGICAL RECEPTORS

A major consequence of the pharmacological classification of ion channels is that they can be regarded as pharmacological receptors. Drugs acting at defined sites that are components of the channel will have discrete structure–activity relationships and will have activator and antagonist properties, and the channels themselves, in common with other pharmacological receptors, should be subject to a variety of homologous and heterologous regulatory influences. Furthermore, it may be

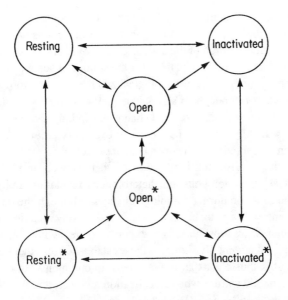

Figure 7. An ion channel depicted as existing in an equilibrium between resting, open, and inactivated states. Interconversion between these states is regulated via a set of time-, voltage-, and ligand-sensitive rate constants. Specific drugs may bind preferentially to one or more of these states (∗).

anticipated that ion channels should be altered in number or expression under pathological conditions and that some disease states may be specifically associated with changes in channel function. All of these expectations have been realized.

However, ion channels may present complexities of drug interaction in addition to those considered in conventional drug–receptor systems. The availability of or access to a drug binding site may be highly dependent upon the state of a channel—resting, open, or inactivated (Figure 7). A drug may exhibit significantly different affinities for these several states, and the binding of a drug may alter the equilibria existing between these states. According to this allosteric model of drug–channel interactions, the apparent affinity of a drug will depend upon the experimental or clinical conditions under consideration: conditions that shift an equilibrium to the favored (high affinity) binding state will increase the apparent affinity of the drug. Additionally, the availability of a drug binding state may depend upon a particular access pathway. Thus, binding sites for hydrophilic drugs may only be available during the open state of a channel, whereas hydrophobic drugs may enjoy continual access through the nonpolar membrane lipid pathway (Figure 7). A drug with preferential affinity for the inactivated state of a channel but which can obtain access to this state only during the open channel state will be a channel antagonist whose apparent affinity is increased during stimulation conditions, increased frequency, that favor the open channel equilibrium. The "modulated receptor" hypothesis of drug–channel interactions has thus proved to be useful to the interpretation of the actions of a number of drug classes including local anesthetics active at Na^+ channels for the control of some arrhythmic states and for the distinction in the pharmacological and therapeutic profiles of the Ca^{2+} channel antagonists (Hondeghem and Katzung, 1984). Such state-dependent behavior occurs also in receptor-operated channels. Binding of drugs such as phencyclidine and MK 801 to the NMDA class of excitatory amino acid receptors is enhanced upon presentation to the open channel state.

A further characteristic of ion channels when viewed as pharmacological receptors is that of homologous and heterologous regulation and alteration of expression and function during pathological states. Although the details of these regulatory events remain to be established, it is likely that ion channels, as membrane proteins, are biosynthesized, exported to the membrane, and regulated by mechanisms fundamentally similar to those established for other pharmacological receptors. It is thus clear that ion channels, in common with other excitable proteins, are dynamic entities whose localization, number, and function are regulated by a variety of influences (Hille, 1984).

SPECIFIC CLASSES OF ION CHANNELS

The direct control of ion channel function by drugs is assuming increased importance as a mode of therapeutic control of cell function. The regulation of Na^+

Table 2. Therapeutic Uses of Ca^{2+} Channel Antagonists

Use	Antagonist		
	Verapamil (I)[a]	Nifedipine (II)	Diltiazem (III)
Angina:			
exertional	+++[b]	+++	+++
Prinzmetal's variant	+++	+++	+++
Paroxysmal			
supraventricular			
tachyarrhythmias	+++	–	+++
Atrial			
fibrillation and flutter	++	–	++
Hypertension	++	+++	+
Hypertrophic cardiomyopathy	+	–	–
Raynaud's phenomenon	++	++	++
Cardioplegia	+	+	+
Cerebral			
vasospasm (post hemorrhage)	–	+	–

[a]Classes I, II, and III as defined by World Health Organization.
[b]Number of plus signs indicates extent of use: +++, being very common; –, not used.

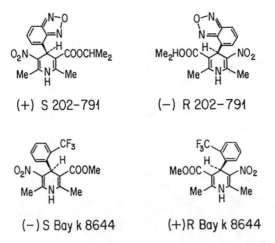

(+) S 202-791   (−) R 202-791

(−) S Bay k 8644   (+)R Bay k 8644

Figure 8. Structural formulas of enantiomeric 1,4-dihydropyridine Ca^{2+} channel antagonists and activators.

channels by local anesthetic and related antiarrhythmic agents and the control of Cl^- channels by benzodiazepines and related anxiolytic and anxiogenic agents are well established examples. More recently, the control of Ca^{2+} channels and the discovery of specific K^+ channel drugs have given greater importance to these latter channel categories (Hille, 1984).

The Ca^{2+} channel antagonists (see Figure 6) are major agents in the control of a number of cardiovascular disorders (Table 2). It is, however, likely that their therapeutic potential with second and third generation agents extends to a multiplicity of disorders from achalasia through migraine to vertigo. Such a wide range of therapeutic applications is totally consistent with the critical roles that cellular Ca^{2+} plays in the maintenance of cell viability and excitability (Janis et al., 1987). The demonstration that nimodipine, a 1,4-dihydropyridine analogue of nifedipine, is effective in animal models of memory enhancement is a further indication of the importance of drugs regulating Ca^{2+} channel function. Current evidence indicates that the L class of Ca^{2+} channels is an oligomeric assembly of five proteins of which a major component, the alpha-1 subunit, carries the drug binding sites arranged as an allosterically interacting assembly. The 1,4-dihydropyridine site is capable of accommodating both potent antagonist and activator 1,4-dihydropyridine molecules (Figure 8); this raises the issue of whether there may exist endogenous regulatory molecules whose functions are mimicked by the synthetic ligands. Interactions of the Ca^{2+} channel antagonists with the Ca^{2+} channel binding sites exhibit prominent use-dependence, and this underlies, at least in significant part, their heterogeneous pharmacological and therapeutic profiles, which exists despite their common interaction at the same channel subunit. 1,4-Dihydropyridine interactions are voltage dependent, activity increasing with the extent of membrane

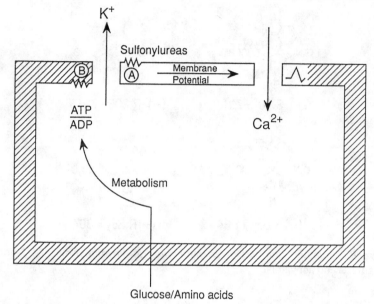

Figure 9. The regulation of ATP-dependent K⁺ channels. Depicted is the K⁺ channel with receptor sites for ATP/ADP and the sulfonylureas that both function as channel antagonists. Depolarization caused by channel closing is sensed by the voltage-dependent Ca^{2+} channels. The ATP/ADP levels are dependent on the metabolic state of the cell.

depolarization. This is consistent with a preferential interaction with an inactivated state of the channel that is favored by persistent depolarization. In contrast, the interactions of verapamil and diltiazem, charged molecules at physiological pH, are demonstrably frequency dependent, consistent with their interacting through an open channel state to produce channel antagonism (see Figure 7).

The L class of voltage-dependent Ca^{2+} channels is but one of several classes of channel. Its electrophysiological characteristics (see Table 1) of large conductance and slow inactivation are consistent with it playing a major role in stimulus–response coupling systems demanding the delivery of significant amounts of Ca^{2+} for prolonged periods of time. In contrast, the T channels, which are not well defined pharmacologically, are activatable only from polarized membrane potentials, inactivate rapidly, are of small conductance, and give rise to transient Ca^{2+} delivery. Such channels may be useful in underlying pacemaker and related rhythmic activities. The N category of voltage-dependent Ca^{2+} channel appears to be restricted to neuronal systems where its activation presumably underlies a variety of neurotransmitter release processes. Although no synthetic drugs are available for this category of channel, it is very sensitive to the polypeptide w-conotoxins secreted from fish-eating molluscs of the *Conus* genus. It may be anticipated that drugs available for the T and N classes of Ca^{2+} channel will expand significantly the therapeutic applications of this broad class of agents.

K^+ channels continue to represent a particularly complex situation (Cook, 1988). Channel classes in excess of a dozen have been characterized by electrophysiological techniques, and the concept of a multiplicity of channel types has been confirmed by molecular biology approaches (see p. 80). Particular interest attaches to K^+ channels characterized by two groups of synthetic drugs. ATP-dependent K^+ channels are associated with pancreatic beta cells and are closed in the presence of elevated levels of ATP. It is postulated that it is the ratio of intracellular concentrations of ATP and ADP that effectively controls the closed/open state of the channel. The K^+ channels of beta cells are maintained closed by insulin secretagogues, glucose and amino acids, which thus serve to depolarize the cell, activate voltage-dependent Ca^{2+} channels, and promote the release of insulin. Conversely, depletion of ATP levels will activate these channels, polarize the cell, and inhibit insulin release (Figure 9). The sulfonylureas, hypoglycemic drugs, including tolbutamide and glibenclamide, serve also as antagonists of ATP-dependent K^+ channels but act at a side distinct from that occupied by adenine nucleotides. The ATP-dependent K^+ channel is associated with other tissues including cardiac, neuronal, and skeletal muscle, and thus it may represent a metabolically sensitive ion channel opening and closing in response to states of energy depletion and repletion, respectively. Such an ion channel is likely to have an important role in the regulation of cellular function during anoxic and ischemic states.

Attention has also been focused on K^+ channels through the actions of a group of drugs referred to as K^+ channel activators (Cook, 1988). These agents, which include cromakalim, nicorandil, and pinacidil and possibly also the vasodilators minoxidil and diazoxide, serve to activate one or more classes of K^+ channels and thus to stabilize cell function. They have therapeutic applications as vasodilators and antihypertensive agents. Of particular interest, the side effects of minoxidil, diazoxide, and pinacidil all include the promotion of hair growth. Whether such growth depends on K^+ channel activity remains to be determined but is a potentially interesting side effect of investigations directed at elucidating channel function through selective drug action.

ION CHANNELS AND DISEASE STATES

Ion channels are subject to a variety of regulatory influences under both experimental and pathological conditions. Because of the therapeutic significance of the Ca^{2+} channel antagonists in cardiovascular diseases, the regulation of these channels has been more intensively studied (Ferrante and Triggle, 1990). Up and down regulation of channel densities and function have been achieved through chronic channel antagonism and activation, respectively. Channel numbers can be regulated experimentally by hormone treatment whereby up and down regulation of cardiac numbers occur in the hypo- and hyperthyroid states. Additionally, patients with hypertrophic obstructive cardiomyopathy shown an increased density of 1,4-dihydropyridine binding sites. The calcium channel antagonists are effective in the clinical treatment of the symptoms of alcohol withdrawal, and it is thus of interest that in experimental animals chronic alcohol administration results in an increased density of neuronal Ca^{2+} channels. In embryonic muscular dysgenic mice a recessive lethal gene results in the loss of 1,4-dihydropyridine binding sites and the associated slow calcium current in skeletal muscle. Microinjection of a plasmid containing complementary DNA encoding the 1,4-dihydropyridine receptor into dysgenic muscle cells restored both receptor sites and channel function.

In cystic fibrosis there is a defective production of respiratory tract fluid attributable to aberrant Cl^- secretion. This defect is at the level of the epithelial Cl^- channel, where aberrant control by β-adrenoceptor–linked adenylate cyclase results in a defective phosphorylation event.

Ion channels are altered under a variety of conditions ranging from chronic use to homologous and heterologous drug and hormone treatment to immune disorders and disease states. It is likely that an increasing number of diseases will be identified that are specifically associated with one or more ion channel defects.

CHANNEL STRUCTURE AND FUNCTION

The general properties of ion channels deduced from electrophysiological and pharmacological analyses indicate that they should be transmembrane proteins, that they should contain sites specifically associated with or linked to chemical- and voltage-sensing functions, and that it is likely that there will exist homologous families of structures representing related classes or types of channels. Structural analyses of a number of ion channels, both voltage- and ligand-dated, have provided general confirmation of these expectations (Barnard et al., 1987; Catterall, 1989).

Sequences of major proteins from Na^+, Ca^{2+}, and K^+ channels are available and reveal substantial homology both in sequence and in putative membrane spanning arrangements of this voltage-gated class. The putative arrangement of the major peptide of the Na^+ channel (approximately 1800–2000 residues according to type) is depicted in Figure 10. The K^+ and Ca^{2+} channels reveal similar arrangements. However, the K^+ channel possesses only one domain of six transmembrane helices, indicating that an oligomeric association of subunits may be involved in the organization of these channels. This arrangement may account for the apparent remarkable diversity of K^+ channels. In addition to the general similarity of organization and overall homology of these voltage-gated channels, there are highly conserved sequences that may indicate conserved function. In particular, the S4 region of each domain constitutes a sequence of one charged residue

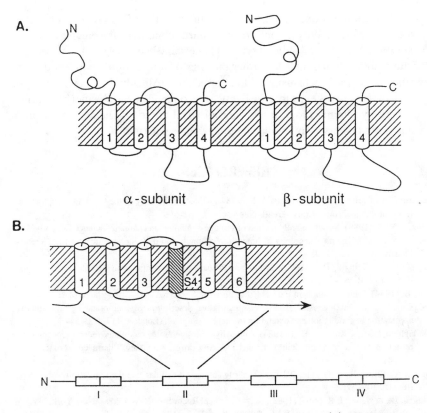

Figure 10. **A.** Organization of the alpha and beta subunits of the GABA$_A$ receptor channel complex. Each unit has four transmembrane helices, and it is suggested that two such structures constitute a functional channel. Not depicted is a third homologous subunit that carries the benzodiazepine receptor and which is substantially homologous to the units depicted here. **B.** Organization of a voltage-dependent Na$^+$ or Ca^{2+} channel. The peptide consists of four homologous domains (I, II, III, IV) each of which contains six transmembrane helices including the highly conserved S4 sequence. The K$^+$ channel consists of only a single domain but otherwise shows substantial homology to the Na$^+$ and Ca^{2+} channels.

followed by two nonpolar residues. These sequences may constitute the voltage sensors.

The sequences and arrangements of at least three ligand-gated ion channels are now available and include the nicotinic acetylcholine receptor, the glycine receptor, and the GABA$_A$ receptor channel complexes. These ligand-gated channels also appear to form one family with considerable homology both between the subunits of each receptor–channel complex and between members of the family. The arrangement of the alpha and beta subunits of the GABA$_A$ complex depicted in

Figure 10 constitute, probably as a tetramer, a functional GABA-gated Cl⁻ channel. Benzodiazepine sensitivity is conferred by a third subunit, but the precise association between these several subunit types is not established.

Future studies will delineate further the sites of drug action, endogenous and exogenous, and the mechanisms of ion permeation. Already, it is clear that two major channel families exist that are likely to accommodate the channel categories recognized by pharmacological and physiological criteria. Whether all channel structures can be thus accommodated remains to be determined.

REFERENCES

Barnard, E.A., Darlison, M.G., & Seeburg, P. (1987). Molecular biology of the GABA$_A$ receptor: the receptor/channel superfamily. Trends Neurosci. 10, 502–509.

Catterall, W.S. (1989). Structure and function of voltage-sensitive ion channels. Science 242, 50–62.

Cook, N.S. (1988). The pharmacology of potassium channels and their therapeutic potential. Trends Pharmacol. Sci. 9, 21–28.

Ferrante, J., & Triggle, D.J. (1990). Drug- and disease-induced regulation of calcium channels. Pharmacol. Revs. 42, 29–44.

Hille, B. (1984). Ionic Channels in Excitable Membranes. Sinauer Assoc., Sunderland, MA.

Hodgkin, A.L., & Huxley, A.F. (1952). A quantitative description of membrane current and its application to conduction and excitation in nerve. J. Physiol. (London) 117, 500–544.

Hondeghem, L.M., & Katzung, B.G. (1984). Antiarrhythmic agents: the modulated receptor mechanism of action of sodium and calcium channel blocking drugs. Ann. Rev. Pharmacol. Toxicol. 24, 387–423.

Janis, R.A., Silver, P., & Triggle, D.J. (1987). Drug action and cellular calcium regulation. Adv. Drug Res. 16, 309–591.

Sakman, B., & Neher, E.R. (eds.) (1983). Single Channel Recording. Raven Press, New York, NY.

Tsien, R.W., Hess, P., McCleskey, E.W., & Rosenberg, R. L. (1987). Calcium channels: mechanisms of selectivity, permeation and block. Ann. Rev. Biophys. Chem. 16, 265–290.

Chapter 5

Molecular Genetics and Evolution of Voltage-Gated Ion Channels

LAWRENCE SALKOFF

INTRODUCTION: THE BRAIN IS AN ELECTRONIC DEVICE DESIGNED TO FUNCTION IN WATER

The brain is an electronic device designed to function in an aqueous environment. Like all of organic evolution, the evolution of biophysical processes in the nervous system took place in the unique environment of molecular water. Water is a special solvent with polar qualities that define, order, and constrain the interactions of

Fundamentals of Medical Cell Biology, Volume 5A
Membrane Dynamics and Signaling, pages 83–98
Copyright © 1992 by JAI Press Inc.
All rights of reproduction in any form reserved.

macromolecules in a way that appears essential for the evolution of life. It is little wonder then that life began in the sea.

Because hydrated ions in water (and not electrons) carry an electrical current very well, the whole environment within and surrounding the brain and nervous system has the properties of a conductor. Special problems are presented for an electronic device required to function in such an environment. To get insight into the problem, the next time you take a bath, take a transistor radio in the tub with you. Of course, you will observe in this experiment that the radio ceases to function immediately. This is because the electrical current is no longer routed through the circuitry of the device. Instead, the current is shorted out instantly at every point and runs into the bath.

Another problem is powering the device (brain). Connecting the brain to batteries, or plugging it into the wall in order to distribute power to all parts of the device, is not possible. Instead, nature has designed each unit of circuitry (each neuron) to function as its own generator and battery. Thus, each individual cell in the nervous system generates and maintains an electrical gradient across its plasma membrane by expending energy in the form of ATP. The generator itself is the sodium/potassium ATPase, an ion pump that creates an ion gradient between the inside and outside of the cell. The ion gradient, principally for potassium ions, generates an electrical gradient of about 80 to 90 mV, with the inside of the cell negative relative to the outside. During the conduction of an action potential from cell to cell, each cell expends energy stored in its electrical gradient in order to propagate the electrical impulse across the length of the cell to the adjacent parts of circuitry. Thus, the brain is powered by internally stored energy distributed throughout the device, and not like a transistor radio, which is powered by a central energy source, a battery.

ION CHANNELS: ORGANIC TRANSISTORS THAT GATE AN ELECTRIC CURRENT IN AN AQUEOUS ENVIRONMENT

A fundamental component of all electronic devices is a rapidly activating switch to turn current flow on or off. In transistor radios and computers the device most commonly used is the transistor, which gates (switches on and off) a current flow carried by electrons. However, ions in solution in the brain, and not electrons, carry an electrical current. Thus, a special switching device is necessary to control this form of electrical current. The ion channel evolved in living systems just for this purpose. In this sense, the ion channel is truly the "transistor" of the brain.

Because extracellular fluids have a high sodium ion concentration, and intracellular fluids are high in potassium ion, these two types of ions carry most of the electrical current in the brain. It is no accident then that the two most common types of voltage-gated ion channels are selectively permeable to either sodium or potassium ions. *Voltage-gating* means that the channels open in response to a change in the voltage gradient across the membrane from the inside to the outside of the cell. This occurs, for example, during the spread of depolarization when the action potential travels along the cell membrane. Sodium channels gate an ion current that flows into the cell, following the concentration gradient for that ion. In contrast, potassium channels gate an electrical current that flows from the inside to the outside of the cell, following the concentration gradient for potassium ions. Both sodium and potassium ions carry a net positive charge. Thus, the effect of the sodium current is to carry positive charge into the cell, depolarizing it (making the inside of the cell more positive). This is the predominate current during the upstroke of the action potential. The potassium current carries net positive charge out of the cell, leaving the inside of the cell more negative. This occurs during the repolarization of the action potential that follows the upstroke of the action potential. The action potential itself is a wave of depolarization, initiated by the sodium current, that spreads along the cell membrane. This is the main mechanism for transmitting electrical signals over distances in the nervous system. The main point to remember is that the action potential is active current flow, powered by the energy stored in every cell. It has to be, because passive current flow, like the current in a copper wire, requires extraordinarily high insulation and, even so, peters out over distance. In the aqueous environment of the nervous system, the action potential mechanism ensures that, like a row of dominos, the signal arrives at the end of the line with undiminished intensity. The fundamental components that make this possible are the voltage-gated ion channels.

MOLECULAR GENETICS HAS SHARPLY ADVANCED THE STUDY OF ION CHANNELS

The study of the protein structures of voltage-gated ion channels has recently been sharply advanced by the application of molecular genetic techniques. The use of these techniques allows the study of the structures of proteins by analyzing the DNA sequence of their genes. The genome (the total DNA) of any living organism is a catalog of genes coding for all of the proteins required to build a complete organism. One of the breakthroughs of modern biology has been the development of techniques enabling the reading of these protein-coding gene "files." The techniques of molecular genetics and nucleic acid chemistry permit the study of protein primary structure in a systematic way that bypasses the most unreliable and difficult aspects of protein biochemistry. Indeed, DNA sequence information is so direct and valuable that projects are now underway to sequence the entire genomes of several animals including the human genome. Because the genetic code is universal among all life forms, gene sequence information integrates all fields of biology and medicine that have a focus on proteins and their functions. Not only does all life share a common genetic code, but the basic molecular machinery of life is also, in large part, shared by all life forms. This commonality is present because the vital innovations of molecular evolution occurred before the great divisions and radiations of vertebrate and invertebrate phyla more than 500 million years ago. Indeed, it was because of the evolution of these molecular processes that such a variation of life forms became possible. The molecular evolution of valuable innovations such as the citric acid cycle enabled life forms to assume energy profligate lifestyles and a myriad of forms. Ion channels also evolved prior to the divergence of vertebrate and invertebrate species and remain conserved today in most species. Exploiting this fact has been a great help in elucidating the molecular structure of these proteins.

Gene cloning and sequencing techniques have, thus, been adopted by neurobiologists to investigate the protein structures of ion channels. Recombinant DNA technology applied to these problems achieved a breakthrough in the early 1980s when the acetylcholine receptor from torpedo electroplax (Numa et al., 1983) was first cloned and sequenced. Soon to follow was the voltage-sensitive sodium channel from electric eel (Noda et al., 1984). Because the messenger RNA, as well as the protein itself, was relatively abundant for both of these channels in the respective tissues chosen, the cloning method of choice involved starting with a cDNA library. (A cDNA library contains the sequences of all the individual mRNAs copied into DNA, which is technically easier to work with.) Information from earlier protein purification and sequencing experiments had provided partial amino acid sequence data, which were used to select the correct cDNAs. The sequencing of these cDNAs yielded the complete deduced amino acid sequence for these channels. The cloning and sequencing of a calcium channel from the sarcoplasmic reticulum of rabbit muscle was achieved in a similar way (Tanabe et al., 1987).

MUTANT ANALYSIS OF THE *DROSOPHILA SHAKER* GENE: CLONING THE POTASSIUM CHANNEL

The approaches discussed in the preceding section worked partially because neurotoxins were available that bound to these channels with high affinity. These toxins aided in the isolation of small amounts of protein, which yielded bits of peptide sequence information. This sequence information was used for synthesizing oligonucleotide probes used in isolating cDNAs for the channels. For other channel types, however, where no toxins were available for channel protein isolation, the task of cloning presented more formidable obstacles. The approach that worked for cloning potassium channels was less direct and involved the exploitation of the fruit fly, *Drosophila melanogaster,* an animal system that can be manipulated genetically.

A variety of behavioral mutants of *Drosophila* had been produced that had the promise of being useful for neurobiological studies (Benzer, 1973; Suzuki, 1974). Some of these had behavioral defects that were most likely due to mutational

alterations in membrane excitability. Physiological studies of one of these mutations, the *Shaker* mutation, suggested that the mutation altered a gene coding for a
potassium channel. *Shaker* mutations produce a behavioral phenotype of poor
coordination and violent shaking upon exposure to ether.

The physiological effects of *Shaker* mutations were first studied at the larval
neuromuscular junction where it was found that the most extreme alleles cause an
abnormally large and asynchronous release of neurotransmitter (Jan et al., 1976).
The abnormal release was found to be due to prolonged depolarization at the mutant
nerve terminal causing a prolonged influx of calcium. Abnormal calcium channels
were ruled out as a cause of the defect, and it was proposed that the repolarization
defect was probably due to an abnormal potassium conductance; the potassium
channel blocker 4-aminopyridine applied to wild-type (normal) neuromuscular
preparations mimics the *Shaker* neuromuscular transmission defect.

Evidence for a repolarization defect in mutant neurons of adult flies came from
intracellular recordings of action potentials from the cervical giant fiber axons
(Tanouye et al., 1981). These recordings showed more directly that the *Shaker*
defect resulted in abnormally broad action potentials. As with larval neuromuscular
preparations, the mutant *Shaker* defect was mimicked in wild-type preparations
when 4-aminopyridine was added.

Direct evidence that *Shaker* mutations affected a potassium current required the
voltage clamp technique. The effects of the *Shaker* gene were most clearly observed
by voltage clamp studies of *Drosophila* flight muscles (Salkoff and Wyman, 1981;
Salkoff, 1983). These studies directly compared the potassium currents in mutant
and wild-type muscles and found that the mutation affected one particular type of
potassium current, a rapidly activating and inactivating potassium current called
the *A-current* (Connor and Stevens, 1971; Neher, 1971).

However, a major question was did the gene code for the channel itself or a
protein that modified the channel? This question was answered by mutant analysis.
If the *Shaker* locus were, indeed, a structural gene for this potassium channel then
three categories of mutant phenotypes, as assayed under voltage clamp, might be
expected to result from mutational alterations of the gene: (1) the complete
elimination of the current by deletions or nonsense mutations in the gene; (2) the
alteration of kinetic properties of the current by missense mutations that changed
the structure of the channel coded by the gene; and (3) the reduction in the amount
of an otherwise normal current by regulatory mutations that limited the amount of
gene expression. *Shaker* mutations were found to produce all three types (Salkoff,
1983). Examples of these mutant phenotypes, as assayed with the voltage clamp
system, are given in Figure 1. This mutant analysis strongly implied that the gene
did, indeed, code for a structural component of the potassium channel.

The subsequent cloning of the *Drosophila Shaker* locus by three independent
groups finally revealed the primary structure of this voltage-gated potassium
channel (Papazian et al., 1987; Kamb et al., 1988; Pongs et al., 1988). These
laboratories employed the chromosome *walking* strategy. The entry point into the

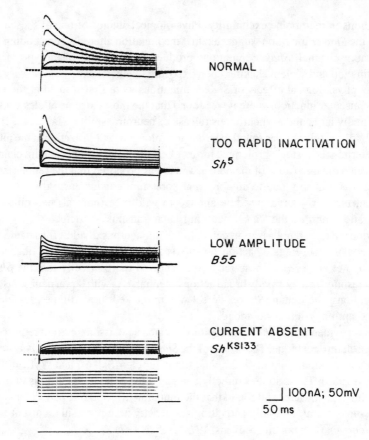

Figure 1. Voltage clamp experiments showing the fast transient potassium current in wild-type (normal) and *Shaker* mutant flies. The potassium current is shown as a deflection of the current trace in the upward direction. In Sh^5 the current is present but inactivates abnormally rapidly. The chromosomal breakpoint *B55* which falls in the *Shaker* region of the salivary chromosome map causes a marked reduction of current. Mutants like Sh^{ks133} completely eliminate the transient potassium current. All records shown are from the 72 h stage of pupal development when the voltage-gated transient potassium current is mature in normal wild-type flies. (See Salkoff, 1983.)

chromosome was a cloned segment of DNA mapping near the *Shaker* gene that was fortuitously available from a project unrelated to *Shaker*. This cloned DNA was used as a hybridization probe to isolate other larger cloned DNA sequences from a DNA library that contains randomly overlapping clones representing the entire *Drosophila* genome. Overlapping clones were then chosen that extended the greatest distance to the left and right along the chromosome. The left and right ends of these new clones were again chosen as new hybridization probes and the selection of new overlapping clones repeated. Thus, the map of cloned DNA was extended farther to the left and right. The process was repeated until the entire genomic region was cloned and the actual *Shaker* gene identified. The identification of the *Shaker* gene was made by the molecular mapping of *Shaker* mutations that fell within the cloned area and by isolating and sequencing cDNAs that mapped close to the mutations.

A SUPERGENE FAMILY INCLUDES BOTH SODIUM AND CALCIUM CHANNELS AS WELL AS POTASSIUM CHANNELS

Molecular and genetic evidence now suggests that most voltage-gated ion channels have a similar protein structure and probably a common evolutionary origin. Before any of the primary sequence data from cloning experiments was known, Bertile Hille suggested (1984) that potassium, calcium, and sodium channels evolved from a common origin, an ancestral cation channel in the earliest eukaryotes; thus he predicted common structural features. The hypothesis of a common evolutionary origin for these voltage-dependent channels is supported by similarities in their biophysical properties. For example; all of these channels are sharply responsive to voltage once a sufficient level of depolarization is reached but show virtually no activation in response to voltage changes that remain near the resting potentials of most cells. Considering similarities like these, it was not unexpected that molecular cloning experiments eventually revealed a common structure and molecular mechanism of voltage dependence for all of these channels.

Cloning and sequencing experiments have shown that both the sodium and calcium channels are composed of a long polypeptide having four internal repeated domains that are homologous but not identical (Figure 2A) (Salkoff et al., 1987; Barchi, 1988; Catterall, 1988). Each of these homology domains in Figure 2 is represented as a pie-shaped wedge. Each homology domain contains six hydrophobic segments, which are proposed to be alpha-helical transmembrane segments. The six transmembrane segments comprising a single homology domain are not indicated in Figure 2. The hydrophilic linker segments that connect the homology domains are postulated to be cytoplasmic whereas the four homology domains are hypothesized to contain all of the portions of the channel residing within the membrane. Thus the channels are composed of $4 \times 6 = 24$ membrane spanning domains surrounding the aperture of the channel.

Potassium channels differ in one important way: the gene product is a smaller polypeptide resembling only a single one of these homology domains (Figure 2B) containing six conserved hydrophobic regions. This has led to the speculation that the tertiary structure of the potassium channel is composed of four of these smaller polypeptides (Stevens, 1987; Agnew, 1988) and is, thus, analogous to the sodium and calcium channels.

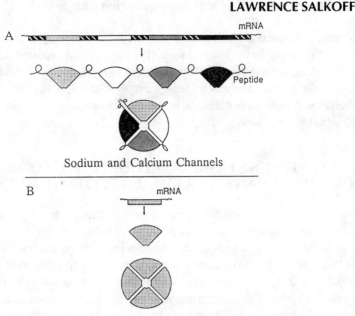

Figure 2. Diagrammatic representation of voltage-gated ion channels. **A.** Sodium and calcium channels. A single mRNA (top) codes for a single long polypeptide (middle) which has four repeated domains which are homologous but not identical. Each of the four domains codes for six hydrophobic regions which are proposed to be transmembrane segments. This single polypeptide folds into a bagel-like structure (bottom) which has an ion selective pore at its center. Thus, both the voltage-gated sodium and calcium channels are composed of 24 membrane spanning segments arranged in a pseudosymmetrical configuration in four groups of six. The structure is said to be pseudosymmetrical (Noda et al., 1984) rather than symmetrical because the four homology domains are not identical. **B.** Potassium channels (proposed). A single mRNA (top) codes for a smaller polypeptide (middle) which resembles a single repeated domain of the sodium or calcium channels. This has led to the speculation that the *Shaker* family potassium channels are composed of four small polypeptides (bottom) (Stevens, 1987; Agnew, 1988; Timpe, et al., 1988) and is, thus, a symmetrical structure.

DIVERSITY OF POTASSIUM CHANNELS: POTASSIUM CHANNEL GENES HAVE SUBFAMILIES

Potassium channels are virtually ubiquitous among the cells of eukaryotes and the most heterogeneous of the voltage-gated cation channels (Hille, 1984; Rudy, 1988). As modulators of many forms of electrical activity, potassium channels are involved in processes as diverse as the mechanisms of learning and memory (Alkon et al., 1982; Seigelbaum et al., 1982) and cardiac pacemaker activity (Giles and

Shibata, 1985). Because it was observed that alternative RNA processing of the *Shaker* gene produced a variety of related peptides, it was initially assumed that this might be the mechanism for producing the great diversity of potassium channel types (Schwarz et al., 1987; Agnew, 1988). Indeed, using the *Xenopus* oocyte expression system, it was found that alternative *Shaker* peptides do produce potassium channels having different kinetics (Timpe et al., 1988; Iverson et al., 1988). The observed kinetic differences, however, are not sufficient to account for the great diversity of potassium channel types seen in *Drosophila* muscle (Salkoff, 1985; Wu and Haugland, 1985; Zagotta et al., 1988) and nerve cells (Solc and Aldrich, 1988; Byerly and Leung, 1988).

More recently it was observed that a greater range of potassium channel diversity results from an extended gene family coding for homologous proteins; the *Shaker* gene is but one of at least four homologous genes in *Drosophila* (Butler et al., 1989; Wei et al., 1990). The proteins encoded by all of these genes, *Shal, Shaw, Shab,* and *Shaker,* share a similar organization in that each resembles one of the four repeated domains of the sodium or calcium channels as in Figure 2B. The cytological locations of these genes are widely scattered in the *Drosophila* genome (Butler et al., 1989).

EXPRESSION OF HOMOLOGOUS GENES SHOWS A WIDE RANGE OF BIOPHYSICAL PROPERTIES

The extended family of *Shaker*-like genes codes for potassium channels with a diverse range of biophysical properties. This was shown by expression of the genes in the *Xenopus* oocyte system as shown in Figure 3. The expressed currents differ with regard to all voltage-sensitive and kinetic properties. The *Shaker* current activates and inactivates very rapidly; the *Shaw* current activates and inactivates very slowly; the *Shal* current is intermediate between the two. Both the *Drosophila Shab* gene and the mammalian *Shab* homologue (Frech et al., 1989) have been expressed, and both have the properties of the delayed rectifier. The delayed rectifier is the first potassium current ever to be characterized (Hodgkin and Huxley, 1952). A more detailed description of the biophysical properties of these expressed channels is given in Wei et al. (1990). It appears that the multigene family of potassium channels has evolved in order to provide a much broader range of voltage sensitivity and kinetic diversity than can be produced by alternative splicing of the *Shaker* gene (see Covarrubias et al., 1991).

MORE EXTENSIVE GENE SUBFAMILIES EXIST IN MAMMALS

In mammals, it appears that gene duplication has produced subfamilies representing all of the voltage-gated ion channels. There appear to be at least three similar sodium channel genes expressed in rat brain, and at least two sodium

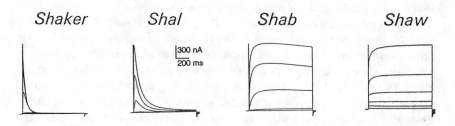

Figure 3. Expression of potassium currents in *Xenopus* oocytes injected with *Shaker*, *Shal*, *Shab*, and *Shaw* RNAs. Outward currents recorded in response to one second voltage step depolarizations ranging between –80 and +20 mV in 10-mV steps from a holding potential of –90 mV. While the *Shaker* and *Shal* currents turn on sharply at about –40 mV, the *Shaw* current has a shallower activation response to voltage and turns on at a more hyperpolarized potential (–60 mV). See Wei, et al., 1990, and Covarrubias et al., 1991. Current calibration is approximate only.

channel genes expressed in skeletal muscle. (There are at least two sodium channel genes expressed in *Drosophila*.) The number of calcium channel genes is unknown but could be at least as large. Presently one calcium channel gene is known to be expressed in skeletal muscle, and a similar but distinct gene is expressed in heart. It is expected that, in the mammalian brain, there are at least three separate calcium channel genes expressed. The number could be much larger.

For potassium channels, the number of similar genes is even greater. Because potassium channels fine-tune almost all aspects of membrane excitability, modifications of their structures are apparently necessary in different cells and tissues. Potassium channels are involved in controlling the height, duration, and rate of repolarization of the action potential. In cells that beat like clocks, potassium channels control the rate of beating. In cells that fire bursts of action potentials at regular intervals, potassium channels are involved in regulating both the interburst interval, as well as the duration of the burst. Apparently extensive gene duplication and specialization of structure occurred to produce the necessary differences.

Current indications now are that each of the *Drosophila* potassium channel genes, *Shaker, Shab, Shaw,* and *Shal* represent a subfamily of genes in mammals. It is not presently known how many members of each subclass there are, but there are at least 10 genes of the *Shaker* class expressed in mammalian brain (see McKinnon, 1989, for a description of two). There are presently known to be four separate genes expressed in brain that have deduced protein features similar to *Drosophila Shaw*, and at least one gene with features distinct to *Drosophila Shab*. The number of homologous potassium channel genes is almost certain to be much larger.

A UNIVERSALLY CONSERVED MECHANISM OF VOLTAGE-DEPENDENT GATING

Voltage-dependent activation of a voltage-gated channel is a property encoded by a specific structural feature that senses and initiates a conformational change in response to changes in transmembrane potential, resulting in the opening of the pore of the channel. These conformational changes are accompanied by a "gating current," a small transient outward current associated with channel opening, which was predicted by Hodgkin and Huxley (1952) and first experimentally measured by Armstrong and Bezanilla (1973). The movement of any charged particles creates an electrical current, even if the current is not carried by ions or electrons. In the case of the gating current, the charged particles that move are parts of the protein structure of the channel.

These physiological observations were the basis for predicting the existence of the positively charged "gating" region of the channel. The primary sequence of voltage-gated channels obtained through cloning and sequencing experiments revealed a region of regularly spaced positive charges common to all voltage-gated channels. Each string consists of a repeated pattern of a positively charged amino acid residue alternating with two uncharged residues (Figure 3). A similar string of charges is present in the fourth membrane spanning region (S4) in each of the four homology domains of the sodium and calcium channels. The charge strings are also present in the S4 region of all potassium channels. As can be seen from Figure 4, the positions of these positive charges are conserved across species.

Empirical and experimental observations suggest that these charges are, indeed, responsible for gating. It is possible that a general rule is that the more positive charges a channel has, the faster the channels open in response to a voltage change. Observations of the activation kinetics of the *Shaw* current, which, as shown in Figure 4, has a lower number of gating charges than either the *Shaker* potassium channel or the sodium channel, have revealed a slower activating current with a shallow activation curve, as predicted for a channel with a low number of gating charges. Site-directed mutagenesis of the charges in the sodium channel (Stuhmer et al., 1989) has also shown that channel gating is affected in a way that strongly implicates these charges in channel gating.

The "sliding helix" model (Guy and Seetharamulu, 1986; Catterall, 1988) is one model that has been proposed for these conserved structures in channel gating. According to this model, these charged strings form membrane-spanning alpha helices, with positively charged residues regularly spaced along the helix facing outward. Presumably, these charged helices are positioned within the channel protein, held in position by salt bridges to neighboring transmembrane segments with complementary positioned negative residues. Each S4 segment is thus held in a transiently stable conformation. In response to a change in the transmembrane electric field, the helix is postulated to rotate such that each positive charge of the helix is shifted into register with another complementary negative charge. A

SODIUM CHANNELS

```
       +          +          +          +          +          +          +          +
1626   R   V   I   R   L   A   R   I   G   R   I   L   R   L   I   K   G   A   K   G   I   R
RAT Brain I

       +          +          +          +          +          +          +          +
1439   R   V   I   R   L   A   R   I   G   R   I   L   R   L   I   K   G   A   K   G   I   R
RAT Muscle I

       +          +          +          +          +          +          +          +
1413   R   V   V   R   V   F   R   I   G   R   I   L   R   L   I   K   A   A   K   G   I   R
FLY

       +          +          +          +          +          +          +          +
1417   R   V   I   R   L   A   R   I   A   R   V   L   R   L   I   R   A   A   K   G   I   R
EEL
```

POTASSIUM CHANNELS

```
       +          +          +          +          +          +          +
 362   R   V   I   R   L   V   R   V   F   R   I   F   K   L   S   R   H   S   K   G   L   Q
FLY Shaker

       +          +          +          +          +          +          +
 362   R   V   I   R   L   V   R   V   F   R   I   F   K   L   S   R   H   S   K   G   L   Q
RAT BK1

                  +          +          +          +
 295   E   F   F   S   I   I   R   I   M   R   L   F   K   V   T   R   H   S   S   G   L   K
FLY Shaw
```

CALCIUM CHANNEL

```
       +          +          +          +          +
 883   K   I   L   R   V   L   R   V   L   R   P   L   R   A   I   N   R   A   K   G   L   K
RABBIT Muscle
```

Figure 4. Conservation of gating charge strings from all voltage-gated ion channels. Positive residues presumed to be gating charges are bold type and labeled with a +. Single letter abbreviations for the amino acid residues are: A, Ala; C, Cys; D, Asp; E, Glu; F, Phe; G, Gly; H, His; I, Ile; K, Lys; L, Leu; M, Met; N, Asn; P, Pro; Q, Gln; R, Arg; S, Ser; T, Thr; V, Val; W, Trp; and Y, Tyr.

rotation of the helix would result in an outward movement of all the charges in the string across the plane of the membrane. The net result would be the outward movement of positive charge detectable as the gating current. The detection of this current, which precedes the ion current through the channel, was an important verification of the fact that part of the structure of the channel moved in response to a voltage change, before the channel passed an ionic current.

GENE DUPLICATION:
NATURE MAKES THE MOST OF A GOOD THING

It has been suspected for some time that membrane proteins mediating transmission of information in the brain would fall into genetically related families (Hille, 1984;

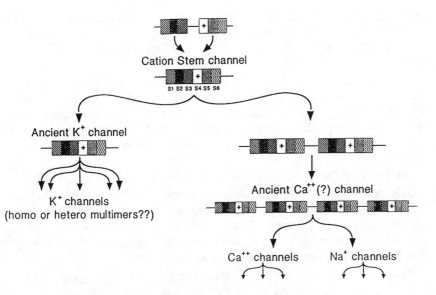

Figure 5. Possible evolutionary descent of voltage-gated cation channels.

Stevens, 1987). This has now been proven to be the case for ligand-gated as well as voltage-gated channels. Thus, the acetylcholine receptor is a member of a superfamily of channels that includes both the GABA as well as the glycine receptor subfamily. It should be pointed out that gene duplication has produced not only the separate channel subfamilies but also the specialized subunits within each subfamily. The nicotinic acetylcholine receptor in muscle is a pentamer composed of the products of four separate, yet homologous, genes (one of the subunits, alpha, is used twice in each channel). In the mammalian brain, the molecular structure of the acetylcholine receptor is distinct in that it is composed of the subunits of homologous genes that are expressed only in the CNS. Thus, in mammals, selective pressure in both muscle and CNS produced separate genes coding for nicotinic acetylcholine receptor subunits. The same pattern of molecular evolution involving gene duplication is seen in the supergene family coding for rhodopsin/beta-adrenergic/muscarinic receptors.

For voltage-gated ion channels, again, a similar evolution has occurred; there is a superfamily of genes that includes sodium, calcium, and potassium, and subfamilies of more closely related genes within each category. Figure 5 summarizes the probable evolutionary line of descent of modern voltage-sensitive channels from a common origin, a primitive nonselective cation channel, which itself probably evolved from earlier genes coding for peptide segments corresponding to regions S1 through S6. In their evolution, sodium and calcium channels possibly underwent two rounds of intragenic duplication; the end result is the modern form

of the genes containing four homology domains within a single large gene. Hille suggested that calcium selectivity may have evolved prior to sodium selectivity because of the vital role that calcium plays in cell metabolism, but this is, by no means, certain.

Potassium channels evolved along a different pathway. It is likely that modern potassium channels resemble the early cation channel—probably a homotetramer. Thus, each potassium channel gene codes for only a single homology domain, but four of the gene products form a complete channel. The formation of hybrid channels among subunits from different genes is restricted (Covarrubias, et al., 1991).

Because both *Drosophila* and mammals have the same highly conserved genes coding for both sodium and potassium (and probably calcium channels), it is almost certain that the genes evolved to virtually their modern form prior to the separation of vertebrates and invertebrates. This means that even the individual members of the potassium channel gene family had evolved their specialized features by the Cambrian period, about 600 million years ago.

Over and over again nature has used gene duplication for expanding and modifying the functional roles of a rare and uniquely valuable protein motif. Examples of convergent evolution in molecular evolution are rare; most enzymes and proteins that perform similar tasks are the products of genes with a common evolutionary origin. Molecular evolution rarely reinvents the wheel.

With regard to the large number of similar, yet distinctive, ion channel genes, a challenge that remains is to piece together an understanding of how all types fit into the pattern of animal behavior and cell metabolism. For more than two decades following the pioneering work of Hodgkin and Huxley (1952) nearly all of our understanding of what ion channels are and how they work was obtained by biophysical investigations. During this period the horizon of our understanding of membrane electrical properties barely extended beyond one single type of sodium and potassium channel. With the great increase in knowledge that molecular genetic techniques has added, a substantial question is presented: how are all these different gene products used in customizing membrane electrical properties in the nervous system, thereby generating the enormously complex and sophisticated patterns of animal behavior?

REFERENCES

Agnew, W.S. (1988). A rosetta stone for K channels. Nature 331, 114–115.

Alkon, D.L., Lederhendler, I., & Showkinas, J.J. (1982). Science 215, 693.

Armstrong, C., & Bezanilla, F. (1973). Currents related to movement of the gating particles of the sodium channels. Nature 242, 459–461.

Barchi, R. (1988). Probing the molecular structure of the voltage-dependent sodium channel. Ann. Rev. Neurosci. 11, 455–495.

Benzer, S. (1973). Genetic dissection of behavior. Sci. Amer. 229, 24–37.

Butler, A., Wei, A., Baker, K., & Salkoff, L. (1989). A family of putative potassium channel genes in *Drosophila*. Science 243, 943–947.

Byerly, L., & Leung, H.-T. (1988). Ionic currents of *Drosophila* neurons in embryonic cultures. J. Neurosci. 8, 4379–4393.

Catterall, W. (1988). Structure and function of voltage-sensitive ion channels. Science 242, 50–61.

Conner, J.A., & Stevens, C. F. (1971). Voltage clamp studies of a transient outward membrane current in gastropod neural somata. J. Physiol. 213, 21–30.

Covarrubias, M., Wei, A., & Salkoff, L. (1991). *Shaker, Shal, Shab* and *Shaw* express independent K^+ current systems. Neuron 7, 763–773.

Frech, G., Vandongen, M., Schuster, G., Brown, A., & Joho, R. (1989). A novel potassium channel with delayed rectifier properties isolated from rat brain by expression cloning. Nature 340, 642–645.

Giles, W.R., & Shibata, E.F. (1985). Voltage clamp of bull-frog cardiac pace-maker cells: a quantitative analysis of potassium currents. J. Physiol. 368, 265.

Guy, B., & Seetharamulu, P. (1986). Molecular model of the action potential sodium channel. Proc. Natl. Acad. Sci. USA 83, 508–512.

Hille, B. (1984). Ionic Channels of Excitable Membranes. Sinauer Associates, Sunderland, MA.

Hodgkin, A.L., & Huxley, A.F. (1952). Currents carried by sodium and potassium ions through the membrane of the giant axon of Loligo. J. Physiol. 116, 449–472.

Iverson, L.E., Tanouye, M.A., Lester, H.A., Davidson, N., & Rudy, B. (1988). A-type potassium channels expressed from Shaker locus cDNA. Proc. Natl. Acad. Sci. USA 85, 5723–5727.

Jan, Y.N., Jan, L.Y., & Dennis, M.J. (1977). Two mutations of synaptic transmission in *Drosophila*. Proc. R. Soc. Lond. B. 198, 87–108.

Kamb, A., Tseng-Crank, J., & Tanouye, M.A. (1988). Multiple products of the *Drosophila Shaker* gene may contribute to potassium channel diversity. Neuron 1, 421–430.

McKinnon, D. (1989). Isolation of a cDNA clone coding for a putative second potassium channel indicates the existence of a gene family. J. Biol. Chem. 264, 8230–8236.

Neher, E. (1971). Two fast transient current components during voltage clamp on snail neurons. J. Gen. Physiol. 58, 36–53.

Noda, M., Shimizu, S., Tanabe, T., Takai, T., Kayano, T., Ikeda, T., Takahashi, H., Nakayama, H.; Kanoak, Y., Minamino, N., Kangawa, K., Matsuo, H., Raftery, M.A., Hirose, T., Inayania, S., Hayashida, H., Miyata, T., & Numa, S. (1984). Primary structure of *Electrophorus electricus* sodium channel deduced from cDNA sequence. Nature 312, 121–127.

Numa, S., Noda, M., Takahashi, H., Tanabe, T., Toyosato, M., Furutani, Y., & Kikyotani, S. (1983). Molecular structure of the nicotinic acetylcholine receptor. Cold Spring Harbor Symp. Quant. Biol. XLVIII, 57–69.

Papazian, D.M., Schwarz, T.L., Tempel, B.L., Jan, Y.N., & Jan, L.Y. (1987). Cloning of genomic and complementary DNA from *Shaker*, a putative potassium channel gene from *Drosophila*. Science 237, 749–753.

Pongs, O., Kecskemrthy, N., Muller, R., Krah-Jentgens, I., Baumann, A., Kiltz, H.H., Canal, I., Llamazares, S., & Ferrus, A. (1988). *Shaker* encodes a family of putative potassium channel proteins in the nervous system of *Drosophila*. EMBO J. 7, 1087–1096.

Rudy, B. (1988). Diversity and ubiquity of K channels. Neuroscience 25, 729–749.

Salkoff, L.B., & Wyman, R.J. (1981). Genetic modification of potassium channels in *Drosophila* Shaker mutants. Nature 293, 228–230.

Salkoff, L. (1983). Genetic and voltage-clamp analysis of a *Drosophila* potassium channel. Cold Spring Harbor Symp. Quant. Biol. 48, 221–231.

Salkoff, L. (1985). Development of ion channels in the flight muscles of *Drosophila*. J. Physiol. Paris 80, 275–282.

Salkoff, L., Butler, A., Wei, A., Scavarda, N., Baker, K., Pauron, D., & Smith, C. (1987). Molecular biology of the voltage-gated sodium channel. TINS 10, 522–527.

Schwarz, T.L., Tempel, B.L., Papazian, D.M., Jan, Y.N., & Jan L.Y. (1988). Multiple potassium-channel components are produced by alternative splicing at the Shaker locus in *Drosophila.* Nature 331, 137–142.

Siegelbaum, S., Camardo, J., & Kandel, E. (1982). Serotonin and cyclic AMP close single K^+ channels in Aplysia sensory neurones. Nature 299, 413–416.

Solc, C.K., and Aldrich, A.W. (1988). Voltage-gated potassium channels in larval CNS neurons of *Drosophila.* J. Neurosci. 8, 2556–2570.

Stevens, C. (1987). Channel families in the brain. Nature 328, 198–199.

Stuhmer, W., Conti, F., Suzuki, H., Wang, X., Noda, M., Yahagi, N., Kubo, H., & Numa, S. (1989). Structural parts involved in activation and in activation of the sodium channel. Nature 339, 597–603.

Suzuki, D. (1974). Behavior in *Drosophila melanogaster,* a geneticist's point of view. Can. J. Gen. Cytol. 16, 713–735.

Tanabe, T., Takeshima, H., Mikanii, A., Flockerzi, V., Takahashi, H., Kangawa, K., Kojima, M., Matsuo, H., Hirose, T., & Nunia, S. (1987). Primary structure of the receptor for calcium channel blockers from skeletal muscle. Nature 328, 313–318.

Tanouye, M.A., Ferrus, A., & Fujita, S. C. (1981). Abnormal action potentials associated with the *Shaker* complex locus of *Drosophila.* Proc. Natl. Acad. Sci. USA 8, 6548–6552.

Timpe, L. C., Schwarz, T.L., Tempel, B.L., Pazian, D.M., Jan, Y.N., & Jan, L.Y. (1988). Expression of functional potassium channels from *Shaker* cDNA in *Xenopus* oocytes. Nature 331, 143–145.

Wei, A., Covarrubias, M., Butler, A., Baker, K., Pak, M., & Salkoff, L. (1990). K^+ current diversity is produced by an extended gene family conserved in *Drosophila* and mouse. Science 248, 599–603.

Wu, C.-F., & Haugland, F.N. (1985). Voltage clamp analysis of membrane currents in larval muscle fibers of *Drosophila*: alteration of potassium currents in *Shaker* mutants. J. Neurosci. 5, 2626–2640.

Zagotta, W. N., Brainard, M. S., & Aldrich, R. W. (1988). Single-channel analysis of four distinct classes of potassium channels in *Drosophila* muscle. J. Neurosci. 8, 4765–4779.

Chapter 6

Evolution of Receptors

GYÖRGY CSABA

THE FUNDAMENTAL RECOGNITION STRUCTURES

Recognition is an inherent property of all living systems, and as such the main prerequisite of viability. It equally operates in intracellular systems, intercellular contacts, and interactions between cell and environment. Although the interactions between receptor and ligand, or receptor and marker, play a decisive role in the mechanism of cellular recognition, molecular recognition had in all probability preceded the evolution of recognition structures in the living systems. Interactions associated with the linking of molecules, with the complementarity of the two strands of DNA, and with the enzyme–substrate reaction equally involve recognition mechanisms, which do not, at this level, seem to presuppose the operation of an independent signal but utilize other cellular potentials. Appearance on the DNA double strand of methyl groups recognizable by the restriction enzymes signifies

Fundamentals of Medical Cell Biology, Volume 5A
Membrane Dynamics and Signaling, pages 99–112

the development of a special signaling function, and the same applies to the development of the signal sequences of polypeptides, which account for entry into the endoplasmic reticulum, and for linking with sugar groups, which determine the further passage of the polypeptide. The elements of this primitive recognition system seem to furnish the basis for the development of the intra- and extracellular receptorial recognition systems, which do themselves involve polypeptide and sugar structures, and certain basic enzymic actions.The glycoprotein receptor and marker system of the intracellular compartments controls not only the transportation of the products but also the contacts between the compartments (and thereby the protection of the cell). Merging of certain elements of the compartments with the cytoplasmic membrane may contribute to the formation of membrane receptors that maintain contact with the environment. Logical as this explanation might seem, it is in all probability valid only for eukaryotic cells, because no intracellular membrane system is present in prokaryotes. Because, however, prokaryotes, too, are capable of recognition functions associated with their cytoplasmic membrane, the hypothesis seems plausible that the limiting membranes of the compartments are internalized details of the cytoplasmic membrane, which represents, in this light, the more primordial structure.

It follows from the preceding considerations that receptors (i.e., molecular structures capable of recognition) are carried by all cells, in varying quality and quantity depending on the internal and external environmental needs of the given cell. Because the quality of a receptor depends largely on the quality and quantity of the signal molecule, the signal–receptor interaction can give rise to development of particular systems, such as the immune system and the endocrine system. The immune system utilizes for differentiation of self from non-self an antigen (or superficial cellular marker) as a signal and the receptors of a coherent cell system as signal receivers, whereas the endocrine system utilizes a hormone (a cell-released marker) as signal and the receptors of different organs, whose functional interrelatedness takes effect only at the level of the macro-system, as receivers.

DEVELOPMENT OF THE SIGNAL MOLECULE–SIGNAL RECEIVER SYSTEM

It thus seems logical that only the cells equipped with a *general* signal receiver potential can acquire the property of *special* signal reception (i.e., phylogenetic development involves a delimitation, yet simultaneous specialization, of the receptor function). The understanding of this process presupposes the experimental study of low living systems, in which specificity is irrelevant, but the potentials of receptor specialization already do exist. Such living systems are the protozoa (Csaba, 1980, 1985, 1986a).

The ciliated unicellulars are primordial organisms, which have existed for about two myriads of years. The ciliated unicellular *Tetrahymena* is able to respond to hormones of higher vertebrates (e.g., to histamine and serotonin by an increased phagocytic activity [Csaba and Lantos, 1973], to thyroxin and its precursors by an increased mitotic rate [Csaba and Németh, 1980], to the sugar metabolism–influencing vertebrate hormones such as insulin [Csaba and Lantos, 1975a] and epinephrine [Csaba and Lantos, 1976] by increased glucose storage), and also shows certain non-specific responses to hormones (Csaba, 1985). Even unicellular plants such as *Acetabularia* (Legros et al., 1975), fungi (Loose and Feldman, 1982; Loose et al., 1983), or *Neurospora* (Fawell et al., 1988; McKenzie et al., 1988) possess receptors capable of specific hormone binding. Primary interaction with the hormone induces in *Tetrahymena* a specific reaction known as *hormonal imprinting* (Csaba, 1980, 1984, 1985), which accounts for a changed—usually increased—response to the hormone at later interactions and persists over several hundreds of offspring generations (Csaba et al., 1982).

Although *Tetrahymena* does itself contain several vertebrate hormones or similar molecules (LeRoith et al., 1980, 1982, 1983), these do not seem to play a role in its regulation but appear to be products of its synthetic activity and cannot, as such, explain the presence at the unicellular level of receptors for active molecules characteristic of vertebrates. A more likely explanation is that the dynamic (fluid mosaic) membrane structure of *Tetrahymena* involves a continuous assembly or reassembly of patterns (Koch et al., 1979) complementary to hormone molecules, and these patterns are transformed to specific receptors by primary interaction with the hormone (Figure 1).

The chemical stimuli acting on the aquatic unicellular in its natural environment represent a wide range of advantageous and adverse influences, whose recognition is vitally important for the protozoa. Recognition of the environment by the dynamic membrane pattern makes possible the nutrition of the unicellular, and its escape from noxious influences as well. *Tetrahymena* recognizes not only vertebrate hormones, but also other exogenous molecules of importance. It follows that the membrane pattern of the unicellular animal is genetically encoded for differentiation in the presence of a wide variety of environmental influences, and imprinting induces a "memory" of the primary interaction, which is transmitted to the progeny generations (Csaba, 1986a). The life span of the unicellular being extremely short, the information acquired by imprinting makes sense only if it is stored and transmitted to daughter cells in due course of time.

The receptor "memory" of the unicellular is in many respects similar to neuronal memory. Extreme stimuli may disturb it, and repetition stabilizes it more efficiently than a single lasting stimulus (Csaba et al., 1984). The material substrates involved in imprinting and receptor memory are not yet known. Evidence is lacking as to whether the receptor memory persists at the receptor, cytoplasmic, or gene level. Although it is known that the normal course of imprinting presupposes an intact cell membrane (Kovács, 1986), partial "peeling" of the cell does not cause loss of memory nor does it prevent its reappearance (Csaba et al., 1986). Transmission of the memory of imprinting may in principle take place via cell division but is

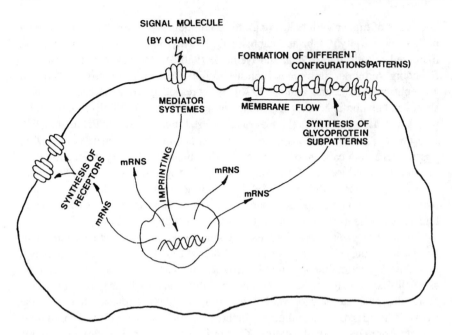

Figure 1. There is a continuous membrane flow of receptor subpatterns helping the formation of different receptor patterns. The signal molecule present in the milieu can bind to the binding site having an adequate configuration. Through mediator systems the information is stabilized in the form of hormonal imprinting, and hence the cell develops a receptor "memory."

mediated by the fluid (nutrient) medium as well. Evidence has in fact been obtained that *Tetrahymena* cells once imprinted transmit the information associated with imprinting not only vertically to daughter cells, but also horizontally, by means of a transfer factor released into the medium, to that part of the population that had not been involved in the interaction (Csaba and Kovács, 1987a and b).

It appears that the existence and function of the unicellulars' receptor system depend on those circumstances that determine the existence of the unicellular organism itself. It follows that the development of multicellular organisms involves in all probability an alteration of the receptor mechanism, to cope with the change in environmental conditions. In multicellular organisms the immediate environment of the cells is furnished by the internal milieu, which is genetically determined, balanced, and largely independent of external influences. In this milieu, the signal molecule is released by a cell that is determined by the same genome as the target cell (Figure 2) and forms an element of the same organism as the latter (Csaba, 1986a and b). This delimits the range of environmental stimuli, and

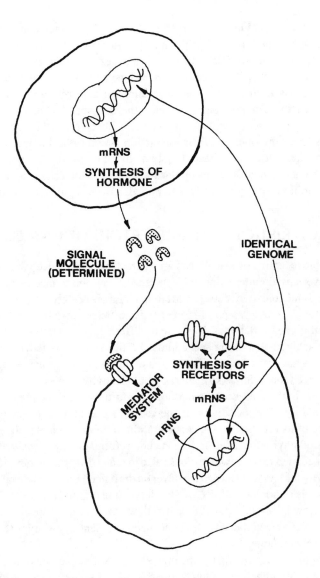

Figure 2. In multicellular organisms, the signal molecule and the receptor are determined by identical genomes in different cells. Nevertheless, the reinforcement of the link between hormone and receptor (hormonal imprinting) is necessary in the perinatal period for the normal development of the receptor.

consequently the receptor function becomes integrated into a system of division of labor. Thus, the genes encoding the receptors furnish not so much the dynamic properties as the stability of the membrane patterns and integration thereof into the diapason of the multicellular organism.

In the course of evolution, the structures functioning as receptor and hormone had equally developed from originally not specific molecules. The establishment of the receptor–hormone relationship presupposes simultaneous presence and structural complementarity of the future hormone and future receptor, as well as the advantageous influence of their interaction on the organism in which they interact. If all these requirements are met, the non-specific membrane pattern transforms to a receiver (receptor), and the molecule carrying the signal potential transforms to a hormone, and their interaction becomes an integral part of the function of the multicellular macro-system. In other words, the open receptor program characteristic of the unicellular transforms to a closed program to cope with the conditions of a genetically determined, strictly regulated environment.

CHARACTERISTICS OF THE RECEPTORS OF UNICELLULARS

Although the receptors of unicellulars are characteristically selective, they do not exclude overlaps of interacting molecules, such as those that occur between agonists and antagonists, or between different members of a given hormone family. The receptor of *Tetrahymena* is scarcely less efficient in this respect than the receptors of higher vertebrates. It is able to differentiate histamine and serotonin from their antagonists (Csaba and Lantos, 1975b) and has practically the same difficulties as have vertebrate cell receptors in differentiating gonadotropins from thyrotropins (Csaba, 1985). As a matter of fact, selectivity applies primarily to those receptors that in the presence of a hormone conform to the principle of structural complementarity. The sensitivity of the receptor can be best characterized by dependence of its function on the concentration of the signal molecule. Although 10^{-9} M diiodothyronine does induce receptor formation in *Tetrahymena* and account for an increased growth response also in many offspring generations, 10^{-18} M diiodotyrosine has no influence whatever on the unicellular (Csaba et al., 1982b). However, cells imprinted with 10^{-9} M diiodotyrosine respond also to 10^{-8} M by an increased growth rate (Csaba, 1985). It follows that receptor specificity can be increased at the unicellular level by a mechanism essentially similar to perinatal imprinting in vertebrates.

It should be pointed out that at the unicellular level not so much the hormonal as the structural properties of the molecule (i.e., its complementarity with a membrane pattern of the unicellular) play a decisive role in the mechanism of interaction. For this reason a hormone precursor may have a greater impact on a given function of the unicellular than the hormone proper (Csaba and Németh, 1980). A characteristic example is provided by the thyroxine series, in which

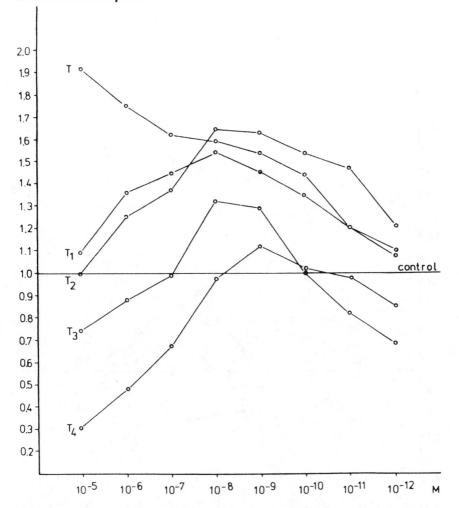

Figure 3. Effect of thyroxine and its precursors on the growth rate of *Tetrahymena*. T = tyrosine, T_1 = monoiodotyrosine, T_2 = diiodotyrosine, T_3 = triiodothyronine, T_4 = thyroxine. The highest peak is given by T_2 at 10^{-8} M. A similar effect is obtained with T and T_1 in this conception. (From Csaba & Németh, 1980.)

diiodotyrosine stimulates the division of *Tetrahymena* to a greater degree than thyroxine (or monoiodotyrosine). Occasionally, even the basic amino acid (tyrosine) may develop a hormone-like action at the unicellular level. Thyroxine and tryptamine enhance cell division to a similar extent as diiodotyrosine and serotonin, respectively, at the same concentration. However, although the dose-response relationship is linear in the case of the amino acids, the hormones proper

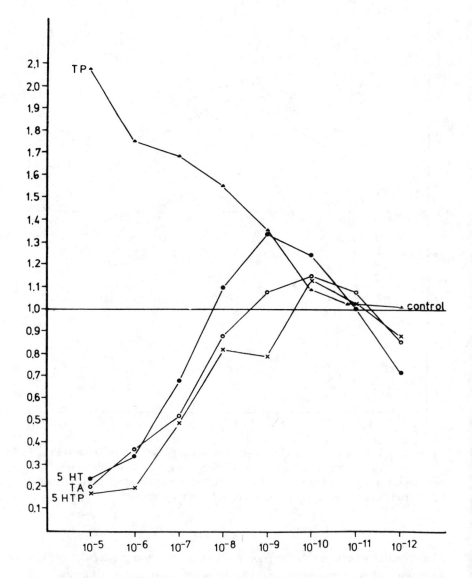

Figure 4. Effect of serotonin and its precursors on the growth rate of *Tetrahymena*. TP = tryptophan, TA = tryptamine, 5HTP = 5-hydroxytryptophan, 5HT = serotonin. The highest peak is caused by 5HT at 10^{-9} M, corresponding to the same rate by TP; however 5HT is toxic in the higher concentrations in opposition to TP, which further increases the growth rate in these concentrations. (From Csaba & Németh, 1980.)

are ineffective in suboptimal concentrations and toxic at supraoptimal ones (Figures 3 and 4).

EVOLUTION OF RECEPTOR–HORMONE CONNECTIONS

Because the basic membrane receptors of unicellulars are in all probability the nutrient receptors (Lenhoff, 1968), there is reason to postulate that hormone receptors arise by transformation of amino acid receptors under favorable conditions. Experimental evidence of induction of imprinting by amino acids supports this hypothesis (Csaba and Darvas 1986, 1987). The imprinting potential of the different amino acids is, naturally, non-uniform; those noted for a powerful imprinting potential (e.g., proline) are important structural elements of vertebrate hormones. The nutrient receptors, on the other hand, display a marked selectivity; they are even able to differentiate between D- and L-amino acids. Of many oligopeptides (e.g., enkephalin substances) tested, those noted for most powerful physiological effects in higher organisms proved to be the most efficient imprinters at the unicellular level (Csaba et al., 1986). This supports the hypothesis that the signal molecules acquire the signal property not so much by chance coincidence as by virtue of certain structural properties, which express receptor formation provoking potentials. This can explain the development of hormone families (e.g., thyrotropin, folliculotropin, luteotropin, and chorionic gonadotropin; vasopressin and oxytocin; gastrin, cholecystokinin, secretin, VIP, and glucagon; ACTH, melatonin, and enkephalin) and the development of amino acid hormones from certain amino acids (tyrosine, tryptophan) as well.

In the light of the theory that phylogenetically ancient cells are genetically able to form a receptor for any ligand, whereas the cells of higher organisms can form only a given specific receptor, it seems self-evident that an established receptor structure may persist for millions of years. For example, the insulin receptor of the hagfish does respond to pig insulin, although the evolutionary line of the two species had separated 500 million years ago (Muggeo et al., 1979a and b). The hagfish insulin differs from porcine insulin in 38% of its components and is 90–95% less active in pig. The hagfish responds more intensively to self- than to porcine insulin but is able to bind the latter, too. Thus, not so much the insulin receptor as the insulin itself had changed in the course of evolution, from which it follows that although the structure of the receptor is highly conserved, that of the ligand is subject to evolutionary changes. This accords well with the previously mentioned greater responsiveness of protozoa to hormone precursors than to the hormone proper, which is also true at higher levels of phylogenesis. A prime example is the pigment cells of *Planaria*. They respond to serotonin, which is present in *Planaria*, and to melatonin which is not present in it but serves as a melatonin precursor in higher organism (Csaba et al., 1980). The greater the effect of serotonin relative to melatonin, the more effective is the precursor molecule tryptamine compared to serotonin.

EVOLUTION OF THE RECEPTOR–MEDIATOR SYSTEM

Reception of a signal involves not only binding of the signal molecule to the receptor, but also mediation of the signal by an adequate effector system. In higher organisms, the adenylate cyclase–cAMP system and/or the Ca^{2+}-calmodulin system, the inositol-trisphosphate and diacylglycerol system serve as signal mediators.

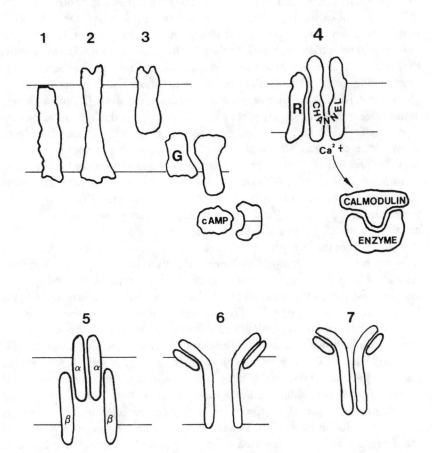

Figure 5. The possible evolutionary forms of receptors: **1** = oncogene receptor; **2** = EGF receptor; **3** = a receptor working through the adenylate cyclase–cAMP system, **4** = Ca^{2+} calmodulin system with receptor linked; **5** = the insulin receptor; **6** = membrane-bound form of the antibody; and **7** = secreted form of antibody.

The cAMP–adenylate cyclase system already appears at the phylogenetic level of bacteria, and release of cAMP represents for the bacterium the hunger signal. Appearance of cAMP in the immediate environment of *Dictyostelium* induces aggregation of the cells (i.e., colony formation). The adenylate cyclase–cAMP and inositol-trisphosphate systems are equally operative in *Tetrahymena* cells and may collaborate in the mediation of signals presented by released molecules or receptors (Kovács,1986). Linking or collaboration of the mediator systems with the receptor may, however, change in the course of evolution.

If the receptor–ligand relationship is regarded as an issue of the enzyme–substrate relationship, one might postulate that one end of the enzyme molecule may function as the receptor and the other as the enzyme. Single-unit receptors, as exemplified by the receptor of the epidermal growth factor (EGF), operate in all probability in this way, in that one limb of it binds the signal (hormone), whereas the other limb (the protein kinase) induces the response of the cell. Although firm proof is lacking, it seems highly probable that evolution proceeds further on in two directions, in that both the mediator limb and the receptor limb become longer.

Elongation of the mediator limb could account for the involvement of either cAMP or Ca^{2+} in the operation of the system, because this would presuppose the collaboration of an additional binding (receptor) protein, which, whether it is a cAMP-binding structure or calmodulin, accounts in any event for a two-step mediation. Prolongation of the receptor limb involves integration of an intermediate protein (e.g., G protein) between receptor and effector (enzyme), and a multi-step course of reception, such as that produced by tetrameric receptors (e.g., the insulin receptor). It is known that the reception of insulin involves an intermediate step of autophosphorylation, from which it follows that elongation of the receptor limb also has an impact on inter-receptor enzymic events. The diversification of receptor structures (function) is obviously an issue of actual need, and in this light it would be hard to differentiate between different levels of receptor development and efficiency (Figure 5).

OTHER ROUTES OF EVOLUTION

Receptor–ligand (signal molecule) mediator systems may of course evolve also by routes other than the one discussed in this review. All genes that determine receptor variability may have a chance for expression in different cells of the same macrosystem and ensure thereby the responsiveness of the organism to an immense variety of signals. This takes place in the immune system, in which a great variety of cells collaborate in the decoding of the signals (antigens), and where the mediator systems initiate cell proliferation in addition to the production and secretion of receptors (antibodies). After secretion into the extracellular space, the soluble receptor recognizes the signal molecule (antigen) in a manner similar to the membrane receptor but lacks the collaboration of a mediator mechanism. The "secreted receptor" is essentially similar to the tetrameric membrane receptor. Evolution has assigned to the receptor a special protective function.

SUMMARY

Cellular receptors are primordial structures that appear already at the unicellular level. Their development and specialization depend greatly on the nature of the environmental molecules, which are "remembered" by the cell. The "memory" of the primary interaction between receptor and ligand is transmitted to the progeny generations. As a structure, the receptor is more conservative than the signal molecule. Although at the unicellular level adaptation to the external environment is furnished by a genetically determined dynamic membrane pattern that is able to interact with an immense variety of signal molecules, at the multicellular level genetically determined receptor structures, characteristic of the different target cells, are developed for adaptation to the internal milieu. The signal molecule potential, more precisely hormone evolution, depends on the formation of adequate receptors (imprinting). Receptors may arise by mechanisms other than hormone evolution and may even be secreted into the extracellular space, as are immunoglobulins.

REFERENCES

Csaba, G. (1980). Phylogeny and ontogeny of hormone receptors: the selection theory of receptor formation and hormonal imprinting. Biol. Rev. 55, 47–63.

Csaba, G. (1984). The present state in the phylogeny and ontogeny of hormone receptors. Horm. Metabol. Res. 16, 329–335.

Csaba, G. (1985). The unicellular *Tetrahymena* as a model cell for receptor research. Intl. Rev. Cytol. 95, 327–377.

Csaba, G. (1986a). Why do hormone receptors arise? Experientia 42, 715–718.

Csaba, G. (1986b). Receptor ontogeny and hormonal imprinting? Experientia 42, 750–759.

Csaba, G., Bierbauer, J., & Fehér, Z.S. (1980). Influence of melatonin and its precursors on the pigment cells of *Planaria*. Comp. Biochem. Physiol. 76, 207–209.

Csaba, G., & Darvas, ZS. (1986). Receptor-level interrelationships of amino acid and the adequate amino-acid type hormones in *Tetrahymena*: a receptor evolution model. BioSystems 19, 55–59.

Csaba, G., Darvas, ZS., Kovács, P., & Madarász, B. (1986). Influence of deciliation and ciliary regeneration on hormonal imprinting in *Tetrahymena*. Expl. Cell. Biol. 54, 49–52.

Csaba, G., & Darvas, ZS. (1987). Hormone evolution studies: multiplication promoting and imprinting ("memory") effects of various amino acids on *Tetrahymena*. Bio-Systems 20, 225–229.

Csaba, G., & Kovács, P. (1987a). Transmission of hormonal imprinting in *Tetrahymena* cultures by intercellular communication. Z. Naturforsch. 42c, 932–934.

Csaba, G., & Kovács, P. (1987b). Taxon dependence of receptor-level cell-to-cell communication in *Tetrahymena*: possible explanation for the transmission of hormonal imprinting. Cytobios 52, 17–22.

Csaba, G., Kovács, P., Török, O., Bohdaneczky, E., & Bajusz, S. (1986). Suitability of oligopeptides for induction of hormonal imprinting—implications on receptor and hormone evolution. BioSystems 19, 285–288.

Csaba, G., & Lantos, T. (1973). Effect of hormones on Protozoa. Studies on the phagocytotic effect of histamine, 5-hydroxytryptamine and indoleacetic acid in *Tetrahymena pyriformis*. Cytobiologie 7, 361–365.

Csaba, G., & Lantos, T. (1975a). Effect of insulin on glucose uptake of Protozoa. Experientia 31, 1097–1098.

Csaba, G., & Lantos, T.(1975b). Specificity of hormone receptors in *Tetrahymena*. Experiments with serotonin and histamine antagonists. Cytobiologie 11, 44–49.

Csaba, G., & Lantos, T. (1976). Effect of epinephrine on glucose metabolism in *Tetrahymena*. Endokrinologie 68, 239–240.

Csaba, G., & Németh, G. (1980). Effect of hormones and their precursors on protozoa—the selective responsiveness of *Tetrahymena*. Comp. Biochem. Physiol. 65B, 387–390.

Csaba, G., Németh, G., & Vargha, P. (1982a). Development and persistence of receptor "memory" in a unicellular model system. Expl. Cell. Biol. 50, 291–294.

Csaba, G., Németh, G., & Vargha, P. (1982b). Influence of hormone concentration and time factor on development of receptor memory in a unicellular (*Tetrahymena*) model system. Comp. Biochem. Physiol. 73B, 307–360.

Csaba, G., Németh, G., & Vargha, P. (1984). Receptor "memory" in *Tetrahymena*: does it satisfy the general criteria of memory? Expl. Cell. Biol. 52, 320–325.

Fawell, S.E., McKenzie, M.A., Greenfield, N.J., Adebodun, F., Jordan, F., & Lenard, J.(1988). Stimulation by mammalian insulin of glycogen metabolism in a wall-less strain of *Neurospora crassa*. Endocrinology, 122, 518–523.

Koch, A.S.,Fehér, J., & Lukovics, I. (1979). Single model of dynamic receptor pattern generation. Biol. Cybernet. 32, 125–138.

Kovács, P. (1986). The mechanism of receptor development as implied from hormonal imprinting studies on unicellulars. Experientia 42, 770–775.

Legros, F., Uitdenhoef, P., Dumont, J., Hanson, B., Jeanmart, J., Massant, B. & Conard, V. (1975). Specific binding of insulin to the unicellular alga *Acetabularia mediterranea*. Protoplasma 86, 119–137.

Lenhoff, H.M. (1968). Behavior, hormones and hydra. Science 161, 432–442.

Le Roith, D., Shiloach, J., Roth, J., & Lesniak, M.A. (1980). Evolutionary origins of vertebrate hormones: substances similar to mammalian insulin are native to unicellular organisms. Proc. Natl. Acad. Sci. USA 77, 6184–6188.

Le Roith, D., Liotta, A.S., Roth, J., Shiloach, J., Lewis, M.E., Pert, C.B., & Krieger, D.T. (1982). Corticotropin and beta endorphin-like materials are native to unicellular organisms. Proc. Natl. Acad. Sci. USA 79, 2086–2090.

Le Roith, D., Shiloach, J., Berelowitz, M., Frohman, L.A., Liotta, A.S., Krieger, D.T., & Roth, J. (1983). Are messenger molecules in microbes the ancestors of the vertebrate hormones and tissue factors? Fed. Proc. 42, 2602–2607.

Loose, D.S., & Feldman, D. (1982). Characterization of a unique corticosterone binding protein in Candida albicans. J. Biol. Chem. 257, 4925–4930.

Loose, D.S., Stover, E.P., Restrepo, A., Stevens, D.A., & Feldman, D. (1983). Estradiol binds to a receptor-like cytosol binding protein and initiates a biological response in *Paracoccidiodes brasiliensis*. Proc. Natl. Acad. Sci. USA 80, 7659–7663.

McKenzie, M.A., Fawell, S.E., Cha, M., & Lenard, J. (1988). Effects of mammalian insulin on metabolism, growth and morphology of a wall-less strain of Neurospora crassa. Endocrinology 122, 511–517.

Muggeo, M., Ginsberg, B.H., Roth, J., Neville, G.M., Meyts, P. de, & Kahn, C.R. (1979a). The insulin receptor in vertebrates is functionally more conserved during evolution than the insulin itself. Endocrinology 104, 1393–1402.

Muggeo, M., Obberghen, E. van Kahn, C.R.,Roth, G., Ginsberg, B.H., Meyts, P. de Emdin, S.O., & Falkmer, S. (1979b). The insulin receptor and insulin of the atlantic hagfish.Diabetes 28, 175–181.

Roth, J., Le Roith, D., Shiloach, J., Rosenzweig, H.L., Lesniak, M.A., & Havrankova, J.(1982). The evolutionary origins of hormones, neurotransmitters, and other extracellular chemical messengers. N. Engl. J. Med. 306, 523–527.

RECOMMENDED READINGS

Barrington, E.J.W. (1979). Hormones and Evolution. Academic Press, New York.
Csaba, G. (1981). Ontogeny and Phylogeny of Hormone Receptors. Karger, Basel.
Csaba, G. (ed.) (1987). Development of Hormone Receptors. Birkhäuser, Basel.
Norman, A.W., & Litwack, G. (1987). Hormones. Academic Press, New York.

Chapter 7

Receptors: Topology, Dynamics, and Regulation

H. STEVEN WILEY

INTRODUCTION

Cells within an organism must be able to recognize and effectively deal with their environment. They must be able to identify other cell types, respond to hormones,

Fundamentals of Medical Cell Biology, Volume 5A
Membrane Dynamics and Signaling, pages 113–142
Copyright © 1992 by JAI Press Inc.
All rights of reproduction in any form reserved.
ISBN: 1-55938-309-7

and selectively incorporate nutritive macromolecules. Cells of the immune system must recognize foreign antigens as well as directionally migrate and identify the appropriate sites of invasion. Most processes of cellular selectivity involve specific receptors. By virtue of their molecular structure and placement on the plasma membrane, receptors are ideally situated to encounter and bind molecules in the extracellular environment. Some receptors can give rise to signals that evoke a cellular response, whereas others allow cells to attach to each other or to the extracellular matrix. Different receptors function to transport large molecules into cells where their component parts can be effectively utilized. Ultimately, cells control all these functions by regulating the number and types of receptors that they display at their surface.

The most fundamental mechanism by which cells regulate the expression of a given receptor is by selective synthesis. If the gene encoding a receptor is not transcribed, then it cannot be expressed at the cell surface. However, the converse is not necessarily true. After receptors are synthesized, their distribution throughout the cell depends on a number of dynamic processes. These processes in turn are highly regulated by the cell. Thus the overall complement of cell surface receptors is dependent on multiple levels of regulation. As one would expect, mutations in receptors that impair their ability to properly localize can inhibit their function.

Role of Receptor Dynamics

The role of receptor dynamics is dependent on the function of the specific receptor system. In the case of the low density lipoprotein (LDL) receptor, internalization of the ligand–receptor complex and LDL delivery to the lysosomes is necessary for cholesterol release and metabolism (Goldstein and Brown, 1985). The iron transport protein transferrin must be internalized into acidic vesicles to release iron into the cells (Klausner et al., 1984). Iron-free transferrin is then rapidly returned to the cell surface still attached to its receptor (Klausner et al., 1983). Many hormones are internalized by their receptors and degraded in the lysosomes as part of a process that regulates information flow between cells (Wiley, 1985). Despite these different roles, the mechanisms that regulate receptor movement and distribution are similar. They all depend on the continual flow of membrane throughout the cell and the ability of receptors to localize to specialized membrane regions.

General Principles

Pathways of Membrane Flow

The ability of receptors to distribute throughout the cell is dependent on the normal flow of membrane. It is well appreciated that the membranes of eukaryotic cells are very dynamic structures undergoing continual turnover. This is especially true for the plasma membrane because of the functional demands that are placed upon it. Any change in a cell's shape will alter its surface:volume ratio, necessitating the removal or insertion of additional membrane. "New" membrane inserted into the cell surface arises partly from biosynthesis, but most is derived from intracellular pools of membrane (Pagano and Sleight, 1985). Only a fraction of the membrane removed from the cell surface is degraded in the lysosomes, whereas the majority is shunted to the intracellular pool for eventual reuse (Koval and Pagano, 1989).

Membrane Biogenesis

Although alterations in cell shape will lead to a induced retrieval/insertion of new plasma membrane, there is always significant steady-state turnover of this structure. Presented in Figure 1A is the general pathway of membrane flow in eukaryotic cells. Although the sites of biogenesis for all membrane lipids are not precisely known, those destined for the plasma membranes are thought to arise in the endoplasmic reticulum (ER); (see Volume 4, Chapter 12). However, exchange studies have shown that lipids in the outer leaflets of cellular membranes can exchange relatively easily with the inner leaflets (Pagano and Sleight, 1985). The ability of a membrane to "flip" a lipid to the relatively inaccessible inner leaflet probably plays a major role in maintaining specialized lipid compositions of the different cellular membranes. Because lipid "flipases" are specialized membrane-associated enzymes primarily synthesized in the ER, this organelle can be considered the site of origin for specialized membrane structures (Backer and Dawidowicz, 1987; Kawashima and Bell, 1987). With only a few exceptions, membrane proteins are essentially trapped in the membrane bilayer into which they were originally inserted. Thus the only significant mechanism by which a protein can travel from one membrane structure to another is by the fusion of the two structures. Recovery of the two initial structures must then be accomplished by membrane division or removal.

Sorting Cycle

The sequence of membrane vesicle fusion followed by fission is the primary mechanism by which membrane proteins and vesicular contents are sorted to their appropriate cellular destination. We can consider this sequence as a *sorting cycle*. Two membrane vesicles of differing compositions encounter and fuse with each other. This step requires a specific recognition of the two vesicles and the physical breaching and joining of the two disparate bilayers (see Chapter 1). After fusion, there must be a segregation of the membrane contents. This segregation occurs both in the plane of the membrane and with respect to luminal content. Segregation of luminal contents can be achieved by creating two regions of membrane with significantly different surface:volume ratios, but other mechanisms are possible (see later). Once segregation occurs, the vesicle divides. This generates two

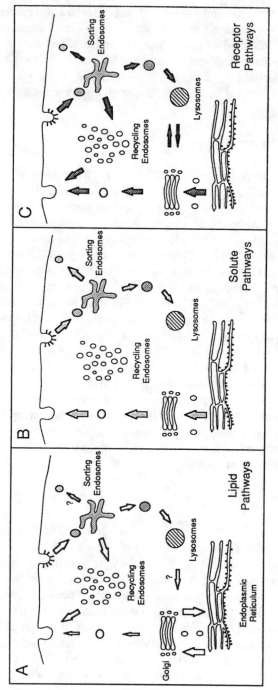

Figure 1. Pathways of membrane flow. Membrane vesicles are composed of lipid, intraluminal contents (solutes), and membrane-associated receptors. Shown separately are the pathways of these three constituents in panels **A**, **B**, and **C** respectively, from their site of synthesis (the endoplasmic reticulum) to the cell surface. Also shown is the flow of these three components back into the cell by endocytosis. The size of the arrows indicates the relative magnitude flow of constituents through the indicated pathways.

vesicles distinct in composition from the original two. Either of the two resultant vesicles may then participate in additional sorting cycles. The vesicles that fuse do not have to be of similar dimensions. Indeed, fusion of small vesicles with larger membrane structures transfers proportionally greater amounts of membrane to the resultant structure.

Asymmetry of Membrane Flow

As illustrated in Figure 1B, the asymmetry between the flow of membrane lipid and vesicular contents in the cell is quite striking. Although there is a rapid transfer of membrane lipid from the ER to the Golgi apparatus and back to the ER, solutes are essentially unidirectional in their transport from the ER to the Golgi (Wieland et al., 1987). Similarly, during endocytosis the bulk flow of solutes is almost entirely to the lysosomes (Silverstein et al., 1977), whereas most membrane lipid is transferred to perinuclear vesicles for eventual recycling (Koval and Pagano, 1989). As shown in Figure 1C, the flow of receptors and membrane proteins through the cell can follow either the membrane lipid or vesicular solute pathway. The precise pathway traversed depends on the specific receptor or membrane protein. The mechanisms responsible for these asymmetric flow processes will be considered in some detail later.

Because the ER is the source of newly synthesized receptors as well as secretory proteins, this can be considered the starting point of membrane flow. Movement from the ER to the Golgi and from compartment to compartment within the Golgi stack is carried out by small transport vesicles (Rothman et al., 1984). There is no apparent selectivity in the bulk flow of membrane and protein from the ER to the Golgi. Instead, selectivity is seen in the retention of certain proteins in the ER (Wieland et al., 1987). Thus the movement of proteins from the ER through the Golgi and then to the cell surface seems to be the "default" or constitutive pathway. The flow of membrane into the cis-Golgi seems to be substantially greater than the flow from the trans-Golgi to other membrane compartments, and so a substantial retrograde movement of membrane lipids must occur from the Golgi back to the ER (Wieland et al., 1987). This may occur either through small transport vesicles or directly through a lipid carrier mechanism.

Endocytosis

Membrane is removed from the plasma membrane by a process known as endocytosis (Silverstein et al., 1977). Although the best known type of endocytosis is mediated by structures known as *coated pits* (Anderson et al., 1977a), some studies indicate that smooth, flask-shaped indentations can also participate in this process (Carpentier et al., 1989). During the process of membrane invagination, which occurs during endocytosis, a bit of extracellular medium is always trapped. Most of these nonspecifically incorporated solutes are directed to the lysosomes.

A fraction of the solutes are very rapidly discharged back into the extracellular environment (Adams et al., 1982) by a process known as *diacytosis*. Whether entry of solute through different endocytic structures dictates the relative extent of lysosomal targeting versus diacytosis is unclear. For example, coated pits could deliver their contents to one type of intracellular structure, whereas smooth pits could target to other compartments (McKinley and Wiley, 1988). Although most internalized solutes are targeted to the lysosomes, the majority of internalized lipid is sorted to small vesicles that eventually recycle to the cell surface (Koval and Pagano, 1989).

Transcytosis

Although the default pathway of solute internalized by cells is usually lysosomal degradation, exceptions to this rule are seen in polarized cells that transport solute from one surface to another, such as kidney epithelial cells (Bomsel et al., 1989). The default pathway in these cells includes both the lysosomal pathway as well as the rapid diacytotic pathway illustrated in Figure 1*B*. The relative fraction of internalized solute that follows either pathway is determined by the surface through which internalization occurs (Bomsel et al., 1989). Entry of material through either the apical or basolateral surface gives rise to a flow of solute through these cells known as *transcytosis*. The mechanisms responsible for the differential compartmentation of solutes through these polarized cells are presently unknown.

Mechanisms of Protein Sorting

It is clear that different membrane proteins follow divergent pathways through the cell. The mechanisms of this specific targeting are generally thought to involve specific sequences in the cytoplasmic tail of the proteins. Evidence for this hypothesis has been provided by studies of proteins that do not follow the default membrane flow pathways but are targeted to specific organelles. The structures of these proteins were systematically modified by site-directed mutagenesis. These experiments showed that specific domains of the cytoplasmic tail were both necessary and sufficient to direct the proteins to their final destinations (Johnson et al., 1987). Similar studies have shown that specific cytoplasmic tail sequences are also important for regulating receptor distribution within cells (Lobel et al., 1989). Luminal proteins destined for export to compartments such as the lysosomes or secretory vesicles are targeted by virtue of their ability to bind to specific receptors. Thus the mechanisms involved in sorting receptors are also responsible for sorting vesicular contents. However, the mechanisms by which these protein domains specify sorting or targeting has remained elusive.

RECEPTOR INTERNALIZATION

Receptor Behavior

Because of their ability to selectively bind specific molecules at vanishingly small concentrations, receptors are one of the best understood classes of membrane proteins. However, most studies of receptor dynamics are actually observations of ligand movements. This is because a ligand can be modified with a radioactive or fluorescent tag before it is given to the cell. Nevertheless, following the fate of the receptor-associated ligand provides significant insight into the dynamic behavior of receptors. Because the behavior of a receptor depends on its function, this is the basis of receptor classification.

Classification of Receptors

Receptors are generally grouped into one of two functional classes (Kaplan, 1981). The class 1 receptors are those whose primary role is to transmit information to cells. This class would include the hormone and growth factor receptors. The class 2 receptors are those whose primary function is to facilitate the transport of large metabolically significant molecules into cells. This class includes the LDL, transferrin, and α2-macroglobulin receptors. The usefulness of this classification scheme is that it emphasizes function. Function ultimately dictates receptor structure, which is quite distinct between the two receptor classes.

Class 1 Receptors

Because of their role in cell-to-cell communication, this class of receptors has been intensively investigated for many years. Particularly well studied are the receptors that bind insulin (McClain et al., 1987), epidermal growth factor (Gill et al., 1987), β-adrenergic agents (Hertel et al., 1985), platelet-derived growth factor (Rosenfeld et al., 1984), and the interferons (Myers et al., 1987). A very distinctive feature of these receptors is that their internalization rates are significantly higher in the presence of ligand than in its absence (Wiley, 1985). An intrinsic consequence of ligand-induced internalization is a relative decrease in surface receptor numbers in the presence of ligand, a process known as endocytic down regulation (see Table 1). The nearly universal occurrence of down regulation in hormone receptors is especially surprising considering their structural diversity as well as differences in signaling mechanisms. For example, the epidermal growth factor (EGF) receptor has only a single membrane-spanning region and generates an intracellular signal by activation of an intrinsic tyrosine-specific protein kinase activity (Gill et al., 1987). The β-adrenergic receptor displays seven membrane-spanning regions and transmits a signal through the interaction with G proteins (Wang et al., 1989). Yet both of these receptors display occupancy-induced

Table 1. Dynamics of Class 1 Receptors

Ligand	k_e (min^{-1})[a]	Down regulated?	Reference
Epidermal growth factor	0.14–0.5	Y	Gill et al. (1988)
Choriogonadotropin	0.02–0.04	Y	Ascoli & Segaloff (1987)
Luteinizing hormone	0.04–0.05	Y	Ascoli & Segaloff (1987)
Glucagon	0.18	Y	Horwitz & Grud (1988)
Insulin	0.05–0.13	Y	Draznin et al. (1984)
Platelet-derived growth factor	0.04	Y	Rosenfeld et al. (1984)
Growth hormone	0.03	Y	Roupas & Herington (1987)
f-met-leu-phe	0.15	Y	Zigmond et al. (1982)
Isoproterenol	0.7	Y	Mahan et al. (1985)
Tumor necrosis factor-α	0.09	Y	Bajzer et al. (1989)
Interferon-a2a	0.05	Y	Bajzer et al. (1989)

[a]Specific internalization rate in the presence of ligand

internalization (Carpenter, 1985; Hertel et al., 1985). Because of its nearly universal occurrence in signal-transmitting receptors, it is clear that ligand-induced internalization plays an important role in the regulation of these receptors.

Class 2 Receptors

By definition, these receptors are involved in the transport of metabolically significant molecules into cells (Kaplan, 1981). Unlike the class 1 receptors, internalization rates of class 2 receptors do not depend on occupancy. In general, their ligand-independent internalization rate is at least as great as that displayed by occupied class 1 receptors, indicating these receptors are always in a conformation that allows them to interact with coated pits (see Table 2). Indeed, electron microscopic studies reveal that members of this class, such as the receptors for LDL and transferrin, are found clustered in coated pits even in the absence of ligand (Goldstein et al., 1985; Watts, 1985).

Although the binding of ligands to both class 1 and 2 receptors is strongly dependent on pH, most class 2 receptors also require the presence of divalent cations (Kaplan, 1981). Recent studies indicate that the low pH encountered in endosomes will dissociate divalent cations from ligand–receptor complexes, effectively stripping the ligand from its receptor (Loeb and Drickamer, 1988). This is important for a recycling receptor system where the ligand is directed to lysosomes and the receptor is recycled back to the cell surface. Because the process of receptor-mediated endocytosis can result in concentrations of solutes up to 10,000-fold greater in endosomes relative to the extracellular environment (McKinley and Wiley, 1988), mechanisms that totally prevent ligand rebinding are required to prevent a significant discharge of ligand back into the medium during receptor recycling. The few cases in which divalent cations are not required for class 2 receptor binding are those in which the ligand is normally recycled back to the cell

Table 2. Characteristics of Class 2 Receptors

Ligand	k_e (min^{-1})	Down regulated?[a]	Divalent cation dependent?	Reference
Low density lipoprotein	0.17	N	Y	Goldstein et al. (1981)
α2-Macroglobulin trypsin	0.23–1.5	N	Y	Novak et al. (1988)
Galactose-terminal glycoproteins	0.21	N	Y	Sharma & Grant (1986)
Vitellogenin	0.08–0.29	N	Y	Opresko & Wiley (1987a)
Mannose-terminal glycoproteins	1.23–1.30	N	Y	Hoppe & Lee (1983)
Transferrin	0.55–0.97	N	N	Wiley (1988)
Mannose-6-phosphate[b]	0.40	N	N	Lobel et al.(1989)
IgG (FcRII-B2 receptor)	0.03	N	N	Miettinen et al. (1989)

[a]Excludes down regulation due to lower receptor synthetic rates
[b]Cation-independent receptor

surface with the receptor (in the case of transferrin) or in which the function of the receptor does not involve the transport of large amounts of ligand to the lysosomes (such as the cation-independent mannose-6-phosphate [Man-6-P] receptor).

Receptor-Mediated Endocytosis

Receptor-mediated endocytosis is the specific internalization of a ligand attached to a receptor. Most cell surface receptors are free to diffuse laterally in the plane of the membrane where they can encounter a coated pit, bind, and be internalized. As shown in Figure 2, coated pit binding of class 2 receptors is independent of ligand, whereas for class 1 receptors, the presence of associated ligand is required (Wiley, 1985). Once internalized, the behavior of a receptor and its associated ligand can be quite different. LDL is rapidly targeted to lysosomes after internalization, while its receptor is rapidly recycled back to the cell surface (Brown et al., 1983). Because the bulk flow of internalized solutes is to the lysosomes, specific mechanisms must exist to redirect internalized receptors into the recycling membrane pathway. Although some ligands may remain attached to their receptors during transit through the cell (notably transferrin), this is the exception rather than the rule.

Mechanism of Ligand-Induced Internalization

Receptor-mediated endocytosis of class 1 receptors is best understood with regard to the epidermal growth factor (EGF) receptor. EGF is a small (~6000 MW) polypeptide mitogen that is thought to play an important role in local tissue processes such as wound healing (Carpenter and Cohen, 1979). The ease with

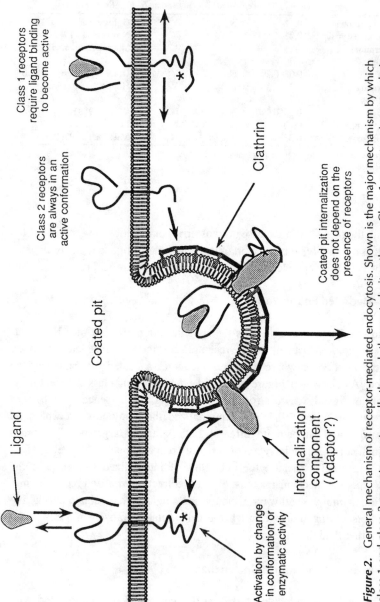

Figure 2. General mechanism of receptor-mediated endocytosis. Shown is the major mechanism by which class 1 and class 2 receptors enter cells through the coated pit pathway. Class 1 receptors increase their affinity for coated pits as a consequence of ligand binding. The affinity of class 2 receptors for coated pits is not altered by ligand binding. Not shown are relatively minor mechanisms of receptor internalization, such as phagocytosis and nonspecific pinocytosis.

Class 1 receptors require ligand binding to become active

Class 2 receptors are always in an active conformation

Clathrin

Coated pit internalization does not depend on the presence of receptors

Coated pit

Ligand

Activation by change in conformation or enzymatic activity

Internalization component (Adaptor?)

122

which EGF can be purified and modified with a variety of fluorescent, radioactive, and electron-opaque markers has greatly facilitated studies of its route of internalization. These investigations reveal that EGF receptors are initially randomly distributed and freely mobile in the plasma membrane (Hillman and Schlessinger, 1982). After ligand binding, these receptors appear to concentrate into coated pit regions where they are subsequently internalized (Haigler et al., 1979a). Ligand-induced clustering in coated pits appears to be due to an increased affinity of the occupied receptors for binding sites in the coated pits themselves (Chen et al., 1989). In the case of normal human fibroblasts, the number of internalization sites is in excess of the number of surface receptors. In the case of cells expressing high numbers of EGF receptors, coated pits can become totally saturated with receptors (Wiley, 1988).

Aggregation of Receptors

It has been proposed that occupancy-induced self-aggregation of receptors is responsible for induced internalization (Zidovetzk et al., 1981; Yarden and Schlessinger, 1987), but there is little evidence to support this hypothesis. Self-aggregation as a rate-limiting step in receptor internalization implies that internalization rates should increase exponentially as a function of surface receptor occupancy. These rates are typically independent of occupancy (Wiley and Cunningham, 1982a; Wiley, 1985). In addition, observed rates of EGF receptor self-aggregation are significantly slower than the rate of receptor internalization (van Belzen et al., 1988) and are most pronounced under conditions that inhibit internalization (Wiley, 1988). LDL receptors will form dimers, but mutated receptors unable to dimerize can still be effectively internalized. In addition, mutations that destroy the ability of the LDL receptor to be internalized by coated pits do not alter their ability to dimerize (van-Driel et al., 1987). Likewise, internalization-defective EGF receptors still effectively form dimers (Glenney et al., 1988). In the case of transferrin receptors, receptors do not dimerize prior to internalization (Hopkins, 1985). Although cross-linking some receptors with exogenous antibodies, such as internalization-defective insulin receptor, facilitates their internalization (Russell et al., 1987), this is probably due to an increase in their normally weak affinity for coated pits. The affinity of any ligand for its receptor increases with valency because a bivalent ligand must dissociate from two receptors simultaneously to escape (Dower et al., 1981). A receptor cross-linked with an antibody could appear as a bivalent ligand with respect to coated pit proteins. Although such a mechanism could facilitate internalization of some ligand–receptor complexes, the preponderance of evidence suggests that a direct alteration of the affinity of a receptor for a coated pit is the primary mechanism of ligand-induced internalization.

Phosphorylation and Endocytosis

Induced internalization of the EGF receptor requires activation of its intrinsic tyrosine kinase activity, but the reason for this is unknown. Lysine 721 in the active site of the EGF receptor tyrosine kinase is essential for its enzymatic activity. If this residue is changed to another amino acid by site-directed mutagenesis, then the resultant EGF receptor binds ligand normally but lacks any occupancy-induced kinase activity (Chen et al., 1987). These receptors are also incapable of undergoing ligand-induced internalization (Glenney et al., 1988; Chen et al., 1989). However, selective truncation of the receptor in the region of the cytoplasmic tail outside of the kinase domain can restore the ability of kinase negative receptors to undergo induced internalization. Conversely, efficient internalization of kinase active receptors requires specific sequences distal to the kinase domain (Chen et al., 1989). Thus kinase activity does not directly mediate ligand-induced internalization of the receptor. Instead, it appears likely that phosphorylation of an accessory protein facilitates binding to specific regions of the cytoplasmic tail of the EGF receptor. Other modifications of the cytoplasmic tail of the receptor could also facilitate binding to these (or other) accessory proteins.

The intrinsic tyrosine kinase activity of the insulin receptor is also required for ligand-induced internalization (Hari and Roth, 1987; McClain et al., 1987; Russell et al., 1987), but many other hormone receptors lacking intrinsic kinase activity also display induced internalization. Because there appears to be little sequence similarity between the cytoplasmic domains of the different class 1 receptors, there is no obvious candidate for a common "internalization domain." Thus the structural determinants of receptors required for binding to coated pits are uncertain.

Binding of Receptors to Coated Pits

The putative accessory proteins that mediate the binding of receptors to coated pits are probably the clathrin assembly proteins (or associated proteins; *APs*). This class of proteins has also been termed *adaptors* (Glickman et al., 1989). The APs are complexes made up of several kinds of polypeptide chains: large (100–115 kDa), medium (45–50 kDa), and small (17–20 kDa). Different families of these proteins have been found in close association with the trans-Golgi and plasma membrane, consistent with their proposed function (Ahle et al., 1988). Recent structural studies have indicated that the APs are found as complexes of two large α and β chains together with single medium and small chains. The relatively constant amino-terminal core domain is proposed to bind to clathrin, which is the primary structural protein of coated pits. The carboxyl-terminal domains are more variable and are thus thought to be responsible for the binding of different types of receptors to coated pits (Kirchhausen et al., 1989). Recent studies have shown that some class 2 receptors can bind to purified APs/adaptors immobilized to affinity matrices, but results from class 1 receptors have not been reported (Glickman et al., 1989).

Constitutive Internalization

The internalization rate of class 2 receptors does not seem to be affected by receptor occupancy and thus can be considered constitutive. One of the most thoroughly studied class 2 receptors mediates the uptake of LDL into cells. Naturally occurring mutants of this receptor were discovered many years ago during an investigation of the causes of familial hypercholesterolemia. These "internalization defective" mutants display a random distribution on the cell surface whereas normal receptors are preferentially associated with coated pits (Anderson et al., 1977b). This led to the hypothesis that the ability of LDL receptors to specifically associate with coated pits was necessary for the efficient internalization of both the receptor and the ligand (Anderson et al., 1977a).

Receptor Domains Specifying Internalization

The specific mutation leading to the internalization-defective phenotype has been identified as a conversion of tyrosine to a cysteine at residue 807 in the cytoplasmic tail of the LDL receptor (Davis et al., 1987b). Detailed molecular mapping studies have confirmed that tyrosine residue 807 is necessary for optimum internalization, but additional structural features may also be required. However, because only the first 22 amino acids of the cytoplasmic domain are needed for rapid internalization, it is clear that little sequence information is necessary (Davis et al., 1987b). Most class 2 receptors possess short cytoplasmic domains as would be expected if little sequence information is required for constitutive internalization through coated pits. It has been reported that substitution of tyrosine for a specific cysteine residue in the cytoplasmic tail of influenza virus hemagglutinin is sufficient to allow this normally non-internalizing membrane protein to enter the coated pit pathway (Lazarovits and Roth, 1988). However, the specific internalization rates displayed by these altered proteins were still quite low. An alternate hypothesis not excluded by those studies is that the normal influenza virus hemagglutinin is immobilized at the cell surface and thus unavailable to coated pits. The tyrosine substitution could release the immobile protein, which in turn allows internalization.

Specificity of Receptor Internalization Domains

Although only a small region of the cytoplasmic domain would seem to be necessary to specify entry of the LDL receptor into the coated pit pathway, it is unlikely that a single sequence is responsible for entry of all proteins. For example, there is no obvious sequence homology between the cytoplasmic domains of the transferrin and LDL receptors required for internalization (Davis et al., 1987b; Rothenberger et al., 1987). In the case of the Class 1 EGF receptor, rapid internalization seems to be specified by extended receptor domains somewhat distal to the transmembrane region (Chen et al., 1989). There are several possible reasons for this lack of a "consensus sequence" specifying rapid internalization. Binding to coated pits could depend only on the general physical properties of the sequence. This is the case for signal sequences that mediate translocation of nascent polypeptide chains across the ER membrane (von Heijne, 1985). In addition, there could be several different sequences that specify binding to the same or different coated pit proteins. Transferrin and EGF receptors, for example, do not compete for entry into the coated pit pathway, indicating separate binding sites/proteins (Wiley, 1988).

RECEPTOR RECYCLING

Recycling of Class 1 Receptors

It is now known that the majority of receptors recycle after internalization regardless of whether they are class 1 or class 2 molecules. Although ligand-induced internalization of class 1 receptors is accompanied by ligand degradation and an acceleration of receptor degradation, the rates of these processes are usually quite disparate (Wiley, 1985). For example, although occupancy of the insulin receptor leads to an increase in its rate of degradation, its rate of internalization is still about 50-fold greater than its rate of degradation (Krupp and Lane, 1982). Thus the average insulin receptor recycles almost 50 times. An exception to the rule of extensive recycling of occupied receptors is seen in the case of the EGF receptor, because in normal human fibroblasts the steady state rate of internalization is similar to the reported rate of degradation (Stoscheck and Carpenter, 1984). However, this is not an intrinsic property of the EGF receptor but instead seems to be a cell-dependent phenomenon. The EGF receptor in hepatocytes undergoes ligand-induced internalization but also undergoes extensive recycling (Dunn and Hubbard, 1984; Gladhaug and Christoffersen, 1987). EGF receptors transfected into mouse cells display a ligand-induced internalization rate very similar to that of fibroblasts, but the rate of receptor degradation is far lower (Lin et al., 1986).

Internalization and Receptor Degradation

The acceleration of receptor degradation that is a consequence of ligand-induced internalization seems to be a result of the shift of receptors from the cell surface to an intracellular pool. Because receptors targeted for degradation are recruited from the intracellular pool, this shift increases the steady-state rate of receptor degradation. However, receptors are still capable of recycling back to the cell surface from

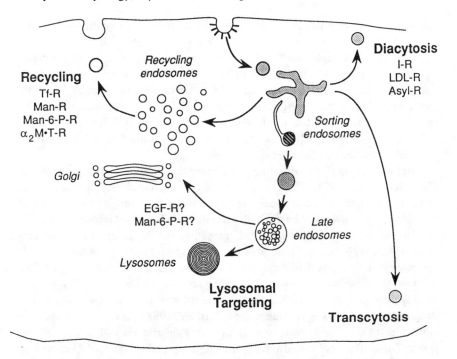

Figure 3. Postendocytic compartmentation of receptors. After endocytosis, receptors can be targeted to a variety of intracellular compartments. The major intracellular pathways are shown together with the receptors that have been shown to utilize these pathways. Abbreviations are: Asyl-R, asialoglycoprotein receptor; α_2M•T–R, α_2- macroglobulin • trypsin complex receptor; EGF-R, epidermal growth factor receptor; I-R, insulin receptor; LDL-R, low density lipoprotein receptor; Man-R, mannose-terminal glycoprotein receptor; Man-6-P-R, mannose-6-phosphate receptor; Tf-R, transferrin receptor.

this pool. There are also multiple intracellular pools of receptors that differ in their ability to undergo recycling or degradation. This situation is perhaps best understood in the case of insulin receptors (Krupp and Lane, 1982; Marshall and Olefsky, 1983; Ezaki et al., 1986; Huecksteadt et al., 1986). However, most receptors are degraded much slower than their rate of internalization and so perhaps only a small fraction enter a compartment targeted for degradation. Because few studies have examined the fate of the receptor instead of its ligand, it is difficult to generalize this aspect of receptor behavior.

Recycling of Class 2 Receptors

It is widely appreciated that most Class 2 receptors rapidly recycle back to the cell surface after internalization. However, the intracellular compartments through which these receptors recycle can be quite divergent. There are at least three different routes through which receptors can recycle (see Figure 3). These routes can exist simultaneously in the same cell type. An additional route is transcytosis where ligands internalized at one face of a polarized cell (such as an intestinal epithelial cell) can be discharged from an opposite face. Transcytosis is a specialized form of diacytosis that exists where its relative extent differs between the apical and basolateral cell surfaces. This results in a net flow of internalized material across the cell (Bomsel et al., 1989).

One of the most rapidly recycling receptors binds asialoglycoproteins (Townsend et al., 1984). These receptors can also recycle slowly, indicating that this receptor type can enter multiple intracellular recycling pathways. Rapid recycling is also displayed by the LDL receptor. No significant intracellular pools of LDL receptors have been described, indicating that this receptor type recycles primarily through the diacytotic pathway. Although it has been shown that LDL receptors lacking a significant cytoplasmic tail are recycled (Davis et al., 1987b), it is not clear whether their route of recycling remains the same. The slowest type of recycling is displayed by the Man-6-P receptor, which traverses the late endosome/lysosomal pathway (Duncan and Kornfeld, 1988). As expected from their slow transit times, these receptors are primarily distributed within intracellular pools. These receptors can also enter the recycling endosome pool (Tanner and Lienhard, 1989). The most well characterized recycling route is that taken by the transferrin receptor (Klausner et al., 1983). Because transferrin remains associated with its receptor during recycling, it has been relatively easy to explore this pathway in detail.

Postendocytic Receptor Sorting

After internalization, the transferrin receptor is rapidly sorted away from other ligands/receptors targeted for the lysosomes (Dunn et al., 1989). Endosomes containing transferrin and other receptors that utilize this route eventually localize to a perinuclear region sometimes associated with the centrioles (Salzman and Maxfield, 1988; Koval and Pagano, 1989). Although these "perinuclear endosomes" cannot be clearly distinguished in all cell types, they have the same functional characteristics and will be referred to as *recycling endosomes*. Specifically, recycling endosomes can be rapidly recruited to the cell surface in response to a number of different hormones and growth factors. Because the recycling endosomes seem to contain most of the recycling pool of membrane lipids (Koval and Pagano, 1989), they probably serve as a ready source of new plasma membrane materials. However, recycling endosomes do not contain previously internalized solutes (Buys et al., 1987). Receptors recycle from this membrane pool at identical rates. Thus recruitment from this pool is independent of vesicular contents (Ward et al., 1989).

Recycling and the Golgi Apparatus

The close association of the recycling endosomes with the Golgi apparatus has led to suggestions that this structure is involved directly in receptor recycling. Antibodies against the transferrin receptor have been reported to be routed to Golgi elements (Woods et al., 1986). Further evidence for the hypothesis that the Golgi sorts both recycled as well as newly synthesized surface receptor proteins comes from studies in which terminal sialic acid residues from surface receptors were removed by enzymatic treatment (Duncan and Kornfeld, 1988). These desialated receptors could then regain their terminal carbohydrate if allowed to be internalized and recycled. Because the Golgi is the only known site where this process occurs, it appears that receptors can reenter the Golgi pathway (Duncan and Kornfeld, 1988). However, the kinetics of reglycosylation are very slow relative to recycling. In addition, other biochemical and ultrastructural studies have indicated that although the recycling endosomes are closely associated with the Golgi, they comprise a distinct organelle (Griffiths et al., 1985; Stoorvogel et al., 1989). It is likely that the majority of recycling receptors do not enter the Golgi during recycling. However, a fraction either derived from the recycling endosomes or recycled from the lysosomes can reenter the Golgi apparatus prior to reappearing at the cell surface (Duncan and Kornfeld, 1988).

Lysosomal Targeting

Although entry of receptors into the lysosomal pathway is normally assumed to lead to degradation, this is not necessarily true. There are a number of receptors that function to deliver lysosomal enzymes and apparently enter and exit this pathway with impunity. An example of this is the Man-6-P receptor. Because Man-6-P receptors are not found in mature lysosomes (Griffiths et al., 1988), an acidified pre-lysosomal compartment must exist in which retrieval and sorting of these receptors occurs. This compartment can be considered a late endosome (or a prelysosome) and is a compartment in which endocytic vesicles destined for lysosomal fusion and Golgi-derived enzymes can apparently mix (Parton et al., 1989). The fate of most receptors that enter the lysosomal pathway is not known, particularly because it is the ligand that is observed and not the receptor. It is clear that some receptors are degraded at an accelerated rate when they enter this pathway, particularly the class 1 receptors (Stoscheck and Carpenter, 1984). However, the fraction that actually enter mature lysosomes rather than late endosomes before recycling is unknown. The pathway involved in this recycling is also unclear, although it may involve an organelle known as the trans-Golgi network.

Mechanisms of Sorting

The specific molecular features of receptors that determine their recycling/sorting patterns and the mechanisms involved in this process are unclear. As discussed earlier, most internalized membrane lipid is sorted to the recycling endosomes while internalized solutes are targeted to the lysosomes. Mechanistically, the relative fates of solute and lipid can be explained by the geometry of the sorting endosome (sometimes referred to as CURL for Compartment for Uncoupling of Receptor and Ligand; Geuze et al., 1983). Morphologically, the sorting endosome consists of a distended vesicle attached to a long tubular structure. It has been estimated that there is approximately 8–10-fold more membrane in the tubular aspect of the sorting endosome than in the cisternal elements (Griffiths et al., 1989). Fusion of the tubular elements derived from the sorting endosomes with recycling endosomes and targeting of the cisternal elements to the late endosomes explains relative membrane and solute sorting but does not provide a mechanism by which this occurs.

Theories explaining receptor sorting have a more difficult task because they must explain why a receptor would prefer one geometric aspect of the sorting endosome over another. Lateral heterogeneity in receptor distribution in sorting endosomes has been observed (Geuze et al., 1987), but there is little direct evidence as to the mechanism involved. One model suggests that receptor occupancy restricts some receptors to the cisternal element (Linderman and Lauffenburger, 1988). A receptor that has no sorting information would partition to the tubular elements because of their greater surface area, whereas an occupied receptor would actively remain in the cisternal element by anchoring to some secondary protein or structure. An implication of this model is that a receptor with no *sorting sequence* or structure for targeting to a particular destination would follow the recycling pathway through the recycling endosomes (or the diacytotic pathway). Although LDL receptors lacking significant cytoplasmic tails recycle, the actual intracellular route they traverse has not been demonstrated (Davis et al., 1987b). Because removing cytoplasmic tails from receptors usually has the effect of severely reducing their internalization rates, it is technically difficult to follow the intracellular routing of these mutants.

Expression of high numbers of transferrin receptors in cells has been reported to slow their rate of recycling, indicating that they compete for a rate-limiting component that mediates sorting and recycling (Rothenberger et al., 1987). This evidence is inconsistent with the notion that the route taken by receptors lacking a sorting sequence involves the recycling endosomes. In addition, inhibiting dissociation of LDL from its receptor effectively blocks receptor recycling (Basu et al., 1981). Again, this implies that an occupied LDL receptor is sorted differently from an empty one and thus a sorting sequence is present. Until we understand the default pathway taken by receptors that enter coated pits, we cannot determine whether any given receptor needs a sorting sequence to dictate its intracellular pathway.

REGULATION OF RECEPTOR DYNAMICS

Receptor Distribution and Dynamics

The overall distribution of receptors within a cell is regulated by their relative transit times though different compartments. For example, if the mean lifetime of a receptor at the cell surface is T_s and the mean lifetime inside the cell is T_i, then the steady-state ratio of receptor distribution between intracellular compartments and the cell surface (In:Sur ratio) is given by the following:

$$In/Sur = T_i/T_s \qquad (1)$$

Thus regulation of the rate at which receptors enter and exit intracellular compartments is fundamental to regulating receptor distribution. Because entry into one compartment is by definition exit from another, we only need to consider mechanisms that regulate the loss of receptors from any given compartment.

Regulation of Receptor Internalization Rates

Internalization of a receptor can be considered loss from the plasma membrane compartment. We have already covered several mechanisms that regulate the internalization rate of class 1 receptors. These involve the selective association of receptors with coated pit structures. Regulation of receptor affinity for coated pits does not seem to occur with class 2 receptors. However, there are other mechanisms for increasing class 2 receptor internalization that are less selective. An increase in either the number or internalization rate of coated pits will also effectively increase receptor internalization (Opresko and Wiley, 1987b). An example of this type of regulation is seen in the case of amphibian oocytes where growth is mediated by the uptake of yolk precursor proteins through the class 2 vitellogenin receptor (Wallace and Misulovin, 1978). Hormonal stimulation increases vitellogenin uptake by increasing the number of coated pits (Wiley and Dumont, 1978). Because this mechanism is nonselective, it leads to a coordinate increase in the receptor-mediated uptake of all macromolecules from the extracellular environment (Opresko and Wiley, 1987b). Thus it is most likely involved in processes where the overall metabolic activity of the cell being regulated.

Endocytic Down Regulation

An intrinsic consequence of ligand-induced internalization is a relative decrease in surface receptor numbers in the presence of ligand (Wiley and Cunningham,

1981). At steady state, the rate at which receptors are inserted into the cell surface will equal their rate of removal. If the removal of receptors is a first-order (random) process,

$$V_r = k_t[R] \qquad (2)$$

where V_r is the velocity at which receptors appear at the cell surface, k_t is the first-order rate constant of removal (by endocytosis), and [R] is the concentration of empty receptors at the cell surface (Wiley and Cunningham, 1981). It thus follows that the concentration of empty receptors at the cell surface at steady state is,

$$[R] = V_r/k_t \qquad (3)$$

Consider the situation where the rate constant of removal of an occupied receptor ([LR]) is significantly greater than that of an empty one. If all of the receptors are occupied then,

$$[LR] = V_r/k_e \qquad (4)$$

where k_e is the rate constant of endocytosis of the occupied receptors. What is the difference between the number of receptors in the presence and absence of ligand? This can be expressed simply as,

$$[LR]/[R] = k_e/k_t \qquad (5)$$

This means that a 10-fold increase in the rate of receptor internalization due to occupancy will result in a 10-fold decrease in the number of surface receptors when they are all occupied (Wiley, 1985). Of course, occupancy of only a fraction of the surface receptors will result in a correspondingly smaller decrease in total receptor number.

Because of the observation that ligand treatment will decrease surface receptor numbers, this phenomenon has been given the unfortunate name down regulation. The term is unfortunate because it implies that the role of this process is to decrease the sensitivity of cells to ligand. However, most evidence indicates that this is not the case. Cells become desensitized to ligands primarily through covalent modifications to the receptors, notably by phosphorylation. The rate of receptor desensitization is also faster than the rate of internalization (Wiley and Cunningham, 1982b; Blake et al., 1987). Finally, preventing ligand-induced internalization does not appear to prevent receptor desensitization, an example being the β-adrenergic receptor (Hertel et al., 1985; Feldman et al., 1986). So what is the role of this process in signal transduction? Paradoxically, ligand-induced internalization most likely serves as a mechanism to *increase* the sensitivity of cells to ligand. Because the rate at which an occupied class 1 receptor is internalized is approximately the same as that for an *un*occupied class 2 receptor, it is more appropriate to consider that unoccupied class 1 receptors are *inhibited* from being internalized. This inhibition of internalization results in an increase in surface receptor number when ligand levels are low or absent. Thus the lower the ambient concentration of hormone, the greater the number of surface receptors, which in turn increases the cell's ability to bind low levels of ligand. This is because the binding rate of hormones to their receptor is a second-order process,

$$V_b = k_a[L][R] \tag{6}$$

where V_b is the velocity of binding, [L] is the ligand concentration, and k_a is the second-order rate constant of ligand binding. The consequence of endocytic down regulation is a reciprocal relationship between ligand concentration and empty receptor number. Thus the product of these two parameters ([L] × [R]) will tend to remain constant, normalizing hormone binding rates over a wide range of concentrations.

Attenuation of Hormonal Signals by Ligand Internalization

The speed at which a cell removes a hormone signal dictates the speed at which it can respond to changes in hormone concentration. A single hormone molecule imparts the same information to a cell regardless of whether it binds once or binds-dissociates-rebinds a hundred times (Berg and Purcell, 1977). Thus the retention of a hormone molecule in the extracellular environment simply interferes with the ability of a cell to perceive new information. The internalization of ligand–receptor complexes can also function in signal termination because internalization is generally rapid relative to ligand dissociation. Receptor systems that mediate rapidly changing signals, such as the acetylcholine receptor system at neuromuscular junctions, have evolved secondary mechanisms to rapidly degrade ligands (Landau, 1978). However, the faster the rate of change in a signal, the greater the amount of energy required to regulate that signal. Because evolution favors more thermodynamically efficient systems, cells will use the slowest rate necessary to accomplish a given task. In the absence of a hormone signal, a slow turnover of a given class 1 receptor is favored. In the presence of a signal, an increase in internalization rates will improve signal regulation.

Negative Regulation of Internalization

The internalization of receptors can also be negatively regulated. For example, activation of the signal mediator protein kinase C (PKC) can block the induced internalization of the EGF receptor. PKC acts to phosphorylate the EGF receptor primarily at an amino acid juxtaposed to the membrane (Hunter et al., 1984; Davis and Czech, 1985). This phosphorylation could thus be involved in the inhibition of receptor internalization. However, PKC-mediated phosphorylation also blocks the intrinsic kinase activity of the EGF receptor (Cochet et al., 1984), making it unclear whether its effect on induced internalization is direct or indirect. Because PKC coordinately blocks EGF receptor activity, decreases its affinity for EGF, and prevents internalization, the effect is to make EGF invisible to the cells. PKC is also known to phosphorylate other receptors, such as the transferrin receptor. However, studies to date have not demonstrated any effect of phosphorylation on either transferrin receptor internalization or recycling (Rothenberger et al., 1987).

A selective inhibition of internalization without loss of receptor kinase activity can increase hormonal responsiveness, which in turn can contribute to a loss of contact inhibition and growth control. For example, cells expressing EGF receptors lacking a domain required for rapid internalization show enhanced sensitivity to EGF and a loss of contact inhibition (Chen et al., 1989). The bovine papillomavirus E5 transforming protein mediates its action in part by preventing ligand-induced internalization of the EGF and colony stimulating factor-1 receptor (Martin et al., 1989). These studies indicate that prevention of normal endocytic down regulation of hormone receptors results in an abnormally high level of activated receptors at the cell surface. Because the ability of cells to regulate activated surface receptors is limited by the supply of regulatory components, abnormally high levels of surface receptors can overwhelm these normally tightly regulated processes (Wiley, 1988).

Regulation of Receptor Recycling Rates

Ligand Dissociation and Receptor Recycling

Much less information is available on the factors that regulate the rate at which a receptor returns to the cell surface. The primary determinant of this process is the pathway followed by the recycling receptor because some pathways have an intrinsically longer transit time. As mentioned earlier, little information is available regarding the determinants of intracellular receptor routing. Prevention of ligand dissociation has been reported to inhibit the return of some (but not all) receptors to the cell surface. For example, the ionophore monensin has been used to prevent the dissociation of LDL from its receptor by raising intravesicular pH. Under these conditions the recycling of the LDL receptor back to the cell surface is strongly inhibited (Basu et al., 1981). Deletion of a segment of the extracellular binding domain of the LDL receptor results in a receptor in which ligand dissociation is pH independent. When occupied, these receptors are internalized normally but are unable to recycle back to the cell surface. Instead, both the ligand and the LDL receptor are shunted to the lysosomes (Davis, et al., 1987a). These studies have been interpreted as showing that ligand dissociation is required for receptor recycling. Agents that raise intravesicular pH have also been reported to inhibit recycling of the insulin receptor (Marshall and Olefsky, 1983). However, photochemically linking insulin to its receptor does not inhibit receptor recycling (Huecksteadt et al., 1986). Other investigators have shown that monensin does not affect overall membrane protein and lipid recycling (Ezaki et al., 1986; Koval and Pagano, 1989). Clearly a number of receptors can recycle despite their state of occupancy, resulting in the discharge of some previously internalized ligand back into the medium (Connolly et al., 1982; Townsend et al., 1984; Korc and Magun, 1985; Ezaki et al., 1986). Therefore, ligand dissociation is not always required for receptor recycling. However, it has been demonstrated that some occupied receptors can block fusion between intracellular vesicles (Opresko, et al., 1980; Opresko and Karpf, 1987). Thus the status of endosomal receptors could dictate the rate of vesicular fusion and recycling.

Translocation of Receptors to the Cell Surface

A more general mechanism that regulates recycling rates is the stimulation of recycling endosome recruitment to the cell surface. This phenomenon was first observed as the ability of EGF to induce a rapid increase of transferrin receptors at the cell surface (Wiley and Kaplan, 1984). It was shown that this was due to an increase in the rate at which the transferrin receptors were recruited to the cell surface from their intracellular recycling pool. EGF will also increase the membrane surface area of cells and induce compensatory endocytosis (Haigler et al., 1979b; Wiley and Cunningham, 1982b). It was subsequently shown that most recycling receptors found in the intracellular pool are also shifted to the cell surface (Ward and Kaplan, 1986; Buys et al., 1987). Many hormones seem capable of shifting intracellular receptor pools to the cell surface (Ward and Kaplan, 1986). This ability has been demonstrated for insulin, platelet-derived growth factor, and insulin-like growth factor (Davis and Czech, 1986) and is most likely related to their ability to raise intracellular calcium.

The ability of hormones to accelerate the rate of receptor recycling is probably related to their ability to increase cellular metabolism and nutrient uptake. For example, one of the primary mechanisms by which insulin increases glucose transport is by increasing the number of glucose transporters at the cell surface (Cushman and Simpson, 1985). These transporters are found in the same intracellular recycling pool as transferrin and Man-6-P receptors (Tanner and Lienhard, 1989). If the perinuclear endosomes are seen as a reservoir of plasma membrane components, then a cell can coordinately regulate nutrient uptake by regulating their fusion back to the cell surface. Note that this process is relatively specific because only receptors found in the perinuclear endosomes and perhaps other specialized organelles are affected. For example, the number of surface LDL or EGF receptors is not affected by hormonal treatment, and internalized solutes are not returned to the cell surface (Buys et al., 1987). Thus cells can selectively alter their membrane composition in response to hormonal signal by regulating the recycling rate of specific intracellular compartments.

SUMMARY

The study of receptor dynamics is in its infancy. Before the dynamics of individual membrane proteins could be determined, it was first necessary to understand the overall flow of membrane in the cells and to develop technologies that allow one to trace the different proteins. Studies of the pathways traversed by membrane proteins show that those without specific targeting sequences will follow the general pathway of membrane flow. The presence of specific sequences in the cytoplasmic tails of receptors allows them to bind to cellular structures that either allow retention in a specific cellular compartments or allow entry into a compartment that is different than the one followed by the bulk of membrane. Current studies are involved in understanding the relative rate of protein transfer between different compartments and the mechanisms that control these processes. Recent advances in molecular and cellular biology have allowed investigators to introduce precise alterations in both receptors and the regulatory proteins involved in their intracellular sorting. This has proven to be a very powerful approach and has shown that receptor distribution and dynamics are regulated at a number of different levels. The two most important processes are regulation of receptor binding to sorting structures (such as coated pits) and regulation of the net amount and flow of membrane components. These processes are involved in regulating cell-to-cell communication as well as cellular metabolism. Dysfunctions in these processes can lead to metabolic diseases as well as contribute to the development of cancer. As we develop a better understanding of receptor dynamics in normal cells, we will improve our knowledge of its role in disease.

REFERENCES

Adams, C.J., Maurey, K.M., & Storrie, B. (1982). Exocytosis of pinocytic contents by Chinese hamster ovary cells. J. Cell Biol. 93, 632–637.

Ahle, S., Mann, A., Eichelsbacher, U., & Ungewickell, E. (1988). Structural relationships between clathrin assembly proteins from the Golgi and the plasma membrane. EMBO J. 7, 919–929

Anderson, R.G., Brown, M.S., & Goldstein, J.L. (1977a). Role of the coated endocytic vesicle in the uptake of receptor-bound low density lipoprotein in human fibroblasts. Cell 10, 351–364.

Anderson, R.G., Goldstein, J.L., & Brown, M.S. (1977b). A mutation that impairs the ability of lipoprotein receptors to localise in coated pits on the cell surface of human fibroblasts. Nature 270, 695–699.

Ascoli, M., & Segaloff, D.L. (1987). On the fates of receptor-bound ovine luteinizing hormone and human chorionic gonadotropin in cultured Leydig tumor cells. Demonstration of similar rates of internalization. Endocrinology 120, 1161–1172.

Backer, J.M., & Dawidowicz, E.A. (1987). Reconstitution of a phospholipid flippase from rat liver microsomes. Nature 327, 341–343.

Bajzer, Z., Myers, A.C., & Vuk-Pavlovic, S. (1989). Binding, internalization, and intracellular processing of proteins interacting with recycling receptors. A kinetic analysis. J. Biol. Chem. 264, 13623–13631.

Basu, S.K., Goldstein, J.L., Anderson, R.G., & Brown, M.S. (1981). Monensin interrupts the recycling of low density lipoprotein receptors in human fibroblasts. Cell 24, 493–502.

Berg, H.C., & Purcell, E.M. (1977). Physics of chemoreception. Biophys. J. 20, 193–219.

Blake, A.D., Hayes, N.S., Slater, E.E., & Strader, C.D. (1987). Insulin receptor desensitization correlates with attenuation of tyrosine kinase activity, but not of receptor endocytosis. Biochem. J. 245, 357–364.

Bomsel, M., Prydz, K., Parton, R.G., Gruenberg, J., & Simons, K. (1989). Endocytosis in filter-grown Madin-Darby canine kidney cells. J. Cell Biol. 109, 3243–3258.

Brown, M.S., Anderson, R.G., & Goldstein, J.L. (1983). Recycling receptors: the round-trip itinerary of migrant membrane proteins. Cell 32, 663–667.

Buys, S.S., Gren, L.H., & Kaplan, J. (1987). Phorbol esters and calcium ionophores inhibit internalization and accelerate recycling of receptors in macrophages. J. Biol. Chem. 262, 12970–12976.

Carpenter, G. (1985). Epidermal growth factor: biology and receptor metabolism. J. Cell. Sci. Suppl. 3, 1–9.

Carpenter, G., & Cohen, S. (1979). Epidermal growth factor. Annu. Rev. Biochem. 48, 193–216.

Carpentier, J.L., Sawano, F., Geiger, D., Gorden, P., Perrelet, A., & Orci, L. (1989). Potassium depletion and hypertonic medium reduce "non-coated" and clathrin-coated pit formation, as well as endocytosis through these two gates. J. Cell. Physiol. 138, 519–526.

Chen, W.S., Lazar, C.S., Lund, K.A., Welsh, J.B., Chang, C.P., Walton, G.M., Der, C.J., Wiley, H.S., Gill, G.N., & Rosenfeld, M.G. (1989). Functional independence of the epidermal growth factor receptor from a domain required for ligand-induced internalization and calcium regulation. Cell 59, 33–43.

Chen, W.S., Lazar, C.S., Poenie, M., Tsien, R.Y., Gill, G.N., & Rosenfeld, M.G. (1987). Requirement for intrinsic protein tyrosine kinase in the immediate and late actions of the EGF receptor. Nature 328, 820–823.

Cochet, C., Gill, G.N., Meisenhelder, J., Cooper, J.A., & Hunter, T. (1984). C-kinase phosphorylates the epidermal growth factor receptor and reduces its epidermal growth factor-stimulated tyrosine protein kinase activity. J. Biol. Chem. 259, 2553–2558.

Connolly, D.T., Townsend, R.R., Kawaguchi, K., Bell, W.R., & Lee, Y.C. (1982). Binding and endocytosis of cluster glycosides by rabbit hepatocytes. Evidence for a short-circuit pathway that does not lead to degradation. J. Biol. Chem. 257, 939–945.

Cushman, S.W., & Simpson, I.A. (1985). Integral membrane protein translocations in the mechanism of insulin action. Biochem. Soc. Symp. 50, 127–149.

Davis, C.G., Goldstein, J.L., Sudhof, T.C., Anderson, R.G., Russell, D.W., & Brown, M.S. (1987a). Acid-dependent ligand dissociation and recycling of LDL receptor mediated by growth factor homology region. Nature 326, 760–765.

Davis, C.G., Van-Driel, I.R., Russell, D.W., Brown, M.S., & Goldstein, J.L. (1987b). The low density lipoprotein receptor. Identification of amino acids in cytoplasmic domain required for rapid endocytosis. J. Biol. Chem. 262, 4075–4082.

Davis, R.J., & Czech, M.P. (1985). Tumor-promoting phorbol diesters cause the phosphorylation of epidermal growth factor receptors in normal human fibroblasts at threonine-654. Proc. Natl. Acad. Sci. USA 82, 1974–1978.

Davis, R.J., & Czech, M.P. (1986). Regulation of transferrin receptor expression at the cell surface by insulin-like growth factors, epidermal growth factor and platelet-derived growth factor. EMBO J. 5, 653–658.

Dower, S.K., Titus, J.A., Delisi, C., & Segal, D.M. (1981). Mechanism of binding of multivalent immune complexes to Fc receptors. 2. Kinetics of binding. Biochemistry 20, 6335–6340.

Draznin, B., Trowbridge, M., & Ferguson, L. (1984). Quantitative studies of the rate of insulin internalization in isolated rat hepatocytes. Biochem, J. 218, 307–312.

Duncan, J.R., & Kornfeld, S. (1988). Intracellular movement of two mannose 6-phosphate receptors: return to the Golgi apparatus. J. Cell Biol. 106, 617–628.

Dunn, K.W., McGraw, T.E., & Maxfield, F.R (1989). Iterative fractionation of recycling receptors from lysosomally destined ligands in an early sorting endosome. J. Cell Biol. 109, 3303–3314.

Dunn, W.A., & Hubbard, A.L. (1984). Receptor-mediated endocytosis of epidermal growth factor by hepatocytes in the perfused rat liver: ligand and receptor dynamics. J. Cell Biol. 98, 2148–2159.

Ezaki, O., Kasuga, M., Akanuma, Y., Takata, K., Hirano, H., Fujita-Yamaguchi, Y., & Kasahara, M. (1986). Recycling of the glucose transporter, the insulin receptor, and insulin in rat adipocytes. Effect of acidtropic agents. J. Biol. Chem. 261, 3295–3305.

Feldman, R.D., McArdle, W., & Lai, C. (1986). Phenylarsine oxide inhibits agonist-induced changes in photolabeling but not agonist-induced desensitization of the beta-adrenergic receptor. Mol. Pharmacol. 30, 459–462.

Geuze, H.J., Slot, J.W., & Schwartz, A.L. (1987). Membranes of sorting organelles display lateral heterogeneity in receptor distribution. J. Cell Biol. 104, 1715–1723.

Geuze, H.J., Slot, J.W., Strous, G.J., Lodish, H.F., & Schwartz, A.L. (1983). Intracellular site of asialoglycoprotein receptor-ligand uncoupling: double-label immunoelectron microscopy during receptor-mediated endocytosis. Cell 32, 277–287.

Gill, G.N., Bertics, P.J., & Santon, J.B. (1987). Epidermal growth factor and its receptor. Mol. Cell. Endocrinol. 51, 169–186.

Gill, G.N., Chen, W.S., Lazar, C.S., Glenny, Jr., J.R, Wiley, H.S., Ingraham, H.A., & Rosenfeld, M.G. (1988). Role of intrinsic protein tyrosine kinase in function and metabolism of the epidermal growth factor receptor. Cold Spring Harb. Symp. Quant. Biol. 53, 467–476.

Gladhaug, I.P., & Christoffersen, T. (1987). Kinetics of epidermal growth factor binding and processing in isolated intact rat hepatocytes. Dynamic externalization of receptors during ligand internalization. Eur. J. Biochem. 164, 267–275.

Glenney Jr., J.R., Chen, W.S., Lazar, C.S., Walton, G.M., Zokas, L.M., Rosenfeld, M.G., & Gill, G.N. (1988). Ligand-induced endocytosis of the EGF receptor is blocked by mutational inactivation and by microinjection of anti-phosphotyrosine antibodies. Cell 52, 675–684.

Glickman, J.N., Conibear, E., & Pearse, B.M.F. (1989). Specificity of binding of clathrin adaptors to signals on the mannose-6-phosphate/insulin-like growth factor II receptor. EMBO J. 8,1041.

Goldstein, B., Wofsy, C., & Bell, G. (1981). Interactions of low density lipoprotein receptors with coated pits on human fibroblasts: Estimate of the forward rate constant and comparison with the diffusion limit. Proc. Natl. Acad. Sci. USA 87, 5695–5698.

Goldstein, J.L., & Brown, M.S. (1985). The LDL receptor and the regulation of cellular cholesterol metabolism. J. Cell Sci. Suppl. 3, 131–137.

Goldstein, J.L., Brown, M.S., Anderson, R.G., Russell, D.W., & Schneider, W.J. (1985). Receptor-mediated endocytosis: concepts emerging from the LDL receptor system. Annu. Rev. Cell. Biol. 1, 1–39.

Griffiths, G., Back, R., & Marsh, M. (1989). A quantitative analysis of the endocytic pathway in baby hamster kidney cells. J. Cell Biol. 109, 2703–2720.

Griffiths, G., Hoflack, B., Simons, K., Mellman, I., & Kornfeld, S. (1988). The mannose 6-phosphate receptor and the biogenesis of lysosomes. Cell 52, 329–341.

Griffiths, G., Pfeiffer, S., Simons, K., & Matlin, K. (1985). Exit of newly synthesized membrane proteins from the trans cisterna of the Golgi complex to the plasma membrane. J. Cell Biol. 101, 949–964.

Haigler, H.T., McKanna, J.A., & Cohen, S. (1979a). Direct visualization of the binding and internalization of a ferritin conjugate of epidermal growth factor in human carcinoma cells A-431. J. Cell Biol. 81, 382–395.

Haigler, H.T., McKanna, J.A., & Cohen, S. (1979b). Rapid stimulation of pinocytosis in human carcinoma cells A-431 by epidermal growth factor. J. Cell Biol. 83, 82–90.

Hari, J., & Roth, R.A. (1987). Defective internalization of insulin and its receptor in cells expressing mutated insulin receptors lacking kinase activity. J. Biol. Chem. 262, 15341–15344.

Hertel, C., Coulter, S.J., & Perkins, J.P. (1985). A comparison of catecholamine-induced internalization of beta-adrenergic receptors and receptor-mediated endocytosis of epidermal growth factor in human astrocytoma cells. Inhibition by phenylarsine oxide. J. Biol. Chem. 260, 12547–12553.

Hillman, G.M., & Schlessinger, J. (1982). Lateral diffusion of epidermal growth factor complexed to its surface receptors does not account for the thermal sensitivity of patch formation and endocytosis. Biochemistry 21, 1667–1672.

Hopkins, C.R. (1985). The appearance and internalization of transferrin receptors at the margins of spreading human tumor cells. Cell 40, 199–208.

Hoppe, C.A., & Lee, Y.C. (1983). The binding and processing of mannose-bovine serum albumin derivatives by rabbit alveolar macrophages. Effect of the sugar density. J. Biol. Chem. 258, 14193–14199.

Horwitz, E.M., & Gurd, R.S. (1988). Quantitative analysis of internalization of glucagon by isolated hepatocytes. Arch. Biochem. Biophys. 267, 758–769.

Huecksteadt, T., Olefsky, J.M., Brandenberg, D., & Heidenreich, K.A. (1986). Recycling of photoaffinity-labeled insulin receptors in rat adipocytes. Dissociation of insulin-receptor complexes is not required for receptor recycling. J. Biol. Chem. 261, 8655–8659.

Hunter, T., Ling, N., & Cooper, J.A. (1984). Protein kinase C phosphorylation of the EGF receptor at a threonine residue close to the cytoplasmic face of the plasma membrane. Nature 311, 480–483.

Johnson, L.M., Bankaitis, V.A., & Emr, S.D. (1987). Distinct sequence determinants direct intracellular sorting and modification of a yeast vacuolar protease. Cell 48, 875–885.

Kaplan, J. (1981). Polypeptide-binding membrane receptors: Analysis and classification. Science 212, 14–20.

Kawashima, Y., & Bell, R.M. (1987). Assembly of the endoplasmic reticulum phospholipid bilayer. Transporters for phosphatidylcholine and metabolites. J. Biol. Chem. 262, 16495–164502.

Kirchhausen, T., Nathanson, K.L., Matsui, W., Vaisberg, A., Chow, E. P., Burne, C., Keen, J.H., & Davis, A.E. (1989). Structural and functional division into two domains of the large (100- to 115-kDa) chains of the clathrin-associated protein complex AP-2. Proc. Natl. Acad. Sci. USA 86, 2612–2616.

Klausner, R.D., Ashwell, G., Van-Renswoude, J., Harford, J.B., & Bridges, K.R. (1983). Binding of apotransferrin to K562 cells: explanation of the transferrin cycle. Proc. Natl. Acad. Sci. USA 80, 2263–2266.

Klausner, R.D., Van-Renswoude, J., Kempf, C., Rao, K., Bateman, J.L., & Robbins, A.R. (1984). Failure to release iron from transferrin in a Chinese hamster ovary cell mutant pleiotropically defective in endocytosis. J. Cell Biol. 98, 1098–1101.

Korc, M., & Magun, B.E. (1985). Recycling of epidermal growth factor in a human pancreatic carcinoma cell line. Proc. Natl. Acad. Sci. USA 82, 6172–6175.

Koval, M., & Pagano, R.E. (1989). Lipid recycling between the plasma membrane and intracellular compartments: transport and metabolism of fluorescent sphingomyelin analogues in cultured fibroblasts. J. Cell Biol. 108, 2169–2181.

Krupp, M.N., & Lane, M.D. (1982). Evidence for different pathways for the degradation of insulin and insulin receptor in the chick liver cell. J. Biol. Chem. 257, 1372–1377.

Landau, E.M. (1978). Function and structure of the Ach receptor at the muscle endplate. Prog. Neurobiol. 10, 253–288.

Lazarovits, J., & Roth, M. (1988). A single amino acid change in the cytoplasmic domain allows the influenza virus hemagglutinin to be endocytosed through coated pits. Cell 53, 743–752.

Lin, C.R., Chen, W.S., Lazar, C.S., Carpenter, C.D., Gill, G.N., Evans, R.M., & Rosenfeld, M. G. (1986). Protein kinase C phosphorylation at Thr 654 of the unoccupied EGF receptor and EGF binding regulate functional receptor loss by independent mechanisms. Cell. 44, 839–848.

Linderman, J.J., & Lauffenburger, D.A. (1988). Analysis of intracellular receptor/ligand sorting in endosomes. J. Theor. Biol. 132, 203–245.

Lobel, P., Fujiimoto, K., Ye, R.D., Griffiths, G., & Kornfeld, S. (1989). Mutations in the cytoplasmic domain of the 275 kd mannose 6-phosphate receptor differentially alter lysosomal enzyme sorting and endocytosis. Cell 57, 787–796.

Loeb, J.A., & Drickamer, K. (1988). Conformational changes in the chicken receptor for endocytosis of glycoproteins. Modulation of ligand-binding activity by Ca^{2+} and pH. J. Biol. Chem. 263, 9752–9760.

Mahan, L.C., Koachman, A.M., & Insel, P.A. (1985). Genetic analysis of beta-adrenergic receptor internalization and down-regulation. Proc. Natl. Acad. Sci. USA 82, 129–133.

Marshall, S., & Olefsky, J.M. (1983). Separate intracellular pathways for insulin receptor recycling and insulin degradation in isolated rat adipocytes. J. Cell. Physiol. 117, 195–203.

Martin, P., Vass, W.C., Schiller, J.T., Lowy, D.R., & Velu, T.J. (1989). The bovine papillomavirus E5 transforming protein can stimulate the transforming activity of EGF and CSF-1 receptors. Cell 59, 21–32.

McClain, D.A., Maegawa, H., Lee, J., Dull, T.J., Ulrich, A., & Olefsky, J.M. (1987). A mutant insulin receptor with defective tyrosine kinase displays no biologic activity and does not undergo endocytosis. J. Biol. Chem. 262, 14663–14671.

McKinley, D.N., & Wiley, H.S. (1988). Reassessment of fluid-phase endocytosis and diacytosis in monolayer cultures of human fibroblasts. J. Cell. Physiol. 136, 389–397.

Miettinen, H.M., Rose, J.K., & Mellman, I. (1989). Fc receptor isoforms exhibit distinct abilities for coated pit localization as a result of cytoplasmic domain heterogeneity. Cell 58, 317–327.

Myers, A.C., Kovach, J.S., & Vuk-Pavlovic, S. (1987). Binding, internalization, and intracellular processing of protein ligands. Derivation of rate constants by computer modeling. J. Biol. Chem. 262, 6494–6499.

Novak, J.M., Ward, D.M., Buys, S.S., & Kaplan, J. (1988). Effects of hypo-osmotic incubation on membrane recycling. J. Cell. Physiol. 137, 235–242.

Opresko, L.K., & Karpf, R.A. (1987). Specific proteolysis regulates fusion between endocytic compartments in *Xenopus* oocytes. Cell 51, 557–568.

Opresko, L.K., & Wiley, H.S. (1987a). Receptor-mediated endocytosis in Xenopus oocytes. I. Characterization of the vitellogenin receptor system. J. Biol. Chem 262, 4109–4115.

Opresko, L.K., & Wiley, H.S. (1987b). Receptor-mediated endocytosis in Xenopus oocytes. II. Evidence for two novel mechanisms of hormonal regulation. J. Biol. Chem. 262, 4116–4123.

Opresko, L.K., Wiley, H.S., & Wallace, R.A. (1980). Differential postendocytotic compartmentation in Xenopus oocytes is mediated by a specifically bound ligand. Cell 22, 47–57.

Pagano, R.E., & Sleight, R.G. (1985). Defining lipid transport pathways in animal cells. Science 229, 1051–1057.

Parton, RG., Prydz, K., Bomsel, M., Simons, K., & Griffiths, G. (1989). Meeting of the apical and basolateral endocytic pathways of the Madin-Darby canine kidney cells in late endosomes. J. Cell Biol 109, 3259–3272.

Rosenfeld, M.E., Bowen-Pope, D.F., & Ross, R. (1984). Platelet-derived growth factor: morphologic and biochemical studies of binding, internalization, and degradation. J. Cell. Physiol. 121, 263–274.

Rothenberger, S., Iacopetta, B.J., & Kuhn, L.C. (1987). Endocytosis of the transferrin receptor requires the cytoplasmic domain but not its phosphorylation site. Cell 49, 423–431.

Rothman, J.E., Miller, R.L., & Urbani, L.J. (1984). Intercompartmental transport in the Golgi complex is a dissociative process: facile transfer of membrane protein between two Golgi populations. J. Cell Biol. 99, 260–271.

Roupas, P., & Herington, A.C. (1987). Receptor-mediated endocytosis and degradative processing of growth hormone by rat adipocytes in primary culture. Endocrinology 120, 2158–2165.

Russell, D.S., Gherzi, R., Johnson, E.L., Chou, C.K., & Rosen, O.M. (1987). The protein-tyrosine kinase activity of the insulin receptor is necessary for insulin-mediated receptor down-regulation. J. Biol. Chem. 262, 11833–11840.

Salzman, N.H., & Maxfield, F.R. (1988). Intracellular fusion of sequentially formed endocytic compartments. J. Cell Biol. 106, 1083–1091.

Sharma, R.J., & Grant, D.A.W. (1986). A differential effect between the acute and chronic administration of ethanol on the endocytic rate constant, k_e, for the internalization of asialoglycoproteins by hepatocytes. Biochim. Biophys. Acta 862, 199–204.

Silverstein, S.C., Steinman, R.M., & Cohn, Z.A. (1977). Endocytosis. Ann. Rev. Biochem. 46, 669–722.

Stoorvogel, W., Geuze, H.J., Griffith, J.M., Schwartz, A.L., & Strous, G.J. (1989). Relations between the intracellular pathways of the receptors for transferrin, asialoglycoprotein, and mannose 6-phosphate in human hepatoma cells. J. Cell Biol. 108, 2137–2148.

Stoscheck, C.M., & Carpenter, G. (1984). Down regulation of epidermal growth factor receptors: direct demonstration of receptor degradation in human fibroblasts. J. Cell Biol. 98, 1048–1053.

Tanner, L.I., & Lienhard, G.E. (1989). Localization of transferrin receptors and insulin-like growth factor II receptors in vesicles from 3T3-L1 adipocytes that contain intracellular glucose transporters. J. Cell Biol. 108, 1537–1545.

Townsend, R.R., Wall, D.A., Hubbard, A.L., & Lee, Y.C. (1984). Rapid release of galactose-terminated ligands after endocytosis by hepatic parenchymal cells: evidence for a role of carbohydrate structure in the release of internalized ligand from receptor. Proc. Natl. Acad. Sci. USA 81, 466–470.

Van Belzen, N., Rijken, P.J., Hage, W.J., De Laat, S.W., Verkleij, A.J., & Boonstra, J. (1988). Direct visualization and quantitative analysis of epidermal growth factor-induced receptor clustering. J. Cell. Physiol. 134, 413–420.

Van-Driel, I.R., Davis, C.G., Goldstein, J.L., & Brown, M.S. (1987). Self-association of the low density lipoprotein receptor mediated by the cytoplasmic domain. J. Biol. Chem. 262, 16127–16134.

Von Heijne, G. (1985). Signal sequences: the limits of variation. J. Mol. Biol. 184, 99–105.

Wallace, R.A., & Misulovin, Z. (1978). Long-term growth and differentiation of Xenopus oocytes in a defined medium. Proc. Natl. Acad. Sci. USA 75, 5534–5538.

Wang, H., Lipfert, L., Malbon, C.C., & Bahouth, S. (1989). Site-directed anti-peptide antibodies define the topography of the beta-adrenergic receptor. J. Biol. Chem. 264, 14424–14431.

Ward, D.M., Ajioka, R., & Kaplan, J. (1989). Cohort movement of different ligands and receptors in the intracellular endocytic pathway of alveolar macrophages. J. Biol. Chem. 264, 8164–8170.

Ward, D.M., & Kaplan, J. (1986). Mitogenic agents induce redistribution of transferrin receptors from internal pools to the cell surface. Biochem. J. 238, 721–728.

Watts, C. (1985). Rapid endocytosis of the transferrin receptor in the absence of bound transferrin. J. Cell Biol. 100, 633–637.

Wieland, F.T., Gleason, M.L., Serafini, T.A., & Rothman, J.E. (1987). The rate of bulk flow from the endoplasmic reticulum to the cell surface. Cell 50, 289–300.

Wiley, H.S. (1985). Receptors as models for the mechanisms of membrane protein turnover and dynamics. Curr. Tops. Membr. Trans. 24, 369–412.

Wiley, H.S. (1988). Anomalous binding of epidermal growth factor to A431 cells is due to the effect of high receptor densities and a saturable endocytic system. J. Cell Biol. 107, 801–810.

Wiley, H.S., & Cunningham, D.D. (1981). A steady state model for analyzing the cellular binding, internalization and degradation of polypeptide ligands. Cell 25, 433–440.

Wiley, H.S., & Cunningham, D.D. (1982a). The endocytotic rate constant: A cellular parameter for quantitating receptor-mediated endocytosis. J. Biol. Chem. 257, 4222–4229.

Wiley, H.S., & Cunningham, D.D. (1982b). Epidermal growth factor stimulates fluid phase endocytosis in human fibroblasts through a signal generated at the cell surface. J. Cell. Biochem. 19, 383–394.

Wiley, H.S., & Dumont, J.N. (1978). Stimulation of vitellogenin uptake in stage IV Xenopus oocytes by treatment with chorionic gonadotropin in vitro. Biol. Reprod. 18, 762–771.

Wiley, H.S., & Kaplan, J. (1984). Epidermal growth factor rapidly induces a redistribution of transferrin receptor pools in human fibroblasts. Proc. Natl. Acad. USA 81, 7456–7460.

Woods, J.W., Doriaux, M., & Farquhar, M.G. (1986). Transferrin receptors recycle to cis and middle as well as trans Golgi cisternae in Ig-secreting myeloma cells. J. Cell Biol. 103, 277–286.

Yarden, Y., & Schlessinger, J. (1987). Epidermal growth factor induces rapid, reversible aggregation of the purified epidermal growth factor receptor. Biochemistry 26, 1443–1451.

Zidovetzki, R, Yarden, Y., Schlessinger, J., & Jovin, T.M. (1981). Rotational diffusion of epidermal growth factor complexed to cell surface receptors reflects rapid microaggregation and endocytosis of occupied receptors. Proc. Natl. Acad. Sci. USA 78, 6981–6985.

Zigmond, S.H., Sullivan, S.J., & Lauffenburger, D.A. (1982). Kinetic analysis of chemotactic peptide receptor modulation. J. Cell Biol. 92, 34–43.

Chapter 8

Electrical Excitability

DAVID J. AIDLEY

INTRODUCTION

It can be argued that the modern sciences of electricity and neurophysiology began in 1791 with Galvani's experiments on the twitching of frogs' legs in thunderstorms. The evidence that nervous messages are electrical impulses was provided by du Bois-Reymond in the mid nineteenth century. In the early years of the twentieth century Gotch, Lucas, and especially Adrian showed that the individual nerve impulse is an all-or-nothing event: its size is independent of the size of the stimulus once the stimulating current crosses the threshold level necessary to produce a response at all. Since then much work has been directed to the crucial question, of what the mechanism of this all-or-nothing event might be.

Fundamentals of Medical Cell Biology, Volume 5A
Membrane Dynamics and Signaling, pages 143–169
Copyright © 1992 by JAI Press Inc.
All rights of reproduction in any form reserved.
ISBN: 1-55938-309-7

THE MEMBRANE THEORY OF NERVOUS CONDUCTION

The idea that the excitability of nerve axons lies in their cell membranes was formulated by du Bois-Reymond and his students Hermann and Bernstein in the latter half of the nineteenth century. They concluded that the inside of nerve and muscle cells is electrically negative to the outside at rest, and that this resting potential is reduced during activity. Thus the membrane is regarded as polarized in the resting state and becomes depolarized on activity. More precisely, Bernstein in 1902 suggested that the nerve axon acts as a concentration cell, with a relatively high concentration of potassium ions inside, and possesses a membrane that is selectively permeable to potassium ions. The action potential, he suggested, is produced by a brief breakdown of this selectivity.

Direct evidence for the membrane theory did not begin to appear until the late 1930s. Then Hodgkin, as we shall see later, obtained experimental results that were in accordance with Hermann's ideas on the mechanism of conduction, and Cole and Curtis began an extensive investigation of the impedance of cells and cell membranes. In the squid axon, Cole and Curtis (1939) concluded that the membrane could be regarded as a resistance and a capacitance in parallel. The

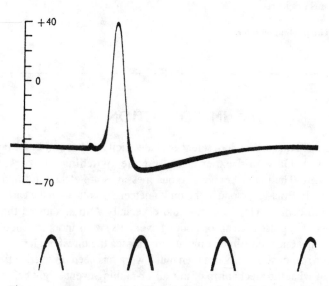

Figure 1. The nerve action potential. One of the first intracellular records of an action potential, recorded from a squid giant axon. The vertical scale shows the potential (in mV) of the internal electrode with respect to the external sea water. The time marker at the bottom shows 2-ms intervals. (From A.L. Hodgkin and A.F. Huxley [1945]. J. Physiol. 104, 176–195.)

membrane resistance fell markedly during the passage of an action potential, to 25 Ω cm^2 from its resting value of about 1000 Ω cm^2. The membrane capacitance, on the other hand, remained essentially unchanged at about 1.1 $\mu F/cm^2$.

The year 1939 was indeed a landmark in electrophysiology. For the first time it was possible to test Bernstein's version of the membrane theory by recording from an electrode inside the cell so as to measure the membrane potential directly. The cell was the squid giant axon, and the experiments were done by Hodgkin and Huxley at Plymouth, England, and by Curtis and Cole at Woods Hole, Massachusetts. Squid giant axons may be up to 1 mm in diameter, so it was possible to insert a fine capillary electrode down the middle of the axoplasm. The results (Figure 1) showed clearly that the membrane potential was in the range of –50 to –60 mV at rest, as expected, and that during the action potential the inside of the membrane became positive to the outside by about 40 to 50 mV. This unexpected "overshoot" was not predicted by Bernstein's theory.

THE IONIC BATTERIES OF THE CELL

How does the resting potential arise? The cytoplasm of squid giant axon (known as the axoplasm) can be squeezed out of the cut end of an axon and subjected to chemical analysis. The results show that the ionic concentrations in the axoplasm are different from those in the extracellular fluid, as is shown in Table 1. Notice particularly that the internal potassium ion concentration is much higher than the external concentration, whereas the reverse is true for sodium ions.

These ionic inequalities provide the "batteries" for transmembrane potentials. Suppose the cell membrane were selectively permeable to potassium ions, for example. A number of potassium ions would flow down the concentration gradient out of the cell, taking positive charge with them. This small depletion of positive charge would make the inside of the cell electrically negative to the outside. The external positive charge would then tend to reduce further outward movements of potassium ions, so that an equilibrium would be reached when the concentration gradient tending to move potassium ions outward would be just balanced by the

Table 1. Ionic Concentrations in Squid Axoplasm and Blood.
(Simplified after Hodgkin, 1964.)

	Concentration in axoplasm (mM)	*Concentration in blood (mM)*
K	400	20
Na	50	440
Cl	40–150	560
Ca	0.4	20
Mg	10	54
Isethionate	250	—
Other organic anions	ca. 110	—

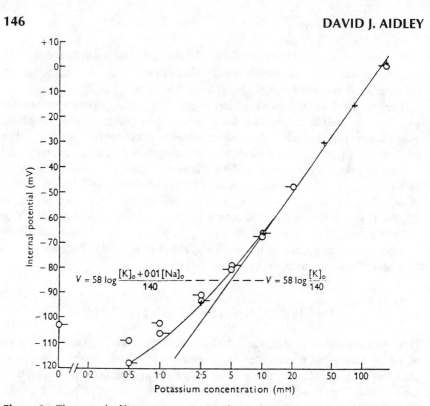

Figure 2. The muscle fiber as a potassium electrode. Circles and crosses show the membrane potential of a frog muscle fiber at different external potassium ion concentrations. (From A.L. Hodgkin and P. Horowicz [1959]. J. Physiol. 148, 127–160.)

electrical gradient tending to move them inward. The potential at this point is called the equilibrium potential or (after the man who first described it) the Nernst potential. Its value, E_K is given by the equation,

$$E_K = \frac{RT}{zF} \log_e \frac{[K]_o}{[K]_i} \qquad (1)$$

where R is the gas constant, T is the absolute temperature, z is the charge on the ion (+1 in this case), F is Faraday's constant, and $[K]_o$ and $[K]_i$ are the potassium ion concentrations outside and inside the cell. At 18 °C, the equation can be written as,

$$E_K = 58 \log_{10} \frac{[K]_o}{[K]_i} \text{ mV}$$

The best way to test the applicability of the Nernst equation is to vary the external potassium concentration and see what effect this has on the resting membrane potential. Equation 2 predicts that the membrane potential should be proportional to the logarithm of the external potassium concentration, with a slope of 58 mV per 10-fold change. Figure 2 shows an experiment of this type done by Hodgkin and Horowicz (1959) on a frog muscle fiber; there is good agreement with equation 2 except at low potassium ion concentrations, where it seems that other factors (such as a small permeability to sodium ions) become important. Similar results have been obtained from nerve axons.

How are the inequalities in ionic distribution set up and maintained? For anions the situation is simple: the cytoplasm contains a number of organic anions that cannot cross the cell membrane, so the concentration of inorganic anions (mainly chloride) is much lower inside the cell than outside. But for cations, which can cross the membrane to varying extents, the situation is more complex.

It turns out that the sodium ion concentration gradient is maintained by an active extrusion of sodium ions; this is commonly known as the sodium pump. The process requires metabolic energy in the form of the energy-rich compound adenosine triphosphate (ATP). Figure 3 shows an experiment that demonstrates this. The axon was injected with radioactive sodium and the rate of its subsequent loss was measured. Inhibition of the mitochondrial ATP-producing reactions by cyanide produced a rapid reduction in the extrusion rate, but this could then be briefly boosted by an injection of ATP. Further experiments showed that sodium extrusion is linked to potassium uptake.

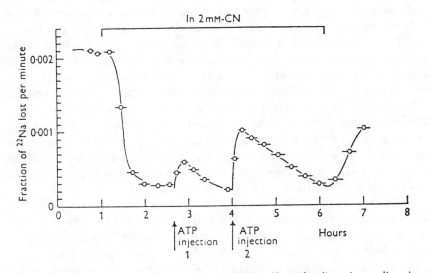

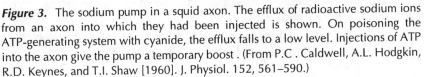

Figure 3. The sodium pump in a squid axon. The efflux of radioactive sodium ions from an axon into which they had been injected is shown. On poisoning the ATP-generating system with cyanide, the efflux falls to a low level. Injections of ATP into the axon give the pump a temporary boost . (From P.C . Caldwell, A.L. Hodgkin, R.D. Keynes, and T.I. Shaw [1960]. J. Physiol. 152, 561–590.)

The sodium pump occurs in a wide range of animal cells. Experiments on red blood cells have shown that for every three sodium ions extruded, two potassium ions are taken up and one ATP molecule is split. In molecular terms the pump is a membrane protein (Na,K-activated ATPase) whose amino acid sequence has now been determined. Its activity can be inhibited by the arrow poison ouabain.

THE SODIUM THEORY OF THE ACTION POTENTIAL

The unexpected finding in 1939 that the axon membrane potential became inside positive for a time during the action potential caused considerable puzzlement, and progress was further delayed by the Second World War. Then Hodgkin and his colleagues decided to investigate the idea that the action potential was produced largely by an inward movement of sodium ions. If this is the case we would expect the peak of the action potential to be near to the sodium equilibrium potential E_{Na}, given by the Nernst equation,

$$E_{Na} = 58 \log_{10} \frac{[Na]_o}{[Na]_i} \tag{3}$$

For the concentrations shown in Table 1, E_{Na} works out at about +55 mV, inside positive.

Equation 3 suggests that, if the sodium theory is correct, a reduction in the external sodium ion concentration should reduce the size of the action potential. This prediction was tested by Hodgkin and Katz (1949), with the results shown in Figure 4. Reduced sodium concentrations lower the peak of the action potential and lower the rate of depolarization during its rising phase, thus providing good evidence for the sodium theory. Notice that the rate of repolarization during the falling phase of the action potential is not affected by changes in sodium concentration, which suggests that sodium ion movements are not involved at this stage.

A more direct approach to ionic movements is provided by measurements of the movements of radioactive ions. It is necessary to measure the fluxes in both directions so as to obtain the net movement of an ion, and to measure fluxes in resting and stimulated axons so as to estimate the net movement per action potential. Keynes (1951) carried out such measurements on cuttlefish giant axons using radioactive sodium and potassium ions. He found that movements of both ions were much increased by stimulation. In one experiment each action potential was accompanied by an influx of 10.3 pmol/cm^2 of sodium and an efflux of 6.6 pmol/cm^2, giving a net inflow of 3.7 pmol/cm^2. For potassium ions there was a net outflow of much the same amount.

Over the whole action potential, therefore, it would seem that the net entry of sodium ions is balanced by a net exit of potassium ions. The two processes must be at least partially separated in time, otherwise there would be no change in

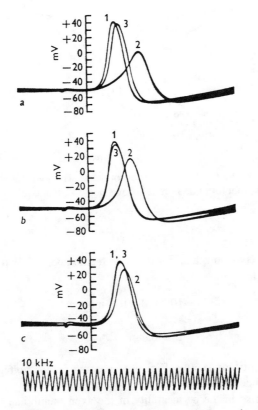

Figure 4. Reducing the external sodium ion concentration reduces the size of the action potential. In each set of records, record 1 shows the response of a squid axon in sea water, record 2 in an experimental solution with a reduced sodium ion concentration, and record 3 in sea water again. Experimental solutions were made by mixing sea water and isotonic glucose solutions, the proportions of sea water being 33% in **a**, 50% in **b**, and 70% in **c**. (From A.L. Hodgkin and B. Katz [1949]. J. Physiol. 108, 37–77.)

membrane potential. It seemed reasonable to suggest that the rising phase was associated with sodium entry and the falling phase with potassium exit.

Are the ionic movements described by Keynes sufficiently large to account for the amplitude of the action potential? The charge Q on a capacitance C is given by,

$$Q = CV$$

where V is the voltage across the capacitance. If this charge is carried by a monovalent ion, then the number of moles of the ion moved from one side of the capacitance to the other is given by,

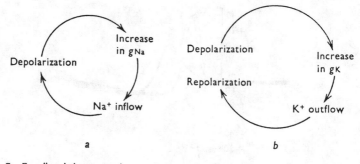

Figure 5. Feedback loops in the excitation of nerve. The loop for sodium (**a**) shows positive feedback, that for potassium (**b**) shows negative feedback.

$$n = CV/F$$

where F is Faraday's constant. For the axon membrane, C is 1 $\mu F/cm^2$ and V is about 100 mV. Hence,

$$n = 10^{-6} \times 0.1/10^5 \text{ mol/cm}^2$$

$$= 1 \text{ pmol/cm}^2 .$$

So Keynes's results are easily sufficient to account for the voltage change seen during the action potential.

Because the rise in the permeability of the axon membrane to sodium ions follows an electrical stimulus that depolarizes the membrane, it seems reasonable to suggest that there is causal link between these two processes. But an inflow of sodium ions will carry positive charge into the axon, producing further depolarization: the system is a positive feedback loop, as is shown in Figure 5A. Keynes's results suggest that the membrane permeability to potassium ions is also increased by depolarization, but here we have a negative feedback loop, because outflow of potassium ions will remove positive charge from the inside of the axon (Figure 5B).

THE CLASSIC VOLTAGE CLAMP EXPERIMENTS

Feedback loops such as those shown in Figure 5 are difficult to investigate experimentally because cause and effect cannot readily be separated. We need some method of cutting the loop so that, for example, ionic movements do not produce changes in membrane potential: only then can we determine just what the effect of membrane potential on ionic movements is. Such a method is provided by the voltage clamp system, initiated by Cole and his colleagues (see Cole, 1968) and developed by Hodgkin, Huxley, and Katz (1952).

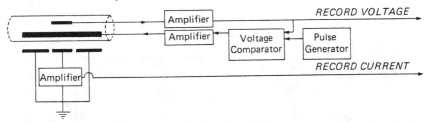

Figure 6. The voltage clamp method for a squid giant axon. One internal electrode serves to measure the voltage across the membrane. This voltage forms one input to a feedback amplifier (shown here as a voltage comparator leading into an amplifier) whose output is passed to the other internal electrode. If the recorded voltage is different from the other input to the comparator, derived from a pulse generator, then the current output will alter so as to reduce the difference; the membrane potential is thus clamped at a value determined by the pulse generator. The current flowing across the membrane during this clamp is monitored by a further amplifier. (From R.D. Keynes and D.J. Aidley [1991]. Nerve and Muscle, 2nd edition. Cambridge University Press, Cambridge.)

The aim of the voltage clamp system is to hold the membrane potential for a time at some predetermined value and to measure the membrane currents that ensue. This is done by using an electronic feedback circuit such as is shown in Figure 6. One input of the feedback amplifier is the membrane potential of a short length of axon, and the other is a command signal voltage determined by the experimenter. The difference between these two is greatly amplified and fed back via the amplifier output to the same stretch of axon membrane. If the membrane potential differs from the command signal, then the output changes so as to reduce that difference. Hence the membrane potential is maintained constant and the current passing through the patch of membrane can be measured.

Before considering the results of Hodgkin and Huxley's voltage clamp experiments, we need to look at a model of the axon membrane that was used by them in analyzing their results (Hodgkin and Huxley, 1952). The model is in the form of an equivalent electric circuit, as shown in Figure 7. It contains three different pathways for ionic current flow: for sodium ions, potassium ions, and a third pathway for leakage current, which is probably largely due to chloride ions.

Each pathway contains a battery in series with a resistance. The batteries represent the Nernst potentials produced by the ionic concentration gradients. Thus the sodium battery E_{Na} is in series with the sodium resistance R_{Na} and produces the sodium current I_{Na}. It is more convenient to use conductance, the reciprocal of resistance, because we can use the sodium conductance g_{Na}, for example, to represent the permeability of the membrane to sodium ions. The sodium and potassium conductances are variable, being affected by the membrane potential E and the time t. In parallel with the ionic pathways is the membrane capacitance, C_m.

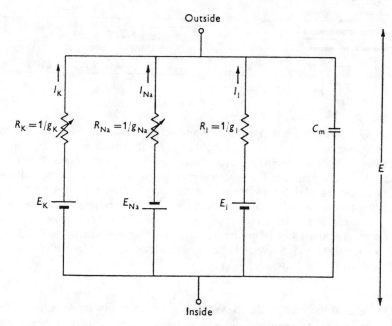

Figure 7. Ionic pathways in a patch of electrically excitable membrane. (After A.L. Hodgkin and A.F. Huxley [1952]. J. Physiol. 117, 500–544.)

The current through each of the ionic pathways is given by Ohm's law (the current through a resistance is proportional to the potential across it) so that,

$$I_{Na} = g_{Na}(E - E_{Na}), \tag{4}$$

$$I_K = g_K(E - E_K), \tag{5}$$

and

$$I_l = g_l(E - E_l)$$

The total ionic current I_i is the sum of the individual ionic currents,

$$I_i = I_{Na} + I_K + I_l$$

and the total membrane current I is the sum of the ionic current and the current through the membrane capacitance, which is itself proportional to the rate of change of membrane potential with time:

$$I = I_i + C_m \frac{dE}{dt}$$

This means that when the membrane potential is held constant (clamped) $dE/dt = 0$, and so the record of current flow gives a direct measure of total ionic flow across the membrane.

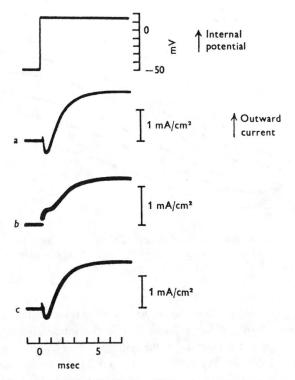

Figure 8. Membrane currents during a voltage clamp of squid giant axon. Traces **a** and **c** show currents with the axon in sea water; for trace **b** the axon was placed in a sodium-free choline chloride solution. Notice that the initial inward current in **a** and **c** becomes outward in **b**. (From A.L. Hodgkin [1958]. Proc. R. Soc. Lond. B 148, 1–37.)

We shall see later that ionic current flow occurs through individual molecular channels, which are essentially pores in the membrane that may be open or closed. In terms of Figure 7, the sodium conductance is proportional to the number of sodium channels that are open, being high when most of them are open and low when most of them are closed. The sodium current is thus the sum of the individual currents through the sodium channels. Similar considerations apply to the potassium channels.

Figure 8A shows the type of record obtained by Hodgkin and Huxley. There are three components to the current flow. First, there is a brief blip of outward current; this is caused by a discharge of the membrane capacitance associated with the change in membrane potential from its resting level to the clamped potential of (in this case) about +15 mV. There are then two phases of ionic current flow: the current is initially inward for about 1 ms, and then it becomes outward and climbs to a steady level, which is maintained for the duration of the clamp.

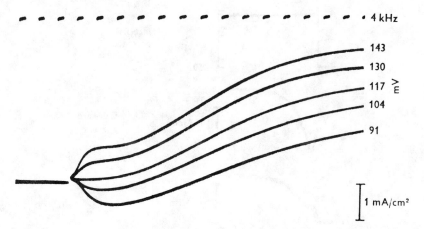

Figure 9. Membrane currents at large depolarizations in a voltage-clamped squid axon. The figures at the right of each record show the change in membrane potential. (From A.L. Hodgkin [1958]. Proc. R. Soc. Lond. B 148, 1–37.)

If we apply the sodium theory to this ionic current flow, we may conclude that the initial inward current is due to sodium ions flowing inward. If this is so it should be altered if we alter the potential of the sodium battery by altering the external sodium ion concentration. Figure 8B shows that this is so; sodium ions have been removed by using choline chloride as a substitute for sodium chloride in the external solution, with the result that the initial current is now outward.

Further evidence for this conclusion is shown in Figure 9, which shows superimposed current records produced by a series of different depolarizations. As the clamped potential approaches the sodium equilibrium potential, the size of the initial inward current decreases; at a depolarization of 117 mV (corresponding to a membrane potential of about +57 mV, which is clearly very near to E_{Na}) it becomes zero, and at greater depolarizations it becomes outward. This is just what we would expect from equation 4; to put it another way, at a depolarization of 117 mV the electrical gradient driving the sodium ions outward just balances the concentration gradient driving them inward, and at greater depolarizations it exceeds it.

The 117 mV trace in Figure 9 contains no sodium current, and therefore (after subtraction of a small constant leakage current) it represents the potassium current. Measurement of the potassium current at other membrane potentials depends on altering the sodium equilibrium potential by changing the external sodium ion concentration. In Figure 10, for example, trace *b* shows the ionic current when the membrane is depolarized by 56 mV in a solution containing only 10% of the normal sodium ion concentration; there is no initial sodium current and so the trace represents the potassium current at this potential. By subtracting trace *b* from that

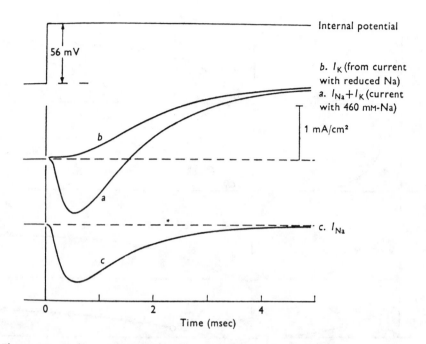

Figure 10. Sodium and potassium currents during voltage clamp in a squid axon. Trace **a** shows the response to a depolarization of 56 mV with the axon in sea water. Trace **b** was obtained with the axon in 10% sea water, 90% isotonic choline chloride: under these conditions the membrane potential during the clamp is equal to the sodium equilibrium potential so there is no sodium current and the trace therefore shows the potassium current. Trace **c** is the difference between **a** and **b** and is therefore the sodium current. (From A.L. Hodgkin [1958]. Proc R. Soc. Lond. B 148, 1–37.)

obtained in normal sea water (trace *a*), we get the sodium current itself, shown in trace *c*.

By applying equations 4 and 5 to traces *c* and *b* of Figure 10 we can see how the sodium and potassium conductances change with time following depolarization. By making similar measurements at different depolarizations, the effects of membrane potential on the ionic conductances can be determined, as is shown in Figure 11. Small depolarizations produce small slow conductance changes, larger ones produce changes that are larger (up to a maximal value) and more rapid. In terms of channels, small depolarizations open a small proportion of the channels, whereas large ones open many more. The sodium conductance increase is always transient, reaching a peak early and falling back to zero within a few milliseconds. Hodgkin and Huxley called the decline in conductance the "sodium inactivation process." The potassium conductance change (potassium activation) is slower than that of sodium but does not show inactivation.

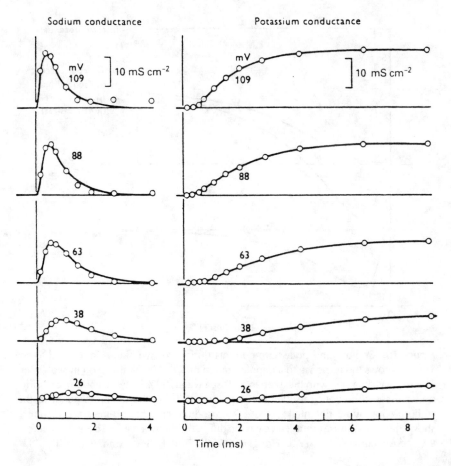

Figure 11. Conductance changes brought about by different clamped depolarizations in squid axon. The circles show values derived from experimental measurements (as in Figure 10), and the curves are drawn according to equations used to describe them. (From A.L. Hodgkin [1958]. Proc. R. Soc. Lond. B 148, 1–37.)

The next step in the analysis was to set up a series of equations to describe how the conductance changes shown in Figure 11 varied with membrane potential and time. These equations were then used to calculate the form of the action potential in an unclamped axon. [Hodgkin (1976) describes how Huxley had to use a hand-cranked calculating machine to do the calculations because the Cambridge University computer was out of action at the time.] The result (Figure 12) was remarkably accurate: there were only slight differences between the calculated action potential and the real one. This provided a striking vindication for the membrane theory of nervous conduction and showed that Hodgkin and Huxley's

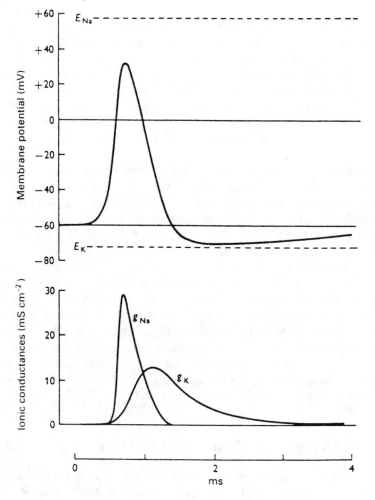

Figure 12. The propagated action potential (upper curve) as calculated from voltage clamp measurements on squid giant axon. The lower curves show the calculated changes in sodium and potassium conductances; they are proportional to the numbers of open sodium and potassium channels. (From D.J. Aidley [1989]. The Physiology of Excitable Cells, 3rd edition. Cambridge University Press, Cambridge. Redrawn after A.L. Hodgkin & A.F. Huxley [1952]. J. Physiol. 117, 500–544.)

description of the ionic conductance changes accounted for the electrical behavior of the nerve axon in a most satisfactory manner.

It is instructive to follow the potential and conductance changes during the time course of the propagated action potential shown in Figure 12. Initially g_K is small but g_{Na} is smaller (i.e., there are just a few of the potassium channels open but almost none of the sodium channels), so that the resting potential is near to the potassium equilibrium potential, E_K. Then, as we shall see later, the presence of an action potential approaching along the axon causes an initial depolarization; notice that this occurs before there is any change in the ionic conductances. When the membrane has been depolarized by about 15 mv, a few sodium channels open so that g_{Na} begins to rise. Hence a relatively small number of sodium ions move down their electrochemical gradient through these open channels. This transfer of charge results in a further depolarization, which causes more sodium channels to open (seen as a further increase in g_{Na}), so more sodium ions cross the membrane, and so on.

The result of this runaway snowball effect is that the membrane potential goes racing up toward the sodium equilibrium potential, E_{Na}. But now the two slower consequences of depolarization, sodium inactivation and potassium activation, begin to take effect: some of the sodium channels close and an increasing number of potassium channels open, so that g_{Na} begins to fall and g_K rises. This means that the rate of sodium ion inflow decreases and the rate of potassium ion outflow increases, so that there is a net transfer of charge out of the axon and the membrane potential begins to fall again. This repolarization closes more of the sodium channels (it also closes potassium channels, but more slowly), so that the membrane potential is brought rapidly back toward its initial inside-negative level. At this point, although the number of sodium channels open is extremely low, there are still quite a lot of potassium channels open, so the membrane potential passes the resting level and moves even nearer to the potassium equilibrium potential, E_K. Finally, as the number of open potassium channels declines to its normal low value, the membrane potential returns to its resting level.

SEPARATING THE IONIC CURRENTS BY DRUG ACTION

Tetrodotoxin is a virulent nerve poison found in the tissues of the Japanese puffer fish. Using voltage-clamped lobster axons, Narahashi, Moore, and Scott (1964) found that tetrodotoxin in the external solution blocks the increase in sodium conductance but has no effect on the potassium conductance. This provides further evidence that sodium and potassium ions flow through different pathways and are controlled by different mechanisms. A similar specific blocking action is produced by saxitoxin, a substance produced by certain marine dinoflagellates.

The dose-response relation of the blocking action suggests that each tetrodotoxin or saxitoxin molecule combines with a single sodium channel. Thus radioactive toxin can be used to count the sodium channel densities in axon membranes. For squid giant axons, densities in the region of 300 channels per μm^2 are found; rather lower values occur in nonmyelinated axons of smaller diameter (Keynes and Ritchie, 1984).

Tetraethylammonium (TEA) ions produce prolongation of the action potential when injected into squid axons. Voltage clamp experiments show that this prolon-

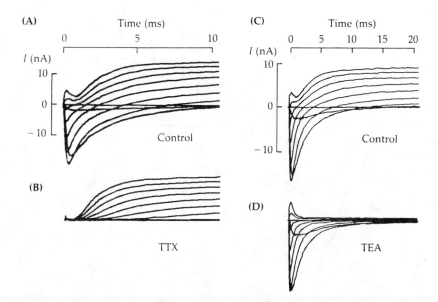

Figure 13. Ionic currents of frog axon (node of Ranvier) separated by specific blocking agents. The records show families of voltage clamp currents produced by depolarizations to membrane potentials ranging from –60 to +60 mV. Trace **A** shows the initial sodium current and later potassium current seen under normal conditions. Trace **B** is from the same node after treatment with tetrodotoxin (TTX); the sodium current has been eliminated. Trace **C** shows control measurements in another node; after treatment with tetraethylammonium (TEA) ions (trace **D**) the potassium current is eliminated. (From B. Hille [1984]. Ionic Channels of Excitable Membranes. Sinauer Associates, Sunderland, Massachusetts.)

gation is caused by a blockage of the potassium channels. Similar effects are produced by 4-aminopyridine and cesium ions.

These drugs can be used to separate the components of the total current flow, as is shown in Figure 13 for frog nerve fibers. When the potassium channels are blocked with TEA ions, the currents produced by clamped depolarizations are inward and transient, whereas when the sodium channels are blocked with tetrodotoxin the currents are entirely outward and maintained. Similar results have been obtained from squid axons; they are clearly in complete accordance with the Hodgkin–Huxley analysis.

SINGLE CHANNEL RESPONSES

Voltage clamp experiments of the Hodgkin–Huxley type can provide us with information about the overall behavior of large numbers of channels, but they

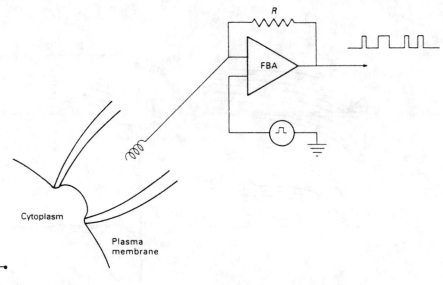

Figure 14. Patch clamp recording. The pipette electrode contains an appropriate saline solution. The voltage clamp circuit consists of a high-gain feedback amplifier (FBA) in which the output (on the right) is fed back via the resistor R to the input. This ensures that the input voltage is kept very nearly equal to the signal voltage derived from the square pulse generator. The output is thus proportional to the current flowing through the patch of membrane in contact with the electrode. (From D.J. Aidley [1989]. The Physiology of Excitable Cells, 3rd edition. Cambridge University Press, Cambridge.)

cannot tell us how the individual channels behave. In order to do this we need to measure much smaller currents from a much smaller area of membrane. The technique for doing this, first used by Neher and Sakmann in 1976 and much developed since then (see Sakmann and Neher, 1983), is known as patch clamping.

Figure 14 gives an indication of how patch clamping is done. A glass pipette microelectrode is polished to produce a smooth tip 1 to 2 μm in diameter, coated with a resin to enhance its insulation, and filled with an isotonic electrolyte solution. It is then pushed against a patch of cell membrane, and suction is applied so as to pull the cell membrane into the tip of the electrode. The result is that a very high resistance seal (a gigaseal) is formed between the membrane and the glass wall of the electrode. This greatly reduces the background noise from the system and so enables the very small currents flowing through the ionic channels to be measured with a voltage clamp circuit.

The first patch clamp records of single sodium channel currents were made by Sigworth and Neher (1980) using myoballs, which are spherical cells produced by growing embryonic rat muscle cells in tissue culture in the presence of colchicine.

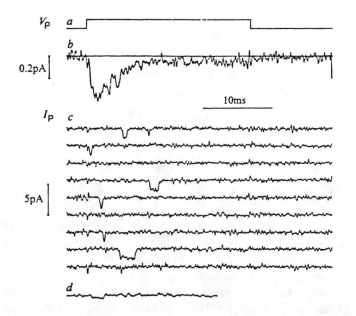

Figure 15. Single sodium channel currents, recorded with the patch clamp technique from cultured rat muscle cells. Trace **a** shows the duration of an imposed depolarization of 40 mV. Trace **b** shows the average current elicited by 300 such pulses at 1-second intervals (compare trace **c** in Figure 10). Nine successive individual records from this set are shown in **c**. Notice that square pulses of inward current (average size 1.6 pA) can be seen in most records: these correspond to the opening of individual sodium channels. Trace **d** shows a record taken when the sodium ion concentration in the patch clamp pipette was reduced to a third; the single channel current is reduced accordingly. (From F.J. Sigworth and E. Neher [1980]. Nature 287, 447–449.)

Embryonic cells are well suited to patch clamp methods because they do not usually have any extracellular material adhering to the cell surface membrane. The patch pipettes contained TEA to block any potassium channels present and also a blocking agent for acetylcholine channels.

The records in Figure 15 show successive responses to a clamped depolarization by 40 mV. Most of the traces show square inward current pulses. These are always the same amplitude but they differ in duration and timing.

How can we be sure that these unitary pulses of current are carried by sodium ions? Sigworth and Neher showed that they were blocked by tetrodotoxin, that they were reduced when the sodium concentration in the pipette was reduced (trace *d* in Figure 15), and that they decreased in size at less negative membrane potentials, disappearing in the region of E_{Na}.

The conductance γ of the open channel can be calculated from the equation

$$i = \gamma \, (E - E_{Na})$$

where i is the single channel current and $(E - E_{Na})$ is the difference between the clamped membrane potential and the sodium equilibrium potential. In the experiment shown in Figure 15,

$$\gamma = 1.6/90 \quad pA/mV$$

$$= 18 \, pS$$

The individual pulses in Figure 15 are much the same size, but they vary considerably in duration and timing. Thus the opening and closing of the channels are stochastic processes, which means that we cannot predict precisely when any particular channel will open or close, but we can in principle determine the probability that it will do so in any one time interval.

We can now give a molecular interpretation of the sodium conductance changes during a voltage clamp of a relatively large area of membrane. At any particular instant a channel is either open or closed; the transitions between the two states are effectively instantaneous. This means that the conductance of an area of axon membrane is proportional to the number of channels that are in the open state. When the membrane is depolarized the probability of any particular channel opening is increased, and so the total number of open channels rises and the overall sodium conductance rises also. During inactivation or after repolarization the probability of any particular channel opening is reduced and so the total number of open channels and the overall sodium conductance fall.

This explanation is a little simplified in that under certain circumstances it is sometimes possible to observe subconductance states, where the current flow through the channel is some fraction of the normal open-channel current. But these are rare events and probably do not make any significant contribution to the macroscopic sodium current (Meves and Nagy, 1989).

The patch clamp technique has also been used to measure the properties of potassium channels in squid axons (Conti and Neher, 1980). The results were in many ways similar to those for sodium channels: the individual channels were either open or shut, and the probability of opening was increased on depolarization. The single channel conductance in the open state was 12 pS. Potassium channels of the type involved in axonal action potential are known as delayed rectifier channels; many other potassium channel types have been revealed by patch clamping and other techniques in recent years (Latorre et al., 1984), but they fall outside the scope of this chapter.

Recent and exciting advances in the molecular biology of the ionic channels, which have given us an understanding of their amino acid sequences and some useful ideas about their molecular structure, are described in Chapter 4.

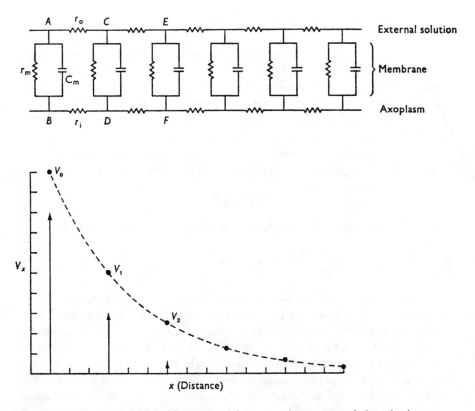

Figure 16. Electrical model of the passive (electrotonic) properties of a length of axon. The graph shows the steady-state distribution of membrane potential when points **A** and **B** are connected to a constant current source. The vertical arrows on the graph are used to indicate that the potential at any point on the membrane rises gradually to its final value when the current is applied. (From D.J. Aidley [1989]. The Physiology of Excitable Cells, 3rd edition. Cambridge University Press, Cambridge.)

THE LOCAL CIRCUIT THEORY

Let us now turn to another aspect of nervous conduction, the mechanism of propagation of the action potential. In order to understand this we need to think about the spread of potential along an electrical circuit representing the passive electrical properties of the axon. Such a circuit is shown in Figure 16; it assumes that each short length of axon membrane possesses a resistance and capacitance in parallel and that these elements are connected together by the longitudinal resistances of the external solution and of the axoplasm. The model is known as the core-conductor model.

If we apply a constant current across the model membrane at one point (across AB in Figure 16, for example), then the voltage at that point will rise rapidly but not instantaneously to a final value V_0; the time taken to reach V_0 depends on the value of the membrane capacitance. The voltage across the next element CD will also begin to rise, but not so rapidly (because its own capacitance has to be charged up as well as that associated with AB), and the final value V_1 is less than V_0 because some of the voltage is dropped across r_0 and r_i. Similarly the voltage across EF also begins to rise, but yet more slowly and to a still lower final value V_2.

Mathematical analysis shows that the distribution of potential along the model at long times (the dotted curve in Figure 16) is an exponential given by,

$$V_x = V_0\, e^{-x/\lambda}$$

where x is distance and λ is a constant (called the space constant) given by

$$\lambda^2 = \frac{r_m}{r_0 + r_i} \tag{6}$$

Such a model describes well the potential changes in an axon produced by hyperpolarizing current or small depolarizing currents. It only accounts for the passive (electrotonic) behavior of the membrane; the membrane resistance remains constant and no channels open or close. But what would happen if we introduce electrical excitability into the model? Suppose the voltage V_0 represents an action potential. Then the threshold depolarization will be about one fifth of V_0, and so it will be crossed at CD as the membrane potential moves toward V_1. But because crossing the threshold changes the properties of the system by opening lots of sodium channels, the voltage at V_1 will now go shooting up toward V_0. That means that the voltage across EF, originally moving toward V_2, will now begin to move more rapidly toward V_1 until it too crosses the threshold and moves rapidly to V_0. So the voltage across AB has moved through CD to EF and will obviously continue propagating at a constant velocity along the length of the model.

Another way of representing this idea is shown in Figure 17, where the currents associated with a propagating action potential are shown. In the active region we have inward currents; we have seen that these are in fact sodium ions flowing through their open channels. These currents spread along the axon in front of the active region and so raise the membrane potential past the threshold at which sodium channels begin to be opened. This idea that propagation occurs as a result of passive current flow in front of the active region is known as the local circuit theory. The outward currents behind the active region are largely the potassium ions moving outward through the potassium channels.

The local circuit theory was first formulated by Hermann in the nineteenth century, but clear evidence for it was not forthcoming until the late 1930s. Then Hodgkin (1937, 1939) managed to demonstrate the existence of the local circuits and also showed that the conduction velocity was affected by the external resis-

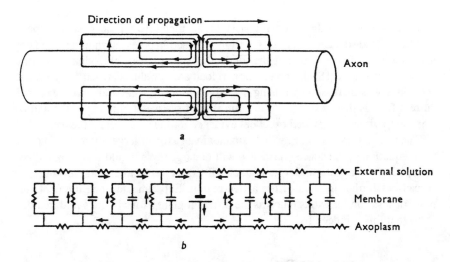

Direction of propagation ⟶

Axon

External solution

Membrane

Axoplasm

a

b

Figure 17. Local circuit currents. **a** shows the currents accompanying a propagating action potential; **b** shows the currents (arrows) set up by the insertion of a battery in the core-conductor model. (From D.J. Aidley [1989]. The Physiology of Excitable Cells, 3rd edition. Cambridge University Press, Cambridge.)

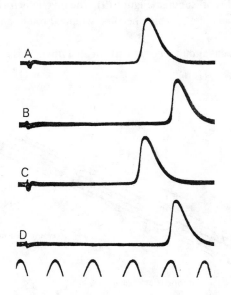

A

B

C

D

Figure 18. Evidence for the local circuit theory. The traces show the action potential in a crab axon following a stimulus, which is seen as an artefact at the beginning of each trace. The axon was immersed in sea water in **A** and **C**, in medicinal paraffin oil in **B** and **D**. Notice that the reduced external resistance produced by the oil has increased the conduction time. The time trace shows ms. (From A.L. Hodgkin [1939]. J. Physiol. 94, 560–570.)

tance. If we increase the external resistance, r_0, then this will decrease the space constant λ, and so the local circuits in front of an action potential should spread out less far along the axon, reducing the conduction velocity. One of Hodgkin's results is shown in Figure 18; the conduction velocity was reduced when the external resistance was increased by raising the axon into medicinal paraffin oil (external current flow is then restricted to the thin film of sea water clinging to the axon). These experiments provided excellent evidence for the local circuit theory.

The internal resistance, r_i, will be smaller in an axon of larger diameter, and so, from equation 6, the space constant λ will be larger. We would therefore expect conduction to be faster in larger diameter fibers as the local circuit currents will spread out further in front of the active region. Experimental measurements on more or less similar axons (such as the different giant axons in the squid mantle) show that this is indeed so.

CONDUCTION IN MYELINATED AXONS

Most vertebrate nerve fibers of large and moderate diameter are myelinated: they are surrounded by a fatty myelin sheath. The sheath is broken at intervals of 1 to 2 mm or so at the nodes of Ranvier (Figure 19). The myelin is formed from the cell membrane of Schwann cells, which become wrapped round and round the axon many times.

If we apply the local circuit theory to myelinated fibers, it is evident that current

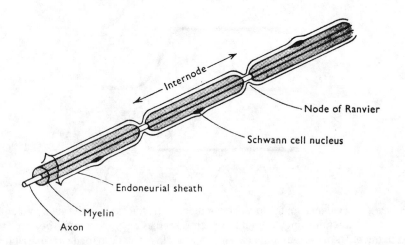

Figure 19. Diagram to show the structure of a myelinated nerve fiber. (From D.J. Aidley [1989]. The Physiology of Excitable Cells, 3rd edition. Cambridge University Press, Cambridge.)

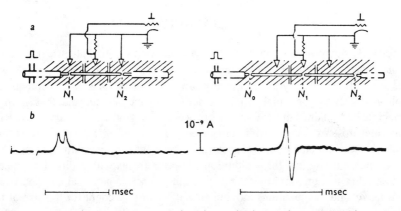

Figure 20. Inward current is restricted to the node during the passage of an action potential in a myelinated fiber. Experiments were done on single frog nerve fibers, using the recording arrangement shown in **a**. The radial currents are shown in **b**, when the middle pool of saline does (right trace) or does not (left trace) contain a node. Inward current is downward. (From I. Tasaki and T. Takeuchi [1942]. Pflügers Archiv. 245, 764–782.)

flow across the membrane will be largely restricted to the nodes. Excitation will therefore jump from one node to the next, and conduction will thus be faster than in a nonmyelinated axon. This idea is known as the saltatory theory of conduction. Good evidence for it was supplied by Tasaki and Takeuchi (1942) and Huxley and Stämpfli (1949), who showed that inward current is indeed restricted to the nodes (Figure 20).

Voltage clamp experiments on myelinated axons require a rather different electrode arrangement than in squid giant axons, because it is not possible to use internal wire electrodes. The results of work on frog fibers, however, accord well with those from squid axons (Frankenhaeuser, 1965; Hille, 1971). As is shown in Figure 13, depolarization produces a transient sodium current that is inward at membrane potentials less than E_{Na} and a maintained outward potassium current. In mammalian fibers, however, there is no potassium current at the nodes (Chiu et al., 1979).

Saxitoxin binding measurements on rabbit central nodes suggest that the sodium channel density there is in the range 400 to 700 channels per μm^2 (Pellegrino and Ritchie, 1984). This value is much higher than axon membranes of nonmyelinated axons. The myelin sheath can be removed by treatment with lysolecithin; measurements on the exposed internodal axon membrane show the presence of potassium currents (Chiu and Ritchie, 1982). There is also some sodium current in the internodal region, but the sodium channel density there is only about 4% of that at the nodes (Shrager, 1987).

SUMMARY

Electrical excitability is seen most clearly in the conduction of electrical impulses in nerve axons. The first stage in understanding the nature of these impulses was the suggestion that they are essentially properties of the plasma membrane: the voltages seen are potential gradients across the membrane, and these can in principle be explained in terms of ionic concentration gradients. Clear evidence for this view arose from the use of internal electrodes in squid giant axons; the resting potential was affected by the concentration gradient for potassium ions, and the action potential by that for sodium ions. Measurements with radioactive tracers showed that sodium ions enter the axon during an action potential and potassium ions leave, and the quantities moving are sufficient to account for the size of the action potential.

The next stage was the use of the voltage clamp method to separate the ionic currents and to determine the relations between depolarization and the sodium and potassium conductances. This enormously successful analysis by Hodgkin and Huxley allowed them to predict the time course of the action potential with remarkable precision. More recently, attention has switched to the responses of single sodium and potassium channels. Such channels (to simplify a little) may be either open or closed; depolarization increases the probability that they will be open.

The current flow through the open sodium channels of an active region spreads along the axon in front of the action potential, where it depolarizes the membrane so as to open more sodium channels. The resulting all-or-nothing action potential is an elegant solution to the problem of sending undistorted electrical messages within the body.

REFERENCES

Chiu, S.Y., & Ritchie, J.M. (1982). Evidence for the presence of potassium channels in the internode of frog myelinated nerve fibres. J. Physiol. 322, 485–501.

Chiu, S.Y., Ritchie, J.M., Rogart, R.B., & Stagg, D. (1979). A quantitative description of membrane currents in rabbit myelinated nerve. J. Physiol. 292, 149–166.

Cole, K.S. (1968). Membranes, Ions and Impulses. University of California Press, Berkeley.

Cole, K.S., & Curtis, H.J. (1939). Electric impedance of the squid giant axon during activity. J. Gen. Physiol. 22, 649–670.

Conti, F., & Neher, E. (1980). Single channel recordings of K^+ currents in squid axons. Nature 285, 140–143.

Curtis, H.J., & Cole, K.S. (1942). Membrane resting and action potentials from the squid giant axon. J. Cell. Comp. Physiol. 19, 135–144.

Frankenhaeuser, B. (1965). Computed action potential in nerve from Xenopus laevis. J. Physiol. 180, 780–787.

Hille, B. (1971). Voltage clamp studies on myelinated nerve fibers. In: Biophysics and Physiology of Excitable Membranes (Adelman, W.J., ed.), pp. 230–246. Van Nostrand Reinhold, New York.

Hodgkin, A.L. (1937). Evidence for electrical transmission in nerve. J. Physiol. 90, 183–232.

Hodgkin, A.L. (1939). The relation between conduction velocity and the electrical resistance outside a nerve. J. Physiol. 94, 560–570.

Hodgkin, A.L. (1964). The Conduction of the Nervous Impulse. Liverpool University Press, Liverpool.

Hodgkin, A.L. (1976). Chance and design in electrophysiology: an informal account of certain experiments on nerve carried out between 1934 and 1952. J. Physiol. 263, 1–21.

Hodgkin, A.L., & Horowicz, P. (1959). The influence of potassium and chloride ions on the membrane potential of single muscle fibres. J. Physiol. 148, 127–160.

Hodgkin, A.L., & Huxley, A.F. (1939). Action potentials recorded from inside a nerve fibre. Nature 140, 710–711.

Hodgkin, A.L., & Huxley, A.F. (1952). A quantitative description of membrane current and its application to conduction and excitation in nerve. J. Physiol. 117, 500–544.

Hodgkin, A.L., Huxley, A.F., & Katz, B. (1952). Measurement of current-voltage relations in the membrane of the giant axon of *Loligo*. J. Physiol. 116, 424–448.

Hodgkin, A.L., & Katz, B. (1949). The effect of sodium ions on the electrical activity of the giant axon of the squid. J. Physiol. 108, 37–77.

Huxley, A.F., & Stämpfli, R. (1949). Evidence for saltatory conduction in peripheral myelinated nerve fibres. J. Physiol. 108, 315–339.

Keynes, R.D. (1951). The ionic movements during nervous activity. J. Physiol. 114, 119–150.

Latorre, R., Coronado, R., & Vergera, C. (1984). K^+ channels gated by voltage and ions. J. Membrane Biol. 71, 11–30.

Meves, H., & Nagy, K. (1989). Multiple conductance states of the sodium channel and of other ion channels. Biochem. Biophys. Acta 988, 99–105.

Narahashi, T., Moore, J.W., & Scott, W.R. (1964). Tetrodotoxin blockage of sodium conductance increase in lobster giant axons. J. Gen. Physiol. 47, 965–974.

Pellegrino, R.G., & Ritchie, J.M. (1984). Sodium channels in the axolemma of normal and degenerating optic nerve. Proc. R. Soc. Lond. B 222, 155–160.

Sakmann, B., & Neher, E. (eds.) (1983). Single-channel Recording. Plenum Press, New York.

Shrager, P. (1987). The distribution of sodium and potassium channels in single demyelinated axons of the frog. J. Physiol. 392, 587–602.

Sigworth, F.J., & Neher, E. (1980). Single Na^+ channel currents observed in cultured rat muscle cells. Nature 287, 447–449.

Tasaki, I., & Takeuchi, T. (1942). Weitere Studien über den Aktionsstrom der markhaltigen Nerfenvaser und über die elektrosaltorische Übertragung des Nervenimpulses. Pflügers Archiv. 245, 764–782.

RECOMMENDED READINGS

Aidley, D.J. (1989). The Physiology of Excitable Cells, 3rd edn. Cambridge University Press, Cambridge.

Hille, B. (1992). Ionic Channels of Excitable Membranes, 2nd edn. Sinauer Associates, Sunderland, Massachusetts.

Hodgkin, A.L. (1964). The Conduction of the Nervous Impulse. Liverpool University Press, Liverpool.

Kuffler, S.W., Nicholas, J.G., & Martin, A.R. (1984). From Neuron to Brain, 2nd edn. Sinauer Associates, Sunderland, Massachusetts.

Chapter 9

Stimulus–Secretion Coupling in Gland Cells

OLE H. PETERSEN

INTRODUCTION

The term stimulus–secretion coupling covers all the processes from interaction of chemical stimulator substances with specific receptor sites on cell membranes to the initiation of the secretory events. Under this heading the innervation of glands will not be discussed nor the mechanisms underlying the control of neurotransmitter secretion. The centerpiece is the gland cell and the processes occurring in the plasma membrane as well as within the cell following a stimulatory challenge.

Fundamentals of Medical Cell Biology, Volume 5A
Membrane Dynamics and Signaling, pages 171–202
Copyright © 1992 by JAI Press Inc.
All rights of reproduction in any form reserved.
ISBN: 1-55938-309-7

It is not the purpose of this chapter to describe the specific details of stimulus–secretion coupling in every type of the very many different gland cells that exist but rather to discuss the main steps that are relevant in most systems. In general an external signal in the form of a hormone, neurotransmitter, or metabolite is received by the gland cells, and this external signal gives rise to an intracellular signal that switches on the secretory process. The secretory event itself may be export of the macromolecules accumulated in secretory vesicles by exocytosis (fusion of vesicle membrane with plasma membrane and opening at the point of fusion) and/or electrolyte and fluid secretion due to opening of ion channels and activation of various ion pumps.

One of the most interesting and complicated topics is the transduction mechanism by which an extracellular message evokes an intracellular signal. The most important intracellular signal for secretion is an increase in the free Ca^{2+} concentration ($[Ca^{2+}]_i$), and this chapter will mainly deal with the various mechanisms for generation of intracellular Ca^{2+} signals and their modulation by other messengers.

OVERVIEW OF SIGNAL TRANSDUCTION MECHANISMS

Figure 1 illustrates the main systems at work. The initial step is always binding of a hormone or neurotransmitter to a receptor. There is of course a well-known overlap between hormones and neurotransmitters (e.g, adrenaline/noradrenaline), and thus the term hormone is meant here to indicate specific chemical stimulator molecules.

Hormone receptors may be defined as those components of target cells that specifically react with hormones and that, as a consequence of such an interaction, initiate events leading to cellular responses. Hormone receptors clearly have two functions: (1) discrimination for a specific hormone, and (2) generation of a response.

Binding of hormone to membranes is generally considered to be the sum of at least two processes, one saturable, normally termed specific binding (to receptors), and the other non-saturable (non-specific). The latter is composed of low-affinity binding to the membrane as well as non-specific adsorption to glass or plastic surfaces in incubation tubes. Because the specific binding sites have a much higher affinity than the non-specific sites, most binding at physiological hormone concentrations occurs to the specific receptor sites. The number of specific binding sites on single cells may vary from 500 to 250,000 (Kahn, 1976). The simplest quantitative analysis of steady-state hormone receptor binding data is based on considering the interaction as a simple reversible bimolecular equilibrium:

$$[H] + [Rec] \rightleftharpoons [H\text{-}Rec]$$

$$k_a/k_d = K_a = [\text{H-Rec}]/[\text{H}] [\text{Rec}]$$

[H] = concentration of free hormone,
[Rec] = concentration of unoccupied receptor,
[H-Rec] = concentration of hormone–receptor complexes,
k_a = association rate constant,
k_d = dissociation rate constant,
K_a = affinity (equilibrium) constant ($K_d = 1/K_a$)

This simple data interpretation appears adequate for a number of hormone–receptor interactions. Most peptide hormones have affinity constants of 10^8 to 10^{10} L/mol. In some cases, however, there may be two or more classes of receptor sites differing in affinity for the hormone. Other complications are site–site interactions (cooperativity). If the binding of the first ligand increases the affinity of the receptor for the second ligand, one talks about positive cooperativity. These kinds of cooperativity are called homotropic. Heterotropic cooperativity indicates a change in affinity between hormone and receptor caused by binding of another type of ligand (e.g., guanyl nucleotides).

The binding of hormones to receptors is very specific; in general a single hormone receptor will only bind one type of hormone (Kahn, 1976). Receptors for peptide hormones and neurotransmitters are at their highest concentration on the plasma membrane. This can be seen by direct studies of labeled hormone binding to intact cells and is clear also from biochemical studies showing that the receptor concentration is markedly enhanced during purification of plasma membrane. Receptors are supposed to face only the external surface of the plasma membrane, and our present concept of the plasma membrane is of a structure composed of lipids and proteins with the lipids arranged in a bilayer configuration. Membrane lipid molecules can move about in the membrane in the fluid state. Integral proteins (i.e., proteins that can only be dissociated from the membrane using strong detergents), such as hormone receptors, associate with the lipid bilayer and appear to be intercalated to various depths in the fluid matrix. Such proteins and glycoproteins are capable of rapid lateral mobility although there are also certain restraining mechanisms preventing free movement and random distribution. In fact, hormone receptors are not always homogeneously distributed (Kahn, 1976). The concentration of hormone receptors is not fixed but depends on continuous synthesis and degradation. A number of factors influence the membrane receptor concentration, and one important factor may be the hormone concentration itself. In the case of insulin receptors, there seems to be an inverse relation between the serum concentration of the hormone and the concentration of receptors.

If every single hormone receptor is linked to an effector, we would expect the response of a target tissue to be directly proportional to the number of receptor sites occupied by hormone (i.e., the response = a [H-Rec]). A maximal response would be obtained when all receptors were occupied and binding and biological response curves should be superimposable. This in fact is not normally observed. Furthermore, Schramm et al. (1977) have been able to dissociate and associate receptors from effectors. By very elegant studies involving fusion of cells with different receptors, but the same effector (adenylate cyclase), Schramm et al. (1977) have clearly shown that it is possible to transfer receptors from one cell to another (mobility of receptors) and for these receptors to couple themselves to another effector. In general one therefore now prefers to think about receptors and effectors as separate molecules floating around in the membrane, preferably interacting when the receptor is bound to hormone.

It is now clear that receptors do not directly interact with effectors, but that special transducing proteins, which are able to bind guanosine triphosphate (GTP), are linking hormone receptors to their effectors (Figure 1). These proteins are normally called G proteins. G proteins are heterotrimers consisting of α, β and γ subunits. The α subunit is specific for a particular G protein and contains the GTP-binding domain. When the receptor is unoccupied by hormone, the α, β, and γ subunits of the associated G protein are joined together and guanosine diphosphate (GDP) is present in the nucleotide-binding region. Occupation of the receptor by the appropriate hormone or neurotransmitter initiates a process by which GDP is exchanged for GTP, and this leads to dissociation of the activated α subunit from

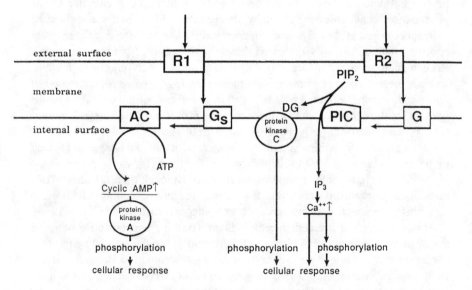

Figure 1. Schematic diagram to explain the general features of signal transduction mechanisms. **R**, receptor; **G**, GTP-binding protein; **AC**, adenylate cyclase; **PIC**, phosphoinositidase C (phospholipase C); **PIP₂**, phosphatidylinositol 4,5-bisphosphate; **DG**, 1,2 diacylglycerol; **IP₃**, inositol 1,4,5-trisphosphate.

the rest of the G protein complex. The α subunit binds to the effector (for example, an enzyme or an ion channel) and causes its activation or inhibition. The process is reversible because GTP can be hydrolyzed to GDP, and in this state the α subunit will dissociate from the effector and rejoin the β and γ subunits, reestablishing the prestimulation situation. A number of different G proteins have been characterized in considerable detail (Gilman, 1987), but a detailed account of this very rapidly moving field is outside the scope of this chapter.

The most important effectors regulated by G proteins that are relevant to the topic of stimulus–secretion coupling are enzymes such as adenylate cyclase (the enzyme that controls the formation of the messenger cyclic adenosine monophosphate [cAMP] from ATP) and phosphoinositidase C (phospholipase C) (the enzyme that controls the formation of inositol trisphosphate [IP_3] and diacylglycerol [DG] from phosphatidylinositol bisphosphate [PIP_2]) (Figure 1) as well as various ion channels.

For gland cells the most important process is the one leading to formation of IP_3 because this is the messenger that releases Ca^{2+} from intracellular stores and therefore initiates most secretory responses. Ca^{2+} can directly bind to regulatory sites on certain ion channels, for example, or can cause phosphorylation of a variety of target proteins. When IP_3 is released there is always an obligatory formation of DG (Figure 1). IP_3 is water soluble, whereas DG is lipid soluble and therefore remains in the plasma membrane where it activates the enzyme protein kinase C that is able to phosphorylate various targets, again including enzymes and ion channels.

The biochemistry of cAMP is particularly well described and a specific G protein, G_s, that activates adenylate cyclase and another called G_i that inhibits adenylate cyclase are known. cAMP evokes phosphorylation of various target proteins including enzymes and ion channels via a cAMP-dependent protein kinase (protein kinase A). It is obvious that there is considerable scope for interaction between the various messenger pathways and there are several known examples of this (Petersen and Bear, 1986).

THE CALCIUM SIGNALING SYSTEM

A Ca^{2+} signal is an increase in the cytoplasmic Ca^{2+} concentration ($[Ca^{2+}]_i$) and can be generated either by release of Ca^{2+} from intracellular stores or by opening of Ca^{2+} channels in the plasma membrane. Intracellular Ca^{2+} release is triggered by the internal messenger IP_3 generated as a result of hormone–receptor interaction-induced breakdown of the membrane inositol lipid phosphatidylinositol bisphosphate (PIP_2) (Figure 1). IP_3 and its phosphorylation product inositol tetrakisphosphate (IP_4) are also involved in regulating Ca^{2+} uptake from the extracellular fluid through pathways that have not yet been well characterized. In some gland cells there is also another Ca^{2+} influx pathway, namely voltage-gated

Ca^{2+} channels. These are channels that open up in response to membrane depolarization. A stimulant that primarily evokes depolarization by causing opening or closing of appropriate ion channels may therefore secondarily evoke Ca^{2+} influx through depolarization-activated Ca^{2+} channels. In some cells both the inositol polyphosphate - Ca^{2+} signaling system and the voltage-gated Ca^{2+} channels operate together.

Control of Ca^{2+} Influx via Voltage-Gated Ca^{2+} Channels

The Voltage-Gated Ca^{2+} Channel

There are a number of different types of voltage-gated Ca^{2+} channels (Tsien et al., 1988), but in gland cells the only type that has been studied in detail is the so-called L type. Figure 2 shows typical L type single-channel currents recorded from an insulin-secreting cell. Because the amplitude of the single-channel currents is relatively small and Ca^{2+} channels mostly coexist with other channel types (particularly voltage-gated K$^+$ channels) that often conduct larger currents, it is generally necessary to block the K$^+$ currents in order to obtain clear single-channel currents from the Ca^{2+} pores. A useful trick is to use Ba^{2+} as the current carrier (Figure 2). This has several advantages: (1) The mobility of Ba^{2+} through Ca^{2+} pores is generally higher than that of Ca^{2+} itself, giving rise to somewhat larger single-channel currents. (2) Ba^{2+} is a very effective blocker of many different K$^+$ channel currents, and (3) intracellular Ba^{2+}, unlike Ca^{2+}, does not cause inactivation of the Ca^{2+} channels. Figure 2 shows the important characteristics of the L type single-channel currents. The channels are opened by membrane depolarization, and a rather large depolarization is required (high threshold channels). During a depolarizing pulse lasting 100 or 200 ms, there are repeated channel openings, and the chance of observing channel openings does not diminish with time within such individual pulses (non-inactivating channels). The single-channel conductance in the type of experiment shown in Figure 2 is about 25 pS, which is higher than for other voltage-gated Ca^{2+} channels of the N or T type (Tsien et al., 1988).

The Ca^{2+} channels are not exclusively controlled by the membrane potential although membrane depolarization is required for their opening. In some gland cells, for example those secreting insulin, the stimulant-evoked depolarization is insufficient by itself to cause any significant degree of Ca^{2+} channel opening. The insulin-secreting cells are stimulated to secrete by an increase in the extracellular glucose concentration, but a glycolytic intermediate, glyceraldehyde, is an equally effective stimulus. As seen in Figure 2, glyceraldehyde changes the relationship between membrane depolarization and Ca^{2+} channel opening so that less depolarization is required for the same degree of opening. The mechanism underlying this effect is not absolutely clear, but it is known that a cell-permeable diacylglycerol analogue can mimic the action of glyceraldehyde (Velasco and

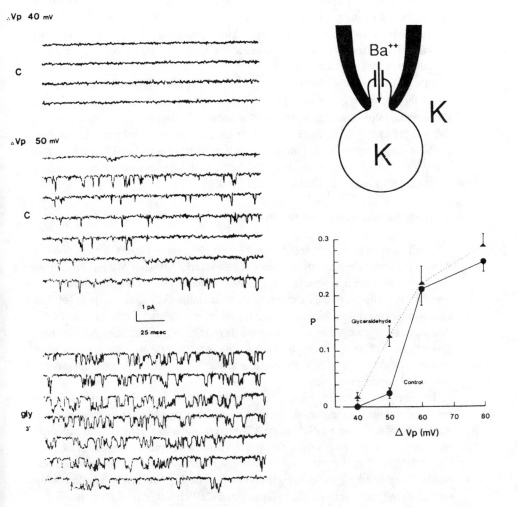

Figure 2. Voltage-gated single-channel Ca^{2+} (Ba^{2+}) currents. The cartoon shows the recording configuration. All traces are from an insulin-secreting cell line derived from rat pancreas (RINm5F). The holding potential was –70 mV (the normal resting potential in these cells), and depolarizing voltage jumps (ΔV_p) lasting 200 ms were applied. **C** denotes that traces were obtained under control conditions (without stimulation). **Gly** indicates that glyceraldehyde (10 mM) had been added to bath solution to stimulate insulin secretion. The graph shows channel open state probability (P) as a function of the magnitude of the depolarizing voltage jump (ΔV_p). (The single-channel current traces are from Velasco et al., 1988.)

177

Petersen, 1989). Diacylglycerol, as shown in Figure 1, may act by activating protein kinase C and could thereby cause specific phosphorylation of appropriate target proteins. Phosphorylation of the Ca^{2+} channels may change the gating behavior toward longer openings. In cardiac cells it is well established that phosphorylation mediated by cAMP-dependent protein kinase increases L type Ca^{2+} channel opening. This effect is normally evoked in the heart cells by noradrenaline acting on beta receptors to generate cAMP.

In a pancreatic acinar cell line (AR 42J) the peptide substance P evokes Ca^{2+} influx through voltage-gated L type Ca^{2+} channels and has been shown to enhance the depolarization-induced inward Ca^{2+} current. Also in this case it would appear that this effect may be mediated via protein kinase C stimulation because activation of this enzyme by various compounds such as phorbol esters and cell-permeable diacylglycerol analogues can generate Ca^{2+} signals similar to those produced by substance P stimulation (Gallacher et al., 1990).

Stimulant-Evoked Depolarization

In all cases where Ca^{2+} influx occurs through voltage-gated Ca^{2+} channels, a depolarization of the plasma membrane is the essential first step. We must therefore consider how such a depolarization can be brought about. There are a number of possibilities. The stimulant (hormone, neurotransmitter, or metabolite) can activate ion channels that allow inward current (in practice, influx of Na^+ or efflux of Cl^-). Another strategy would be to close channels through which outward currents pass (essentially outflux of K^+). A third possibility is to activate an electrogenic carrier system allowing influx of cations (in practice Na^+). All these mechanisms are employed in various glands.

In adrenal chromaffin cells, acetylcholine (ACh) interaction with nicotinic receptors causes opening of relatively non-selective cation channels, and the main depolarization is due to Na^+ influx. These channels are actually the nicotinic receptors as is the case in the end-plates of skeletal muscle. Opening of such channels causes a depolarization sufficient to activate voltage-dependent Na^+ channels, again like those in muscle and nerve, causing yet further depolarization sufficient to finally activate the voltage-gated Ca^{2+} channels (Petersen, 1980).

In insulin-secreting cells, the primary effect of the main carbohydrate secretagogues is to close resting K^+ channels. These K^+ channels are sensitive to changes in the intracellular adenosine triphosphate (ATP) and adenine diphosphate (ADP) concentrations, and it is likely that glucose acts by increasing the cytosolic ATP:ADP concentration ratio, thereby decreasing the channel open state probability (Petersen and Findlay, 1987). The closure of the resting K^+ channels reduces the membrane potential, and this helps to promote Ca^{2+} channel opening although as mentioned earlier an additional stimulus directly to the Ca^{2+} channels is also needed. A neuropeptide, vasopressin, that also stimulates insulin secretion can close the same nucleotide-sensitive K^+ channels, possibly directly via a G protein (Martin et al., 1989). Substances that inhibit insulin secretion, for example another neuropeptide galanin, act by opening the nucleotide-sensitive K^+ channels, thereby hyperpolarizing the cell membrane, and thus make Ca^{2+} channel opening less likely (Dunne et al., 1989).

Amino acids can evoke depolarization of insulinoma cells by allowing linked Na^+ influx, via, for example, an alanine–Na^+ cotransport system (Petersen and Findlay, 1987). This depolarization can in some cells be sufficient to cause opening of voltage-gated Ca^{2+} channels.

Ca^{2+} Signals Evoked by Activation of Receptors Linked to Inositol Lipid Hydrolysis

In this section we are concerned with Ca^{2+} signals evoked by hormone or neurotransmitter activation of receptors linked via G proteins to the enzyme phosphoinositidase C (also referred to as phospholipase C). Binding of neurotransmitters or hormones to such receptor types causes breakdown of the membrane-bound inositol lipid phosphatidylinositol 4,5-bisphosphate (PIP_2), producing initially two separate messengers, namely the lipid-soluble diacylglycerol (DG) and the water-soluble inositol 1,4,5-trisphosphate (IP_3), (Berridge, 1987) (Figure 1). In 1983 it was shown for the first time that IP_3 releases Ca^{2+} from an intracellular non-mitochondrial store (Streb et al., 1983), and soon thereafter it became clear that this Ca^{2+} store was in the endoplasmic reticulum (Berridge, 1987).

The Ca^{2+} signals often oscillate (Figure 3) and because receptor-mediated repetitive Ca^{2+} spikes can be observed in the absence of extracellular Ca^{2+} (Berridge & Irvine, 1989), it would appear that pulsatile release of Ca^{2+} from intracellular stores (Figure 3) is primarily responsible for the oscillations.

Pulsatile IP_3 Formation Is Not Required for Pulsatile Intracellular Ca^{2+} Release

From the simplest possible model concept shown in Figure 3, it seems that pulsatile IP_3 formation could provide a straightforward explanation for the oscillating Ca^{2+} signal. At this point in time it is not possible to measure IP_3 concentrations in single cells with a high time resolution, and we therefore do not know whether IP_3 levels oscillate during receptor activation. It is, however, possible to address the central question in a different manner by asking whether a constant concentration of IP_3 can evoke pulsatile Ca^{2+} release.

Intracellular injections of IP_3 into *Xenopus* oocytes have been shown to evoke repetitive membrane depolarizations due to Ca^{2+} activation of Cl^- channels (Figure 3). In such experiments a constant intracellular IP_3 concentration cannot be

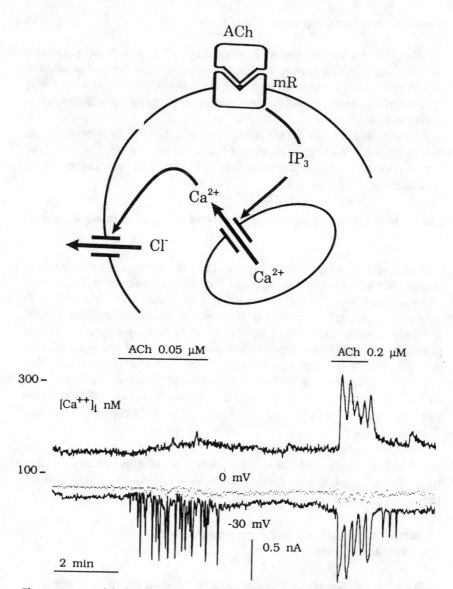

Figure 3. Acetylcholine (ACh)-evoked intracellular Ca^{2+} signals in a single mouse pancreatic acinar cell. The cartoon illustrates the general scheme for the ACh action. The traces shown in the lower half of the figure were all obtained from the same cell. The top trace shows $[Ca^{2+}]_i$ (nM) measured by fura-2 fluorescence. The bottom trace shows the Ca^{2+}-dependent Cl^- current at -30 mV (patch clamp whole-cell recording). At 0 mV (middle trace) ACh has virtually no effect as the Cl^- concentration in this experiment was the same inside and outside ($E_{Cl} \sim 0$). ACh in a low concentration (0.05 μM) causes Ca^{2+} spikes near the cell membrane (Ca^{2+}-dependent Cl^- current) and at a higher concentration (0.2 μM) causes synchronous Ca^{2+} spikes near the membrane and in the cell at large. (The original traces are from Osipchuk et al., 1990.)

achieved, but similar findings have been obtained when mouse pancreatic acinar cells are internally perfused with a constant IP_3 concentration in patch clamp experiments where the cytoplasmic Ca^{2+} fluctuations have been assessed by measuring the Ca^{2+}-dependent Cl^- current (Figure 3) (Wakui et al., 1989). Even such experiments can be criticized because it cannot be ruled out that metabolism and particularly Ca^{2+}-dependent phosphorylation of IP_3 to IP_4 (Berridge and Irvine, 1989) occurs. To solve this problem it is therefore necessary to use a non-metabolizable IP_3 analogue with Ca^{2+}-releasing activity. Inositol 1,4,5-trisphosphorothioate (IPS_3) is an effective releaser of Ca^{2+} from intracellular stores but is not metabolized by phosphatase or kinase pathways. IPS_3 perfused continuously via a pipette into single pancreatic acinar cells evokes regular and repetitive spikes of Ca^{2+}-dependent Cl^- current, showing that a constant messenger concentration can cause pulsatile intracellular Ca^{2+} release (Wakui et al., 1989). Although these experiments cannot exclude that receptor activation results in pulsatile IP_3 formation, it is an unnecessary complication to postulate that such a phenomenon occurs. The most economical hypothesis is that a constantly elevated IP_3 concentration elicits pulsatile Ca^{2+} release in normal intact cells when stimulated by appropriate concentrations of hormones or neurotransmitters.

Ca^{2+}-Induced Ca^{2+} Release Is Important for the Generation of Ca^{2+} Oscillations Initiated by IP_3

Because IP_3 at a constant concentration can evoke repetitive spikes of intracellular Ca^{2+} release (Wakui et al., 1989) and because the simplest suggestion would be that IP_3 primarily evokes a constant movement of Ca^{2+} from the IP_3-sensitive Ca^{2+} pool, it is pertinent to ask whether a small constant flow of Ca^{2+} into the cytosol can evoke repetitive transport of Ca^{2+} from another pool.

Ca^{2+}-induced Ca^{2+} release was originally discovered in skinned cardiac muscle cells where a small increase in the Ca^{2+} concentration of the fluid in contact with the sarcoplasmic reticulum causes a large Ca^{2+} release (Endo, 1977). In the presence of the drug caffeine, skinned muscle fibers exhibit spontaneous contractions when bathed in a solution with a relatively low Ca^{2+} concentration (Endo, 1977). If Ca^{2+}-induced Ca^{2+} release is directly responsible for the cytoplasmic Ca^{2+} oscillations evoked by stimuli generating IP_3, then intracellular Ca^{2+} infusion should be able to mimic the effect of IP_3. Figure 4 shows the result of an experiment in which Ca^{2+} was infused into a single cell while $[Ca^{2+}]_i$ was assessed both by microfluorimetry using the fluorescent dye fura-2 and by measurement of Ca^{2+}-dependent Cl^- current (Osipchuk et al., 1990). In this type of experiment the electrical current trace monitors $[Ca^{2+}]_i$ in the immediate vicinity of the inner surface of the plasma membrane whereas the microfluorimetrical recording reports the average $[Ca^{2+}]_i$ in the cell. As seen in Figure 4 the intracellular Ca^{2+} infusion results in a gradual rise in $[Ca^{2+}]_i$, and this increase is reversed when the Ca^{2+} infusion is

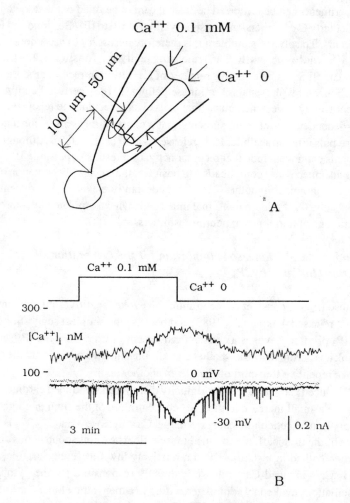

Ca^{++} 0.1 mM

Ca^{++} 0

100 μm 50 μm

A

Ca^{++} 0.1 mM

Ca^{++} 0

300 –

[Ca^{++}]$_i$ nM

100 –

0 mV

-30 mV

0.2 nA

3 min

B

Figure 4. Ca^{2+}-induced repetitive Ca^{2+} release from stores close to the cell membrane in single mouse pancreatic acinar cell. The cartoon in **A** shows the arrangement used to exchange solution at the tip of the patch clamp pipette. Application of 200 mm Hg pressure to the tube containing 0.1 mM Ca^{2+} resulted in a slow rise of [Ca^{2+}]$_i$ and Ca^{2+}-dependent Cl$^-$ current (bottom trace in **B**) and also in repetitive short-lasting spikes of Ca^{2+}-dependent Cl$^-$ current (bottom trace in **B**. Application of pressure to the tube with Ca^{2+}-free solution (also containing the Ca^{2+} chelator EGTA) reversed the effects. (Adapted from Osipchuk et al., 1990.)

182

stopped and a Ca^{2+} chelator applied. In the cytoplasmic space near the plasma membrane there are additional short-lasting repetitive Ca^{2+} spikes that are particularly pronounced just before the major sustained $[Ca^{2+}]_i$ rise and in the phase after discontinuation of the Ca^{2+} infusion. These results indicate that Ca^{2+} can induce pulses of Ca^{2+} release primarily from pools very close to the cell membrane. IP_3 infusion also evokes Ca^{2+} spikes near the cell membrane, which are mostly not reflected in the average $[Ca^{2+}]_i$ recording, although in small cells it is possible to detect IP_3-evoked Ca^{2+} oscillations synchronously with both microfluorimetry and electrophysiology (Osipchuk et al., 1990). Low doses of ACh also predominantly evoke Ca^{2+} spikes seen only at the cell membrane, whereas larger doses cause slightly broader Ca^{2+} signals seen throughout the cell (see Figure 3) (Osipchuk et al., 1990). The conclusion from these experiments is therefore that Ca^{2+} infusion can mimic the action of IP_3.

In the course of normal signal transduction, IP_3 most likely evokes a small steady flow of Ca^{2+} into the cytosol that subsequently causes repetitive pulses of Ca^{2+} release primarily from stores close to the cell membrane. According to this model, two different types of Ca^{2+} channels must exist in intracellular Ca^{2+}-storing organelles, namely IP_3-activated and Ca^{2+}-activated Ca^{2+} release channels (see Figure 5).

IP_3 and Ca^{2+} Activate Two Separate Types of Ca^{2+}-Release Channels

Table 1 summarizes some of the properties of the two functionally very different Ca^{2+} release channels from intracellular Ca^{2+}-storing organelles. The pharmacology is of particular significance. The IP_3-activated Ca^{2+} channel is inhibited by heparin, whereas this substance has no effect on the Ca^{2+}-induced Ca^{2+} release channel. The Ca^{2+}-induced Ca^{2+} release channel is inhibited by ruthenium red and activated by caffeine, substances that have no effect on the IP_3-activated Ca^{2+} channel. The Ca^{2+}-induced Ca^{2+} release channel, often referred to as the ryanodine receptor, because it has a very high affinity for the plant alkaloid ryanodine, has been characterized mainly in muscle tissues. In skeletal muscle this channel normally functions by releasing Ca^{2+} from the sarcoplasmic reticulum in response to depolarization of the T-tubular cell membrane. A junctional complex has been described consisting of the voltage-sensing dihydropyridine receptor in the T-tubule membrane and the ryanodine receptor Ca^{2+} channel (foot) complex in the sarcoplasmic reticulum. In cardiac cells it would appear that the dihydropyridine receptor functions as a voltage-sensitive Ca^{2+} channel, and the Ca^{2+} entering from the outside during membrane depolarization then evokes Ca^{2+} release through the ryanodine receptor channel (Fleischer and Inui, 1989). In general this channel can therefore be activated either by electromechanical transduction, in a way not completely understood, or by Ca^{2+} binding to a site on the cytoplasmic side of the

Table 1. Characteristics of the Two Ca^{2+} Release Channels from
Non-Mitochondrial Intracellular Ca^{2+} Stores

	IP₃-activated Ca²⁺ channel	*Ca²⁺activated Ca²⁺ channel*
Structure	Probably homotetramer (each of the 4 polypeptides with M_T 250,000) with square shape (Maeda et al.,1990). The amino acid sequence has been determined (Furuichi et al., 1989).	Homotetramer (each of the 4 polypeptides with M_T 400,000) forming four-leaf clover "feet" (Lai et al., 1988). The amino acid sequence has been determined (Takeshima et al., 1989).
Activator	IP₃ [only (1,4,5) and (2,4,5) but not (1,3,4) isomer] (Ferris et al. 1989)	Ca²⁺ (Lai et al., 1988)
Potentiator		Caffeine (Rousseau et al., 1988), ATP, and ATP analogues (Meissner, 1984; Smith et al., 1985)
Inhibitor	Heparin (Ferris et al. 1989)	Ruthenium red and Mg²⁺ (Meissner et al., 1986)
Single-channel conductance	10 pS (Ehrlich & Watras, 1988)	~100 pS (Smith et al., 1986; Lai et al., 1988; Ehrlich & Watras, 1988)

sarcoplasmic reticulum membrane, thereby inducing the conformational change presumably required for the increase in open state probability (Fleischer and Inui, 1989). It is clear that the concentrations of ATP and Mg^{2+} in particular are important, as ATP promotes opening whereas Mg^{2+} has the opposite effect (Table 1). When the skeletal muscle ryanodine receptor is expressed in Chinese hamster ovary cells there is no depolarization-evoked Ca^{2+} release, but caffeine effectively releases Ca^{2+} (Penner et al., 1989). In pancreatic acinar cells surface membrane depolarization does not cause Ca^{2+} release (Matthews et al., 1973), but Ca^{2+}-induced Ca^{2+} release occurs (Osipchuk et al., 1990) and caffeine can evoke cytoplasmic Ca^{2+} oscillations under certain conditions (Petersen and Wakui, 1990). It is therefore clear that the Ca^{2+}-induced Ca^{2+} release channel does not necessarily have to be functionally linked to a plasma membrane voltage sensor, and in electrically non-excitable cells, for example those found in epithelia, the absence of such a coupling is probably normal.

Two Intracellular Non-mitochondrial Ca²⁺-Pools

The model shown in Figure 5 requires two separate Ca^{2+} pools with Ca^{2+} channels having different characteristics. The existence of at least two non-mitochondrial intracellular Ca^{2+} pools has been proposed on the basis of studies where it can be shown that IP₃ is only able to release Ca^{2+} from a minor compartment of the endoplasmic reticulum Ca^{2+} pool (Berridge and Irvine, 1989). In pancreatic acinar cells the two Ca^{2+} pools seem to have different types of Ca^{2+} uptake mechanisms. The IP₃-insensitive pool has a normal Ca^{2+} ATPase pump,

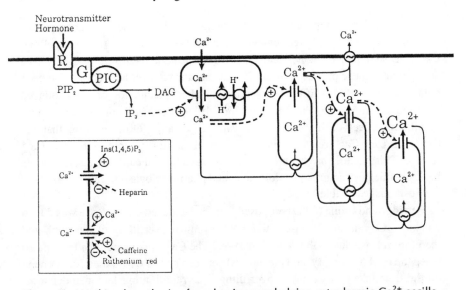

Figure 5. Working hypothesis of mechanisms underlying cytoplasmic Ca^{2+} oscillation evoked by receptor activation. The basic concept is based on the model by Berridge and Irvine (1989). This present version of the model emphasizes the different characteristics of the two types of Ca^{2+} channels in intracellular Ca^{2+} storing organelles that are now known and also highlights the progression of the Ca^{2+} signal from the surface cell membrane to the cell interior. (This idea is based on the data from Osipchuk et al. 1990.)

whereas the IP_3-sensitive pool accumulates Ca^{2+} via a machinery consisting of a H^+ ATPase pump and a Ca^{2+}-H^+ exchanger (Figure 5).

The IP_3 binding protein has now been identified as the IP_3-activated Ca^{2+} channel and has been localized by immunohistochemical techniques to intracellular particles associated with the endoplasmic reticulum. In exocrine acinar cells there is evidence indicating that the major part of the internal Ca^{2+} release evoked as a consequence of the processes initiated by receptor activation primarily comes from stores very close to the plasma membrane. Cytoplasmic Ca^{2+} oscillations evoked by low doses of ACh, as well as by only just suprathreshold IP_3 or Ca^{2+} stimuli applied internally, are detected near the cell membrane but not in the cell at large (Figures 3 and 4). Caffeine potentiates these responses and makes them easily detectable in the whole cytoplasm (Osipchuk et al., 1990). This suggests that the Ca^{2+}-induced Ca^{2+} release, which appears to be the major fraction of the Ca^{2+} signal, primarily occurs close to the cell membrane although Ca^{2+}-induced Ca^{2+} release can be evoked throughout the cell interior with stronger stimulation or when Ca^{2+}-induced Ca^{2+} release is potentiated by caffeine.

The two-pool model shown in Figure 5 is based on a considerable amount of experimental evidence. The IP_3-activated Ca^{2+} channel has been isolated and sequenced and most importantly its physiological and pharmacological properties investigated (Table 1). Direct evidence for Ca^{2+}-induced repetitive intracellular Ca^{2+} release in pancreatic acinar cells and evidence that the resulting Ca^{2+} spikes are similar to those produced by surface membrane receptor activation have been obtained (Figure 4). Caffeine, a well-established potentiator of Ca^{2+}-induced Ca^{2+} release in muscle (Fleischer and Inui, 1989), markedly enhances the Ca^{2+}-mobilizing effect of IP_3 and ACh (Osipchuk et al., 1990), showing that the Ca^{2+}-activated Ca^{2+} release channel is important for receptor-activated Ca^{2+} release. This channel has been isolated from skeletal and cardiac muscle, and its basic chemical, structural, physiological, and pharmacological properties are now well-established (Table 1).

Berridge and Irvine (1989) proposed that Ca^{2+}-induced Ca^{2+} release was due to uptake of Ca^{2+} in an IP_3-insensitive pool, eventual overfilling of the pool, and therefore release due to the "Ca^{2+} pressure." The Ca^{2+} spike would be terminated when the pool was empty, and the interval between spikes would therefore depend on the time taken for renewed overfilling. The rate-limiting factor under most circumstances would be the IP_3-evoked Ca^{2+} flow that provided the Ca^{2+} to be taken up into the Ca^{2+}-sensitive pool. According to this model concept Ca^{2+} release would be activated by the "Ca^{2+} pressure" inside the store. The Ca^{2+}-induced Ca^{2+} release (ryanodine) channel is, however, activated by Ca^{2+} acting on the outside (cytoplasmic side) of the channel, and it is therefore simpler to suggest that the Ca^{2+} released by IP_3 into the cytoplasm acts directly on the ryanodine channel to cause opening. Whether the Ca^{2+}-sensitive Ca^{2+} pools are actually emptied during individual spikes is unknown. An alternative explanation for the oscillatory behavior of the system may be negative feedback by Ca^{2+} so that a rise in $[Ca^{2+}]_i$ at a critical site above the level that opens the ryanodine channels would switch them off. The concept of repetitive emptying and filling of Ca^{2+}-sensitive pools is appealing from the point of view of explaining the increasing frequency of spikes with increasing agonist dose observed in some cell types but seems unattractive energetically. The amounts of Ca^{2+} released at low intensities of stimulation, apparently only generating Ca^{2+} spikes close to the cell membrane (Osipchuk et al., 1990), may be quite small.

Repetitive release of short-lasting Ca^{2+} pulses, which under physiological conditions may be restricted to the crucial cytoplasmic area immediately under the plasma membrane (Osipchuk et al., 1990), is an effective and precise, but also energy-saving, signaling mechanism. The sequential operation of two types of Ca^{2+} pools as shown in Figure 5 allows signal amplification in a simple manner. The possibility of the Ca^{2+} waves spreading by Ca^{2+}-induced Ca^{2+} release throughout the cell at higher intensities of stimulation (Osipchik et al., 1990) is also useful because Ca^{2+} may serve as a messenger for many different types of events not all necessarily controlled at the cell membrane. By distributing the sensitivity of

Ca^{2+} release to Ca^{2+} in particular ways it may be possible to control the spreading of Ca^{2+} signals at different intensities of stimulation.

Control of Ca^{2+} Influx by IP_3 and IP_4

Up to now the receptor-activated cytoplasmic Ca^{2+} oscillations have been dealt with as if they were completely independent of extracellular Ca^{2+}, and the model shown in Figure 5 does not have an element of regulated Ca^{2+} inflow. This is reasonable in the sense that oscillations in $[Ca^{2+}]_i$ have been observed in many systems in the absence of external Ca^{2+}, but in other studies receptor-activated Ca^{2+} spikes disappear when external Ca^{2+} is removed and rapidly reappear after Ca^{2+} readmission (Berridge and Irvine, 1989).

The mouse pancreatic acinar cells are particularly interesting because extracellular Ca^{2+} dependency is present or absent depending on the precise experimental situation. In intact single cells where the Ca^{2+} oscillations are studied by microfluorimetry after loading with fura-2 acetoxymethyl ester (Yule and Gallacher, 1988), ACh evokes oscillations in $[Ca^{2+}]_i$ that in the initial phase (1–2 minutes) are independent of external Ca^{2+}. In the sustained phase of stimulation, however, the removal of Ca^{2+} immediately causes cessation of the oscillations and these reappear after external Ca^{2+} readmission. Surprisingly the same cells isolated in exactly the same way in the same laboratory behave in a somewhat different manner when investigated in the patch clamp whole-cell recording configuration. In such experiments both ACh- and IP_3-evoked Ca^{2+} oscillations are completely unaffected by removal and readmission of external Ca^{2+} for about 7–8 minutes after start of stimulation (Wakui et al., 1989), but external Ca^{2+} dependency thereafter gradually develops and finally becomes absolute. These experiments show that the oscillation mechanism is not primarily linked to control of Ca^{2+} influx through the surface cell membrane but indicate that during prolonged periods of stimulation intracellular Ca^{2+} deprivation occurs and reloading from the extracellular Ca^{2+} compartment becomes necessary. It therefore seems justified to focus on the intracellular transport events with regard to understanding the basic mechanisms underlying the cytoplasmic Ca^{2+} oscillations (Figure 5). Nevertheless there is regulation of Ca^{2+} influx, and in some preparations IP_3 alone cannot give a sustained Ca^{2+} signal but requires the presence also of IP_4 (Figure 6).

IP_3 is metabolized via two routes. IP_3-5 phosphatase converts its substrate to IP_2, a compound that does not release Ca^{2+}, but there is also an IP_3-3 kinase that phosphorylates IP_3 to IP_4. Subsequently IP_4 is broken down to the (1,3,4) IP_3 isomer that has no effect on Ca^{2+} homeostasis. IP_3-3 kinase is a Ca^{2+}-regulated enzyme, which means that Ca^{2+} released by the action of IP_3 will help to promote formation of IP_4. In many cell types it has been shown that both IP_3 and IP_4 formation occur very rapidly (within seconds) upon receptor activation.

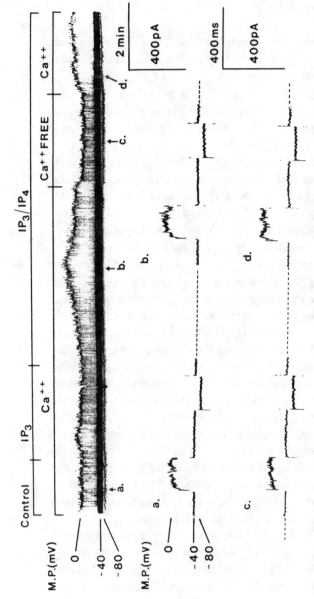

Figure 6. The effects of IP$_3$ and IP$_4$. Whole-cell current (patch clamp) recordings from a single mouse lacrimal acinar cell. The effect of intracellular application of IP$_3$ and the combination of IP$_3$/IP$_4$ (both 10 μM) on the Ca^{2+}-dependent outward K$^+$ current is demonstrated in the presence and absence of extracellular Ca^{2+}. The upper part represents a continuous record obtained with a slow time base and shows the small and transient IP$_3$-evoked increase in the Ca^{2+}- and voltage-dependent outward K$^+$ current. IP$_3$/IP$_4$ evokes a larger response with a sustained component that is acutely dependent on the presence of Ca^{2+} in the bath solution. The lower part shows four oscilloscope photographs demonstrating the outward current evoked by a depolarizing pulse from −40 mV to 0 as well as the very small inward current evoked by a hyperpolarizing pulse to −80 mV before IP$_3$ stimulation (**a**), during IP$_3$/IP$_4$ internal perfusion (**b**), after external Ca^{2+} has been removed during continued IP$_3$/IP$_4$ stimulation (**c**), and finally after readmission of external Ca^{2+} still in the presence of internal IP$_3$/IP$_4$ (**d**). (Adapted from Morris et al., 1987.)

188

Following preliminary and somewhat inconclusive experiments on sea urchin eggs, it was proposed that IP_4 together with IP_3 controls Ca^{2+} influx, but hard evidence was difficult to come by until experiments on internally perfused lacrimal acinar cells showed that, at least in that system, IP_4 had reproducible effects (Petersen, 1989).

The effects of IP_4 in contrast to those of IP_3 are likely to depend on an intact plasma membrane. It had therefore not been easy initially to develop a good system for testing the actions of IP_4 in a controlled manner, because this requires access to the cell interior under conditions where transport functions of the cell membrane or the consequences of such functions can be assessed. However, voltage clamp (patch clamp) studies on internally perfused single cells (Hamill et al., 1981) (Figure 3) are suitable for this type of work, because in such experiments the cell interior is equilibrated with the solution used to fill the patch clamp pipette (i.e., known concentrations of inositol polyphosphates can be introduced into the cell). Monitoring Ca^{2+}-dependent currents can, as previously mentioned, provide a qualitative or semiquantitative measure of $[Ca^{2+}]_i$ (Figures 3 and 4).

The mouse lacrimal acinar cells employed in the study of Morris et al. (1987) possess, unlike the previously mentioned mouse pancreatic acinar cells, but like most other mammalian exocrine gland acinar cells, Ca^{2+}- and voltage-activated high-conductance K^+-selective channels localized in the basolateral membrane (Petersen, 1986; Petersen and Gallacher, 1988). The ACh-evoked increase in the outward K^+ current was shown to be sustained in the presence of external Ca^{2+} but to be transient in its absence. The sustained K^+ current response was abolished by introducing into the cell a solution with a high concentration of the Ca^{2+}-chelator EGTA (Morris et al., 1987). It would therefore appear that agonist–receptor interaction evokes an initial release of stored Ca^{2+} followed by entry of the divalent cation from the extracellular fluid.

Figure 6 shows a typical result from an internally perfused mouse lacrimal acinar cell in which membrane depolarization evokes an outward Ca^{2+}-sensitive K^+ current. When IP_3 is introduced after an initial control period of several minutes' duration, there is an increase in the magnitude of the outward (K^+) current, but despite continuing dialysis of the cell with the IP_3-containing pipette solution, the K^+ current response is only transient. Thereafter a combination of IP_3 and IP_4 is introduced, evoking a renewed increase in the K^+ current, which is now sustained until Ca^{2+} is removed from the bath (external) solution. Thereafter the K^+ current quickly returns to the prestimulation level. Ca^{2+} readmission to the external (bath) solution in the continued presence of internal IP_3/IP_4 stimulation restores the sustained response. Both IP_3 and IP_4 are needed in order to evoke a sustained increase in the voltage-sensitive K^+ current that is acutely dependent on the presence of external Ca^{2+}.

The protocol illustrated in Figure 6 has been utilized to investigate the specificity of the action of a number of different inositol polyphosphates alone and in

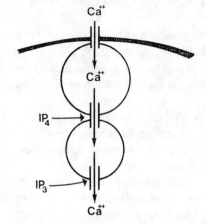

Figure 7. Simple model concept showing that whereas IP_3 opens channels allowing Ca^{2+} to move from within stores to the cytosol, IP_4 connects different Ca^{2+} stores, thereby also establishing a connection to extracellular compartment. (Adapted from Changya et al., 1989b.)

combination. A number of conclusions can be drawn from these experiments: (1) Sustained activation of the Ca^{2+}-sensitive K^+ current can only be achieved by a combination of an inositol tris- and tetrakisphosphate, and this effect is acutely dependent on the presence of external Ca^{2+} (Figure 6). (2) All the inositol trisphosphates capable of evoking release of Ca^{2+} from internal stores (transient effects in the absence of external Ca^{2+}) can also support IP_4 to evoke Ca^{2+} influx (sustained effect dependent on external Ca^{2+}). (3) The inability of IP_3 to evoke a sustained response cannot be explained by rapid metabolism of the inositol trisphosphate because the stable analogue IPS_3 also fails to evoke sustained responses.

The IP_3 antagonist heparin inhibits the ACh-evoked K^+ current response and also blocks the IP_3-evoked transient as well as the sustained K^+ current increase evoked by stimulation with internal IP_3 and IP_4. When during stimulation with IP_3 and IP_4 one of the inositol polyphosphates is removed, the K^+ current declines, but whereas removal of IP_3 results in an immediate termination of the response, removal of IP_4 only causes a gradual and slow reduction. IP_4 is therefore not an acute controller of Ca^{2+} release into the cytosol but modulates Ca^{2+} release induced by IP_3 by an unknown mechanism (Petersen, 1989).

With regard to the action of IP_4 specific binding sites distinct from those for IP_3 have been demonstrated, but there is no information about the intracellular localization of these sites. The precise identification of the target(s) for the action of IP_4 and the mechanism of its action are major tasks for the future. It should be borne in mind that IP_4 need not necessarily act only at the level of the plasma membrane. The work of Changya et al. (1989a) shows that IP_4, when added to the internal solution already containing IP_3, evokes a transient Ca^{2+} release in the absence of external Ca^{2+}, suggesting that IP_4 may also have a role in augmenting IP_3-evoked Ca^{2+} release, possibly by linking up different intracellular Ca^{2+} pools.

Figure 7 shows a much simplified model indicating how IP_4 may act together with IP_3 to promote Ca^{2+} influx. It should be emphasized that a role for IP_4 has not been demonstrated in all cellular systems investigated. One reason may be that the action of IP_4 is to establish semi-stable links between intracellular pools. In some cells these may be much more long-lasting than in others thereby explaining the difficulties of observing IP_4 effects in some systems.

MESSENGER ACTIONS IN CONTROL OF SECRETION

The two important secretory processes are exocytosis of macromolecules and fluid formation. There is now evidence available showing that regulation of ion channels is important for control of fluid secretion, and some of the functionally best defined messenger-activated molecules in secretory cells are Ca^{2+}-activated ion channels. Of these, K^+-selective channels are the ones studied in most detail.

Ca^{2+}-Activated Channels

By inserting the tip of a fine glass microelectrode into an acinar cell, it is possible to record the intracellular electrical potential relative to that in the extracellular fluid. Values vary between different glands and range from –30 to –75 mV (cell interior always negative with respect to outside). It is possible to make glass microelectrode tips selectively permeable for K^+ and in this way measurements can be obtained of the intracellular K^+ activity. Values obtained from resting acinar cells vary between 90 and 120 mM. The resting extracellular K^+ activity (concentration multiplied by activity coefficient [about 0.75 in physiological saline solutions]) is about 3 mM, and using the Nernst equation we can find the electrical potential difference across the acinar cell membrane that is required to maintain thermodynamic equilibrium for K^+:

$$E_K = RT/F \ln (aK^+o/aK^+i)$$

where E_K is the K^+ equilibrium potential, R is the gas constant, T the absolute temperature, F the Faraday constant, aK^+o the extracellular K^+ activity, and aK^+i the intracellular K^+ activity. At a temperature of 37 °C and using units for R and F giving the result in millivolts (mV) one obtains:

$$E_K = 61.5 \log (3/100) \text{ mV} = -93\text{mV}$$

It can be seen that the actual resting membrane potential (–30 to –75 mV) is always less negative than E_K and there must for this reason be a tendency for K^+ to move out of the cell. In the steady state this continuing K^+ loss down the

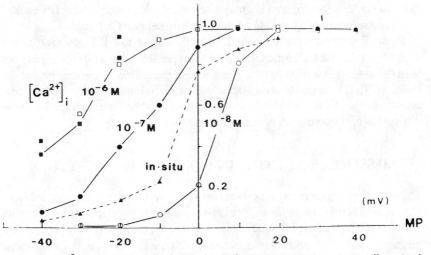

Figure 8. Ca²⁺-activation of single K⁺ channels. Pig pancreatic acinar cell excised inside-out membrane patch. Curves showing the relationship between open-state probability of K⁺ channel and membrane potential at three different levels of [Ca²⁺]ᵢ: O, 10^{-8} mol/L; ●, 10^{-7} mol/L; □, ■, 10^{-6} mol/L. All the experiments were carried out in the presence of normal Na⁺/K⁺ gradients. The open-state probability was determined by examining 30–40-s long continuous stretches of recordings at each potential level. The triangles (dashed line) represent results from an experiment on a cell-attached patch. [Ca²⁺]ᵢ in the intact cell was apparently between 10^{-7} and 10^{-8} mol/L. (Adapted from Maruyama et al., 1983.)

electrochemical gradient has to be balanced by an energy-requiring uptake against the gradient.

The most direct method available for investigating ion channels in cell membranes is patch clamp single-channel current recording. Instead of inserting a microelectrode into a cell, a microelectrode tip is pressed onto the surface of a cell, effectively isolating a small patch of membrane (1–10 μm²). With appropriate low noise amplifiers, the pico-ampere currents flowing through single channels when they open can be measured directly. The seal between the tip of a microelectrode and the outer surface of the cell membrane has, under suitable conditions, not only a high electrical resistance but also is mechanically very stable. The electrically isolated patch membrane can be pulled off the cell (excised) in such a way that the inside of the plasma membrane faces the bath solution (inside out) or alternatively so that the inside faces the solution in the micropipette (outside out). In the excised inside-out membrane patches taken from the basolateral side of pancreatic or salivary gland acini, the relationship between the membrane potential, [Ca²⁺]ᵢ, and the fractional open time (open state probability, p) of the channel can be investigated (Figure 8). When [Ca²⁺]ᵢ is increased 10-fold from 10^{-8} M to 10^{-7} M, the open state probability markedly increases at the negative membrane potentials. The K$^+$ channel can therefore be controlled by changing the membrane potential and/or by changing [Ca^{2+}]$_i$. This type of channel is referred to as a Ca^{2+}- and voltage-activated K$^+$ channel, and it is this channel that dominates the electrical characteristics of the basolateral plasma membrane in a large number of mammalian exocrine acinar cells, including the human pancreas and human submandibular gland (Petersen, 1986; Petersen and Findlay, 1987; Petersen and Gallacher, 1988). In the normal resting acinar cell the open state probability of the channel is low, but it is the only available exit pathway for K$^+$.

In isolated intact cells, single-channel currents due to opening of the high-conductance K$^+$ channels can be observed at the spontaneous resting potential, but in the acinar cells from both the pig pancreas and the mouse submandibular gland, the open state probability is very low with a value of only about 0.002. Depolarization of the plasma membrane evokes a marked increase in the fractional open time of the channels. The electrical resistance of the plasma membrane, which is dominated by the high-conductance K$^+$ channels, is therefore highly voltage dependent (Petersen and Gallacher, 1988). In order to quantify the total membrane K$^+$ conductance, it is not enough to know the single-channel conductance properties and the average open state probability, it is also necessary to obtain information on the channel density. This can only be done by comparing single-channel current data with whole-cell current recordings (Petersen, 1986).

In the resting pig pancreatic acinar cells, where this approach to channel quantification was first used, the total number of operational channels appears to be quite small, only about 50 per cell (Petersen and Maruyama, 1984). In ensemble fluctuation analysis of whole-cell currents it is possible to estimate the number of channels per cell (N), the single-channel current amplitude, and the open state probability (p) from measurements of current variance and mean current as a function of membrane potential, and such an analysis has been carried out on parotid acinar cells. The single-channel current–voltage relationship obtained from this analysis is virtually identical to that obtained directly from single-channel current recording, and the estimate of p as a function of membrane potential from the ensemble fluctuation analysis also agrees closely with the direct single-channel current measurements. The estimate, from the ensemble noise analysis, of N was scattered between 30 and 108 with a mean value of 76 (Petersen and Gallacher, 1988). Because the open state probability (p) in the resting intact cells is only about 0.002, this means that on average only 0.15 channel will be open in a single acinar cell. However, in the intact salivary glands and pancreas, the acinar cells are in fact extremely well coupled within units consisting of 100–500 cells (Petersen, 1980). Using the conservative estimate of 100 cells per acinar unit, there would be 7500 operational channels in the resting situation, and with $p = 0.002$ the average number of open channels would be 15 per unit. Because the resting value of p is so very low, the membrane has an enormous reserve K$^+$ conductance, which can be

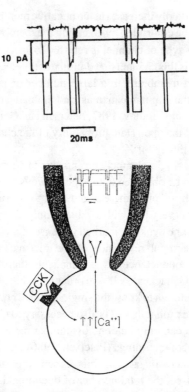

Figure 9. Schematic diagram illustrating messenger-mediated activation of an ion channel by the hormone cholecystokinin (CCK). Single-channel current trace displayed in the pipette tip shows transition between complete channel closure and repeated inward current steps due to openings of a K^+ channel in the isolated patch membrane, following application of CCK to the cell outside the isolated patch area. The interpretation is that CCK–receptor interaction releases calcium into the cytoplasm, evoking an increase in intracellular Ca^{2+} concentration ($[Ca^{2+}]_i$). Ca^{2+} diffuses into the patch area, activating the channel. The single-channel recording is from a pig pancreatic acinar cell. (Adapted from Petersen and Findlay, 1987.)

mobilized either by depolarizing influences or by hormones and neurotransmitters evoking intracellular Ca^{2+} release (Petersen and Gallacher, 1988).

The most direct demonstration of the messenger-mediated effects of neurotransmitters and hormones on the activity of ion channels is obtained in the cell-attached recording mode. The cells are intact, there is no disturbance of cellular function, and ion channels are characterized in their most natural environment (i.e., bathed on their intracellular aspect by cytosolic fluid). Any effect of agonists on intracellular electrolyte (including pH) composition will also be manifest. In this situation the agonists are applied to the solution bathing the acinar cells. The neurotransmitters or hormones thus have no direct access to the isolated patch membrane area and any effect of the agonist must therefore be mediated by the generation of some intracellular mediator(s) (Figure 9).

The ability of Ca^{2+}-mobilizing agonists to activate K^+ currents in intact cells has now been demonstrated in human and rodent salivary acinar cells, in pig pancreatic acinar cells, and also for lacrimal acinar cells (Petersen, 1986). In each of these tissues patch clamp recordings from excised patches of basolateral membranes are dominated by currents in the voltage- and calcium-activated K^+ channel, and in each case it is this channel that is activated *in situ* by Ca^{2+}-mobilizing agonists. The effects of the agonists are mimicked in all these tissues by the calcium ionophore A23187 (Petersen and Gallacher, 1988).

In salivary acinar cells the Ca^{2+}-dependent effects of neurotransmitters on ion channels have been extensively investigated. Acetylcholine stimulation results in a dramatic increase in the frequency and duration of opening of K^+ channels in the cell-attached patches. In all of these experiments the solution in the recording pipette was Ca^{2+} free and the only source of Ca^{2+} to activate channels in the patch is via the cytosol. In the absence of extracellular Ca^{2+}, ACh stimulation results in an increase in the single-channel current amplitude but a fall in the open probability of the channels in the patch. These effects, in Ca^{2+}-free solutions, are explained as due to the release of intracellular Ca^{2+}, activation of K^+ channels in the membrane area outside the patch, and consequently acinar cell hyperpolarization. The cell-attached recording configuration has been used to demonstrate that there is no voltage-dependent Ca^{2+} influx pathway. In the pig pancreatic acinar cell, sustained cholecystokinin (CCK) activation of channels *in situ* was dependent on the presence of Ca^{2+} in the recording pipette itself. In this tissue the K^+ channel in the cell-attached patches is only activated when hormonal stimulation promotes entry of Ca^{2+} across the area of membrane covered by the patch pipette. CCK stimulation must then be associated with the generation of some intracellular regulator of Ca^{2+} influx, most likely IP_3 and/or IP_4 (see Figure 7).

Role of Ion Channels and Pumps in Secretion

The main electrolytes in the isotonic primary acinar fluid in both salivary glands and pancreas are Na^+ and Cl^- although there are also HCO_3^- and K^+ ions present (Petersen, 1980; Schulz, 1987). In the salivary glands, the primary secretion is the only source of the fluid in the final secretion (Young et al., 1987) whereas in the pancreas a bicarbonate-rich fluid is added in the duct system (Schulz, 1987). The proteins secreted are mainly derived from the acinar cells, and the mechanism underlying the release is exocytosis (Baker and Knight, 1986).

Figure 10 summarizes some of the most important electrolyte transport events in the acinar cells. Different aspects are, for the sake of clarity, shown separately in the five cells depicted. In cell *a,* neurotransmitters or hormones act by releasing

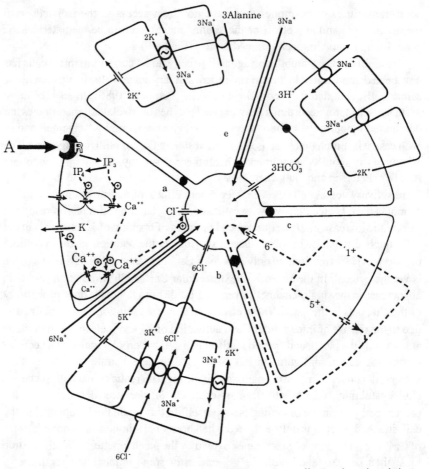

Figure 10. Summary of transport events in exocrine acinar cells. For the sake of clarity different aspects are shown in separate cells labeled **a,b,c,d**, and **e** for discussion in the text. The plasma membrane is divided into two parts by the tight junctions placed close to the lumen. All the transport events shown to take place across the basal membranes undoubtedly also occur in the lateral membranes. Abbreviations: A, agonist; R, Receptor; IP$_3$, inositol 1,4,5-trisphosphate; IP$_4$, inositol 1,3,4,5-tetrakis-phosphate. (Adapted from Petersen and Gallacher, 1988.)

intracellular Ca^{2+}, evoking an increase in $[Ca^{2+}]_i$. The released Ca^{2+} activates K^+ channels in the basolateral plasma membrane. The Ca^{2+}-activated Cl^- channels are not nearly so well known. They may be present in the luminal cell membrane because after stimulation the lumen becomes more negative with respect to the interstitial fluid (Lundberg, 1958). It is possible that the Cl^- channels are in part derived from the secretory granules and inserted into the luminal membrane during exocytosis. Cell *b* shows the ion transport events underlying NaCl secretion. The basolateral plasma membrane contains three transport proteins: the Ca^{2+}- and voltage-activated K^+ channel (blocked by tetraethylammonium or Ba^{2+}), the Na^+-K^+-$2Cl^-$ cotransporter (blocked by loop diuretics such as bumetanide, piretanide, and furosemide), and the energy-requiring Na^+-K^+ pump (blocked by ouabain) (Petersen, 1986). In the steady secreting state, K^+ recirculates via pump and cotransporter. The only net transport is that of Cl^- uptake, and Cl^- leaves the cell via the luminal Cl^- channels. The lumen negativity allows Na^+ to move between the cells into the narrow intercellular spaces and through the so-called tight junctions placed at their luminal end. The net result of all these transport events is transcellular NaCl transport. Water follows along by osmosis. Cell *c* in Figure 10 presents the overall electrical circuit due to the transport events shown in cell *b*. The outward current across the basolateral membrane is mostly due to K^+ exit through the Ca^{2+}-activated channels, but it is also partly due to the imbalance of ion movements through the Na^+-K^+ pump. This outward current is matched by the inward current through the luminal Cl^- channels. To complete the circuit, current has to pass between the cells and through the so-called tight junctions. This "paracellular" current is carried mainly by Na^+ moving from the interstitial fluid into the acinar lumen. The low transcellular specific resistance (much lower than the specific transmembrane resistance) that has now been described for many epithelia was in fact first demonstrated in salivary acini and is explained by the leakiness of the so-called tight junctions (Petersen, 1980). The initial K^+ release to the blood side is due to Ca^{2+} activation of K^+ channels, but K^+ has to be accompanied by something else, namely Cl^-. In the initial period after start of stimulation, a large part of the paracellular cation flow is made up of K^+, explaining the "K^+ transient" in the secreted juice. The release of cellular KCl creates a more favorable overall electrochemical gradient for Na^+-K^+-$2Cl^-$ via the cotransporter, and the Na^+ inflow raises the intracellular Na^+ concentration, which stimulates the Na^+-K^+ pump (Petersen, 1986). Both transport proteins help to reaccumulate K^+, and in the secreting steady state this exactly balances the K^+ lost via the channels. When stimulation is discontinued, $[Ca^{2+}]_i$ sharply decreases (Osipchuk et al., 1990) and both K^+ and Cl^- channels close. There is now net KCl reuptake through the Na^+-K^+ pump and Na^+-K^+-$2Cl^-$-cotransporter. Thereafter the overall electrochemical gradient for Na^+-K^+-$2Cl^-$-cotransport gradually becomes less favorable for uptake, and when Na^+ inflow is reduced the activity of the Na^+-K^+ pump is also brought back to the prestimulation level, and the gland is again in a true resting state.

Cell *d* illustrates the transport apparatus needed for the secretion of $NaHCO_3$. The Na^+/H^+ exchanger may be activated by stimulation via diacylglycerol (see Figure 1). The only transport protein that has not been identified here is the postulated HCO_3^- channel in the plasma membrane. This could be the Cl^- channel already depicted in cells *b* and *c* with some HCO_3^- permeability, it could be a combination of this Cl^- channel with a Cl^--HCO_3^- exchanger, or it might be a

Na$^+$-dependent bicarbonate transporter with a stoichiometry of 3 HCO$_3^-$:1 Na$^+$. In order to save space the electrical circuit of the bicarbonate transport apparatus is not shown but could easily be produced in a way similar to that done for NaCl secretion in cell c.

Cell e shows the transporters involved in amino acid uptake; an essential prerequisite for protein synthesis and therefore protein secretion. Only sodium–alanine cotransport (with a 1:1 stoichiometry) is shown here as this is the only amino acid transporter that has been characterized in detail in electrophysiological experiments (Petersen and Findlay, 1987).

The Ca^{2+}- and voltage-activated K$^+$ channel has a central role in many important aspects of secretion from the acinar cells because it is not only essential for chloride and bicarbonate secretion but also for amino acid uptake and therefore protein synthesis and secretion. The Ca^{2+}-activated K$^+$ channel is the key to the understanding of the Ca^{2+} dependence of fluid secretion as well as the stimulant-evoked K$^+$ release and membrane hyperpolarization described many years ago (Petersen, 1986).

Figure 10 deals exclusively with channel openings mediated by internal Ca^{2+}. However, cAMP may also be involved in mediating fluid secretion. In the salivary glands, for example, excitation of adrenergic beta-receptors increases the cellular cAMP concentration and evokes a small increase in fluid secretion in addition to the substantial protein secretory response (Petersen, 1986). It is known that isoproterenol depolarizes the parotid acinar cell membrane (Na$^+$ and Cl$^-$ dependent) and evokes a slight increase in membrane conductance. Because isoproterenol also evokes K$^+$ uptake (without preceding K$^+$ release), which is Cl$^-$ as well as Na$^+$ dependent and can be blocked by the loop diuretic furosemide (Petersen, 1986), the acinar cell model shown in Figure 10 may also apply to this situation except that cAMP may have very little effect on the K$^+$ channels in most exocrine acinar cells and therefore would mainly act to open the Cl$^-$ conductance pathway. This could explain why there is no initial stimulant-evoked K$^+$ loss but instead a stimulant-evoked K$^+$ uptake and also explain why isoproterenol-evoked fluid secretion in the salivary glands is very small compared with that evoked by cholinergic or alpha-adrenergic stimulation because the lack of K$^+$ channel activation would severely limit the basolateral Cl$^-$ uptake.

In the exocrine pancreas it is also known that cAMP is the internal messenger that switches on the anion-selective channels that are responsible for the bicarbonate-rich fluid secretion evoked by stimulation with the peptide hormone secretin (Schulz, 1987).

The acinar cell model shown in Figure 10 appears to be compatible with the results so far obtained in all the mammalian salivary and lacrimal glands investigated as well as with the data from the human and pig pancreas. The mouse and rat pancreatic acinar cells do not, however, possess Ca^{2+}-activated K$^+$-selective channels but have instead Ca^{2+}-activated non-selective cation channels (Petersen and Maruyama, 1984). The rat pancreatic acinar cells secrete fluid in response to stimulation with ACh and CCK, and this process is Ca^{2+} dependent (Petersen, 1980). Stimulant-evoked increase in membrane Cl^- conductance (Petersen, 1980) as well as Na^+- and Cl^--dependent K^+ transport (Petersen, 1986; Wakui et al., 1989) has been shown in mouse pancreatic acinar cells. Many but not all of the elements shown in Figure 10 are therefore present also in the mouse and rat pancreatic acinar cells, but more work is clearly needed to establish the precise fluid secretion mechanism in these cells.

Messenger Actions in the Control of Exocytosis

Secretion of macromolecules often depends on an external signal, which via an intracellular messenger evokes movement of secretory granules to the surface cell membrane. Fusion of the granule and plasma membranes with an opening developing at the point of fusion then occurs, establishing continuity between the interior of the secretory granule and the lumen through which the stored proteins can escape. This mode of secretion has been termed exocytosis. The granule membrane inserted into the plasma membrane is subsequently recaptured and can be reutilized (membrane recycling) (Baker and Knight, 1986).

All organelles and therefore also the secretory granules are embedded in the cytoskeleton and in order to make contact with the plasma membrane they have to pass through the actin filament network. Actin filaments are connected by a number of Ca^{2+}-dependent cross-linking proteins such as caldesmon, and there is evidence suggesting that the actin network prevents exocytosis until the relevant signal is received. Agents shown to evoke disassembly of the actin network promote secretion, and the disruption of the actin network occurring in response to hormones or neurotransmitters mobilizing internal Ca^{2+} may involve regulation of caldesmon and depolymerization of the actin filaments (Burgoyne, 1990).

The details of the molecular mechanisms involved in the late stages of the secretion process, namely the movement of the secretory granules to the inner surface of the cell membrane, docking of the vesicles on the membrane, and finally the fusion and fission events underlying exocytosis, are still mostly unknown.

One interesting aspect of exocytosis is that it is very tight (i.e., the fusion of lipid bilayers, although it involves considerable molecular rearrangements, causes no damage to cell or vesicle membranes). Apparently the only connection formed is between vesicle lumen and extracellular space. High-resolution electrophysiological studies have revealed the initial conductance of the fusion pore (200–300 pS), and this may correspond to an inner diameter of about 2 nm. Thereafter pore dilation occurs. The initial opening of the fusion pore could be Ca^{2+}-activated. As with ion channels in general the fusion pore opens and closes reversibly (flickering) (Almers, 1990).

SUMMARY

Stimulus–secretion coupling is normally initiated by an extracellular signal in the form of a hormone or a neurotransmitter binding to a specific receptor site on the outer surface of the gland cell membrane. Several membrane transduction mechanisms exist, but with regard to stimulus–secretion coupling those involved in the generation of intracellular Ca^{2+} signals are the most important. Receptor activation can lead to opening of voltage-gated Ca^{2+} channels by causing membrane depolarization. Such membrane depolarization can be mediated by, for example, closure of K^+ channels due to receptor–channel linkage via a GTP-binding (G) protein. Receptor activation can also, again via a G protein, switch on the enzyme phosphoinositidase C (phospholipase C), generating inositol trisphosphate (IP_3). IP_3 acts on an intracellular Ca^{2+} store in the endoplasmic reticulum to open Ca^{2+} channels and thereby releases the messenger ion into the cytosol. Ca^{2+} acts on another Ca^{2+} store also in the endoplasmic reticulum to open a different type of Ca^{2+} channel, sensitive to caffeine, thereby causing further Ca^{2+} release. The release of Ca^{2+} from the latter pool is often pulsatile, causing oscillations in the cytoplasmic Ca^{2+} concentration. The intracellular Ca^{2+} signal can activate exocytosis by inducing the formation of a pore linking the interior of a secretory vesicle with the outside of the cell membrane. The initial pore is small but soon dilates, allowing export of the secretory product. Intracellular Ca^{2+} can also activate ion channels in both basolateral and luminal cell membranes, thereby initiating fluid and electrolyte secretion in a variety of exocrine glands.

REFERENCES

Almers, W. (1990). Exocytosis. Ann. Rev. Physiol. 52, 607–624.

Baker, P.F., & Knight, D.E. (1986). Exocytosis: control by calcium and other factors. Br. Med. Bull. 42, 399–404.

Berridge, M.J. (1987). Inositol trisphosphate and diacylglycerol: two interacting second messengers. Ann. Rev. Biochem. 56, 159–193.

Berridge, M.J., & Irvine, R.F. (1989). Inositol phosphates and cell signalling. Nature 341, 197–205.

Burgoyne, R.D. (1990). Secretory vesicle-associated proteins and their role in exocytosis. Ann. Rev. Physiol. 52, 647–659.

Changya, L., Gallacher, D.V., Irvine, R.F., Potter, B.V.L. & Petersen, O.H. (1989a). Inositol 1,3,4,5-tetrakisphosphate is essential for sustained activation of the Ca^{2+}-dependent K^+ current in single internally perfused mouse lacrimal acinar cells. J. Membr. Biol. 109, 85–93.

Changya, L., Gallacher, D.V., Irvine, R.F., & Petersen, O.H. (1989b). Inositol 1,3,4,5-tetrakisphosphate and inositol 1,4,5-trisphosphate act by different mechanisms when controlling Ca^{2+} in mouse lacrimal acinar cells. FEBS Letts. 251, 43–48.

Dunne, M.J., Bullett, M.J., Li, G., Wollheim, C.B., & Petersen, O.H. (1989). Galanin activates nucleotide-dependent K^+ channels in insulin-secreting cells via a pertussis toxin-sensitive G-protein. EMBO J. 8, 413–420.

Ehrlich, B.E., & Watras, J. (1988). Inositol 1,4,5-trisphosphate activates a channel from smooth muscle sarcoplasmic reticulum. Nature 336, 583–586.

Endo, M. (1977). Calcium release from the sarcoplasmic reticulum. Physiol. Rev. 57, 71–108.

Ferris, C.D., Huganir, R.L., Supattapone, S., & Snyder, S.H. (1989). Purified inositol 1,4,5-trisphosphate receptor mediates calcium flux in reconstituted lipid vesicles. Nature 342, 87–89.

Fleischer, S., & Inui, M. (1989). Biochemistry and biophysics of excitation-contraction coupling. Ann. Rev. Biophys. Biophys. Chem. 18, 333–364.

Furuichi, T., Yoshikawa, S., Miyawaki, A., Wada, K., Maeda, N. & Mikoshiba, K. (1989). Primary structure and functional expression of the inositol 1,4,5-trisphosphate-binding protein P_{400}. Nature 342, 32–38.

Gallacher, D.V., Hanley, M.R., Petersen, O.H., Roberts, M.L., Squire-Pollard, L.G., & Yule, D.I. (1990). Substance P and bombesin elevate cytosolic Ca^{2+} by different molecular mechanisms in rat pancreatic acinar cell line. J. Physiol. 426, 193–207.

Gilman, A.G. (1987). G proteins: transducers of receptor-generated signals. Ann. Rev. Biochem. 56, 617–649.

Hamill, O.P., Marty, A., Neher, E., Sakmann, B., & Sigworth, F.J. (1981). Improved patch-clamp techniques for high resolution current recordings from cells and cell-free membrane patches. Pflugers Arch. 391, 85–100.

Kahn, C.R. (1976). Membrane receptors for hormones and neurotransmitters. J. Cell Biol. 70, 261–286.

Lai, F.A., Erickson, H.P., Rousseau, E., Liu, Q.-Y., & Meissner, G. (1988). Purification and reconstitution of the calcium release channel from skeletal muscle. Nature 331, 315–319.

Lundberg, A. (1958). Electrophysiology of salivary glands. Physiol. Rev. 38, 21–40.

Maeda, N., Niinobe, M., & Mikoshiba, K. (1990). A cerebellar Purkinje cell marker P_{400} protein is an inositol 1,4,5-trisphosphate (InsP3) receptor protein. Purification and characterization of InsP3 receptor complex. EMBO J. 9, 61–67.

Martin, S.C., Yule, D.I., Dunne, M.J., Gallacher, D.V., & Petersen, O.H. (1989). Vasopressin directly closes ATP-sensitive potassium channels evoking membrane depolarization and an increase in the free intracellular Ca^{2+} concentration in insulin-secreting cells. EMBO J. 8, 3595–3599.

Maruyama, Y., Petersen, O.H., Flanagan, P., & Pearson, G.T. (1983). Quantification of Ca^{2+}-activated K^+ channels under hormonal control in pig pancreas acinar cells. Nature 305, 228–232.

Matthews, E.K., Petersen, O.H., & Williams, J.A. (1973). Pancreatic acinar cells: acetylcholine-induced membrane depolarizations, calcium efflux and amylase release. J. Physiol. 234, 689–701.

Meissner, G. (1984). Adenine nucleotide stimulation of Ca^{2+}-induced Ca^{2+} release in sarcoplasmic reticulum. J. Biol. Chem. 259, 2365–2374.

Meissner, G., Darling, E., & Eveleth, J. (1986). Kinetics of rapid Ca^{2+} release by sarcoplasmic reticulum. Effects of $Ca^{2+}+$, Mg^{2+} and adenine nucleotides. Biochemistry 25, 236–244.

Morris, A.P., Gallacher, D.V., Irvine, R.F., & Petersen, O.H. (1987). Synergism of inositol trisphosphate and tetrakisphosphate in activating Ca^{2+}-dependent K^+ channels. Nature 330, 653–655.

Osipchuk, Y.V., Wakui, M., Yule, D.I., Gallacher, D.V., & Petersen, O.H. (1990). Cytoplasmic Ca^{2+} oscillations evoked by receptor stimulation, G-protein activation, internal application of inositol trisphosphate or Ca^{2+}: simultaneous microfluorimetry and Ca^{2+}-dependent Cl^- current recording in single pancreatic acinar cells. EMBO J. 9, 697–704.

Penner, R., Neher, E., Takeshima, H., Nishimura, S., & Numa, S. (1989). Functional expression of the calcium release channel from skeletal muscle ryanodine receptor cDNA. FEBS Letts. 259, 217–221.

Petersen, O.H. (1980). The Electrophysiology of Gland Cells. Academic Press, New York.

Petersen, O.H. (1986). Calcium-activated potassium channels and fluid secretion by exocrine gland. Am. J. Physiol. 251, G1–G13.

Petersen, O.H. (1989). Does inositol tetrakisphosphate play a role in the receptor-mediated control of calcium mobilization? Cell Calcium 10, 375–383.

Petersen, O.H., & Bear, C.E. (1986). Two glucagon transducing systems. Nature 323, 18.

Petersen, O.H., & Findlay, I. (1987). Electrophysiology of the pancreas. Physiol. Rev. 67, 1054–1116.

Petersen, O.H., & Gallacher, D.V. (1988). Electrophysiology of pancreatic and salivary acinar cells. Ann. Rev. Physiol. 50, 65–80.

Petersen, O.H., & Maruyama, Y. (1984). Calcium-activated potassium channels and their role in secretion. Nature 307, 693–696.

Petersen, O.H., & Wakui, M. (1990). Oscillating intracellular Ca^{2+} signals evoked by activation of receptors linked to inositol lipid hydrolysis: mechanism of generation. J. Membr. Biol. 118, 93–105.

Rousseau, E., LaDine, J., Liu, Q.-Y., & Meissner, G. (1988). Activation of the Ca^{2+} release channel of skeletal muscle sarcoplasmic reticulum by caffeine and related compounds. Arch. Biochem. Biophys. 267, 75–86.

Schramm, M., Orley, J., Eimerl, S., & Korner, M. (1977). Coupling of hormone receptors to adenylate cyclase of different cells by cell fusion. Nature 268, 310–313.

Schulz, I. (1987). Electrolyte and fluid secretion in the exocrine pancreas. In: Physiology of the Gastrointestinal Tract (2nd ed.) (Johnson, L.R., ed.), pp. 1147–1171. Raven, New York.

Smith, J.S., Coronado, R., & Meissner, G. (1985). Sarcoplasmic reticulum contains adenine nucleotide-activated calcium channels. Nature 316, 446–449.

Smith, J.S., Coronado, R., & Meissner, G. (1986). Single channel measurements of the calcium release channel from skeletal muscle sarcoplasmic reticulum. J. Gen. Physiol. 88, 573–588.

Streb, H., Irvine, R.F., Berridge, M.J., & Schulz, I. (1983). Release of Ca^{2+} from a nonmitochondrial intracellular store in pancreatic acinar cells by inositol 1,4,5-trisphosphate. Nature 306, 67–69.

Takeshima, H., Nishimura, S., Matsumoto, T., Ishida, H., Kangawa, K., Minamino, N., Matsuo, H., Ueda, M., Hanaoka, M., Hirose, T., & Numa, S. (1989). Primary structure and expression from complementary DNA of skeletal muscle ryanodine receptor. Nature 339, 439–445.

Tsien, R.W., Lipscombe, D., Madison, D.V., Bley, K.R., & Fox, A.P. (1988). Multiple types of neuronal calcium channels and their selective modulation. Trends Neurosci. 11, 431–437.

Velasco, J.M., Petersen, J.U.H., & Petersen, O.H. (1988). Single-channel Ba^{2+} currents in insulin-secreting cells are activated by glyceraldehyde stimulation. FEBS Letts. 231, 366–370.

Velasco, J.M., & Petersen, O.H. (1989). The effect of a cell-permeable diacylglycerol analogue on single Ca^{2+} (Ba^{2+}) channel currents in the insulin-secreting cell line RINm5F. Q.J. Exp. Physiol. 74, 367–370.

Wakui, M., Potter, B.V.L., & Petersen, O.H. (1989). Pulsatile intracellular calcium release does not depend on fluctuations in inositol trisphosphate concentration. Nature 339, 317–320.

Young, J.A., Cook, D.I., VanLennep, E.W., & Roberts, M. (1987). Secretion by the major salivary glands. In: Physiology of the Gastrointestinal Tract (2nd ed.) (Johnson, L.R., ed.), pp. 773–815. Raven, New York.

Yule, D.I., & Gallacher, D.V. (1988). Oscillations of cytosolic calcium in single pancreatic acinar cells stimulated by acetylcholine. FEBS Letts. 239, 358–362.

RECOMMENDED READINGS

Bronner, F. (ed.) (1990). Intracellular Calcium Regulation. Wiley-Liss, New York.

Fleischer, S., & Fleischer, B. (eds.) (1989). Methods in Enzymology. Vol. 171, Biomembranes part R; Transport theory: Cells and Model Membranes. Academic Press, New York.

Forte, J.G. (ed.) (1989). Handbook of Physiology, Section 6 The Gastrointestinal System, Volume III Salivary Gastric, Pancreatic and Hepatobiliary Secretion. Am. Physiol. Soc., Bethesda, Maryland.

Thorn, N.A., Treiman, M., & Petersen, O.H. (eds.) (1988). Molecular Mechanisms in Secretion. Munksgaard, Copenhagen.

Chapter 10

Pacemaker Periodicity in Secretory Systems

E. KEITH MATTHEWS

INTRODUCTION

Periodicity or rhythmic oscillations have long been recognized as an essential feature of normal electrical and mechanical activity in cardiac and smooth muscle cells, but the discovery of oscillatory electrical behavior in secretory cells is of more recent origin (see Dean and Matthews, 1968; Matthews, 1985). This and subsequent work established that many endocrine cells share with their counterparts in nerve and muscle not only a common developmental origin but the distinctive property of membrane electrical excitability. For example the β-cells of the pancreatic islets of Langerhans generate voltage-dependent changes in ionic conductance across the plasma membrane in a rhythmic fashion when the cells are exposed to glucose. These bursts of action potentials increase in frequency as the glucose concentration is raised and initiate a corresponding increase in insulin

Fundamentals of Medical Cell Biology, Volume 5A
Membrane Dynamics and Signaling, pages 203–221
Copyright © 1992 by JAI Press Inc.
All rights of reproduction in any form reserved.
ISBN: 1-55938-309-7

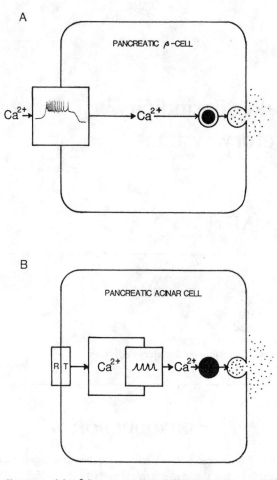

Figure 1. Oscillations of $[Ca^{2+}]_i$ in secretory cells determined by (**A**) the entry of calcium ions across the cell membrane from the extracellular environment (membrane oscillator) or (**B**) the release of calcium from sequestered stores (cytoplasmic oscillator).

secretion. In contrast, the neighboring acinar cells of the exocrine pancreas secrete amylase and other enzymes in response to an agonist but in the complete absence of action potentials because, as in most epithelia, these cells are electrically inexcitable. Yet both types of cells release their secretory product by a similar process of exocytosis triggered primarily by an increase in the cytoplasmic concentration of the ionized calcium ion, $[Ca^{2+}]_i$. However, the time-dependent fluctuations in $[Ca^{2+}]_i$ are derived essentially from a different source for each cell type (Figure 1) (i.e., from the extracellular space via the plasma cell membrane [β-cell] or from discrete subcellular stores [acinar cells]). On this basis, it might be suggested that *either* oscillators located at the plasma membrane *or* intracellular cytosolic oscillators govern the response of different cells to their extracellular command signals (e.g., neurotransmitters or hormones). In reality, these mechanisms are not necessarily mutually exclusive and, although one mechanism may predominate in any given cell type, both can be present and closely linked to ensure a coordinated and well-controlled output of secretory product in response to a wide variety of different input signals converging on the target secretory cell. In recent years, a combination of powerful electrophysiological techniques used in parallel with calcium-sensitive indicators and digital image analysis has enabled us to begin to unravel the common mechanisms that underlie these oscillatory phenomena and extend across a wide spectrum of cells secreting a diversity of product from polypeptide hormones to hydrolytic enzymes, electrolytes to steroids. The operational characteristics of each type of secretory cell will first be described in detail and the functional significance of the periodic signals is discussed in the later part of this chapter.

PANCREATIC β-CELLS

In the absence of glucose, or with basal concentrations below some 3 mM the pancreatic β-cell is electrically quiescent and maintains a stable resting membrane potential of the order of –60 mV. When exposed to concentrations of glucose > 4 mM dynamic oscillations in potential (i.e., bursts of action potentials) are seen with a periodicity dependent on glucose concentration. At the glucose concentration that is approximately half-maximal for insulin release, 11.1 mM, each burst of electrical activity has a duration of about 5 s (Figure 2A) and is followed by a silent period of approximately the same length. This cycle can be continued with a remarkably regular periodicity of approximately 6 min^{-1} for several hours. Increasing the glucose concentration lengthens the duration of the burst and progressively shortens the interburst "silent" interval, which is eventually obliterated at concentrations of glucose > 20 mM when spiking becomes continuous. Further electrophysiological investigation including patch clamp analysis of single membrane channels has confirmed that the inward current responsible for the individual spikes is carried by the Ca^{2+} ion. It is the transmembrane influx of calcium by this mechanism that is largely instrumental in discharging insulin from the cell initially in proportion to the extent of electrical activity. The influx of calcium may also act in a feedback manner to limit the changes in membrane conductance and modulate, in turn, the periodic bursting process. Figure 2B outlines the way in which various interacting ionic processes contribute to the control of bursting behavior and insulin secretion.

D-Glucose or the orally active hypoglycemic sulfonylurea compounds are believed to initiate a depolarization of the β-cell via an ATP-sensitive channel, which acts as a kind of metabolic sensor and decreases outward membrane K^+ flux.

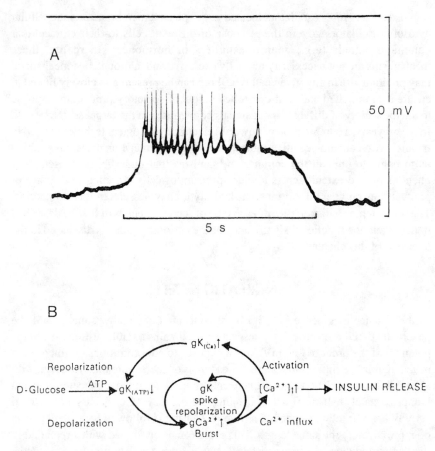

Figure 2 Pancreatic islet β-cells and insulin release. **A.** Electrical activity induced by D-glucose, 11.1 mM, in a single β-cell of the mouse islet. **B.** Relationship between changes in ionic conductance, electrical activity, $[Ca^{2+}]_i$, and insulin release induced by D-glucose or sulfonylureas. (From Matthews et al., 1973; Matthews, 1986.)

Once the membrane potential is decreased by this process below a threshold of <–40 mV, voltage-gated Ca^{2+}-channels in the membrane open and calcium enters the cell down its concentration gradient to activate insulin release. Individual spikes repolarize by rapid activation of a potassium permeability (as in nerve and muscle). The burst of activity is then terminated by the accumulation of $[Ca^{2+}]_i$, which activates an outward Ca^{2+}-dependent K^+ permeability, repolarizing the cell. Additional factors may contribute to controlling this phase of the burst, including Ca^{2+} channel inactivation or relaxation of ATP channel control, all of which may be Ca^{2+} dependent (see later). It is highly likely that the depolarization–repolarization cycle induced by D-glucose is necessary to maintain the sensitivity of the islet cell

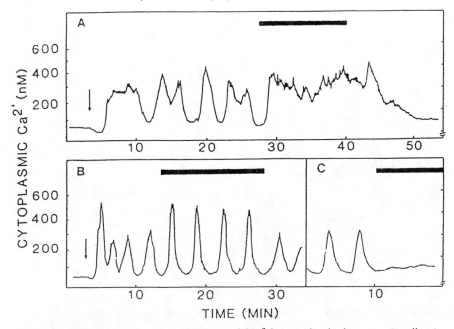

Figure 3. Glucose-induced oscillations of $[Ca^{2+}]_i$ in individual mouse β-cells. **A.** Glucose increased from 3 to 11 mM at the arrow, and to 30 mM (black bar) followed by a return to 3 mM. **B.** External Ca^{2+} concentration, $[Ca^{2+}]_o$ increased from 1.28 to 3.84 mM (black bar) after exposure to 11 mM glucose at the arrow. **C.** Action of methoxyverapamil, 10 μM (black bar), on response of $[Ca^{2+}]_i$ to glucose, 11 mM. (From Grapengeisser et al., 1989.)

in a delicately balanced state optimal for an integrated response to the wide variety of input signals modulating insulin release (Matthews, 1985). The overall process may be affected by neurotransmitter action. For example, activation of muscarinic receptors increases the frequency of glucose-induced bursting and decreases the repolarization phase between bursts. It is therefore particularly intriguing to note that measurements of cytoplasmic calcium concentration reveal oscillations in the ionized intracellular calcium level $[Ca^{2+}]_i$, which vary in pattern depending upon the stimulant used. Glucose-induced oscillations in $[Ca^{2+}]_i$ are dose dependent (Figure 3A) and sensitive to extracellular calcium, $[Ca^{2+}]_o$ (Figure 3B), and agents blocking membrane calcium current (Figure 3C), whereas responses to the cholinergic agonist carbachol, although dose dependent (Figure 4), were not eliminated by removing external calcium. This suggests a correlation of changes in $[Ca^{2+}]_i$ with glucose-induced membrane electrical activity (though the matching is not precise) and, in the case of carbachol, Ca^{2+} cycling across internal stores (see Figures 1B and 4). Both a membrane oscillator and a cytoplasmic oscillator may therefore be involved in matching insulin output more precisely to converging input

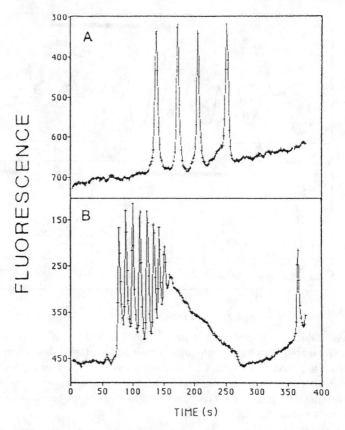

Figure 4. Effect of carbamylcholine concentration on latency and frequency of $[Ca^{2+}]_i$ oscillations in clonal pancreatic β-cells. Exposure at time zero to 25 μM (**A**) and 200 μM (**B**) carbamylcholine. (From Prentki et al., 1988.)

signals, a dominant glucose-activated membrane oscillator being responsive not only to direct negative feedback but also to the operation of the modulatory neurotransmitter-activated cytosolic oscillator. Mechanisms controlling the cytoplasmic Ca^{2+} oscillator are well expressed in other cells and will be considered in the following sections.

PANCREATIC ACINAR CELLS

The acinus of the exocrine pancreas is, like that of the salivary or lacrimal gland, composed of polarized epithelial cells arranged in clusters. The pancreatic acinar cells secrete hydrolase enzymes, including amylase, and fluid in response to neurotransmitter or peptide agonists (e.g., acetylcholine [ACh] or cholecystokinin

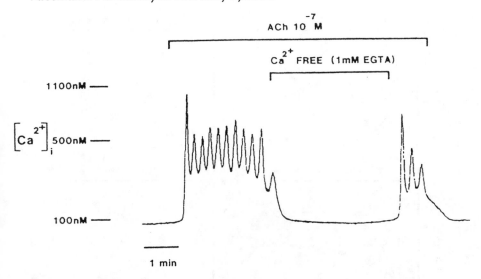

Figure 5. Measurement of $[Ca^{2+}]_i$ in a single pancreatic acinar cell superfused at 30 °C with acetylcholine (ACh) in the presence or absence of external calcium ions. (From Yule and Gallacher, 1988.)

[CCK]. These exocrine cells can be readily isolated and their electrophysiological profile, cytosolic Ca^{2+}, and secretory characteristics studied under precisely controlled conditions in vitro. This has allowed critical examination of the effects of $[Ca^{2+}]_o$ and temperature in relation to oscillatory behavior. As pointed out already, the acinar cells are not electrically excitable in the sense that they do not display voltage-gated action potentials, but they do depolarize in response to stimulants and they possess an ensemble of membrane ionic channels that are linked to changes in cellular activity. Agonists like ACh and CCK interact with their specific receptors and via a common signal transduction pathway (see Figures 1 and 7) elevate $[Ca^{2+}]_i$ by mobilizing free calcium ions from cytoplasmic storage sites. Increased $[Ca^{2+}]_i$ not only activates the exocytosis of zymogen granules, and hence enzyme secretion from the apical aspect of the cell, but by its action on large-conductance Ca^{2+}-dependent K^+ channels increases the basolateral efflux of K^+. A local increase in $[K^+]_o$ in turn enhances the turnover of a K^+-Na^+-Cl^- cotransporter, leading to the apical secretion of a NaCl-rich fluid into the lumen.

Cellular activation with ACh evokes an increase in cytoplasmic calcium, which oscillates in a manner dependent upon the presence of extracellular calcium (Figure 5). These oscillations are marked at lower temperatures (e.g., 0.1 Hz at 30 °C), becoming only transient at higher temperatures superimposed on a more sustained and persistent increase in $[Ca^{2+}]_i$. This effect is clearly seen in parotid acinar cells (Figure 6). In these cells also two calcium-dependent currents, one an outwardly directed membrane K^+ current and the other an inwardly directed Cl^- current,

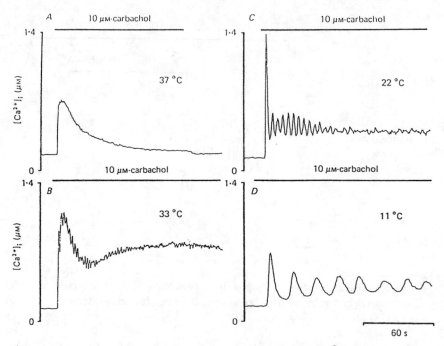

Figure 6. Parotid acinar cell: temperature dependence of [Ca²⁺]. Oscillations in response to carbachol. (From Gray, 1988.)

oscillate in phase with the cytoplasmic Ca^{2+}. Although the temperature and time constants of the various components will affect the overall process, a simplified version of the ionic events applicable to a stimulated acinar cell is illustrated graphically in Figure 7. In all the different types of acinar cells so far examined, the signal transduction across the membrane from receptor to cell interior is mediated by a G protein linked to phospholipase C, which, by an action on phophatidylinositol 4,5-bisphosphate (PIP₂), produces two second messengers, inositol 1,4,5-triphosphate (IP₃) and diacylglycerol (DAG). DAG acts by stimulating phosphokinase C in a calcium-dependent manner, and IP₃ releases calcium from internal stores. This basic signaling system may then account for the oscillating mobility of intracellular calcium. In the simplest operating system of this type (Figure 8A), a negative feedback loop operates by a reversible phosphorylation of the G protein transducer. Alternatively, reciprocal coupling by positive feedback is possible between the two messengers (Figure 8B) followed by rapid mitochondrial sequestration of calcium. Of more consequence in acinar cells would be a negative feedback of calcium levels on the endoplasmic reticulum (Figure 8C). An additional refinement of the model, for which there is growing evidence, involves a calcium-induced calcium release from an IP₃-insensitive pool of calcium and incorporates a component of [Ca²⁺]ᵢ through entry across the cell membrane

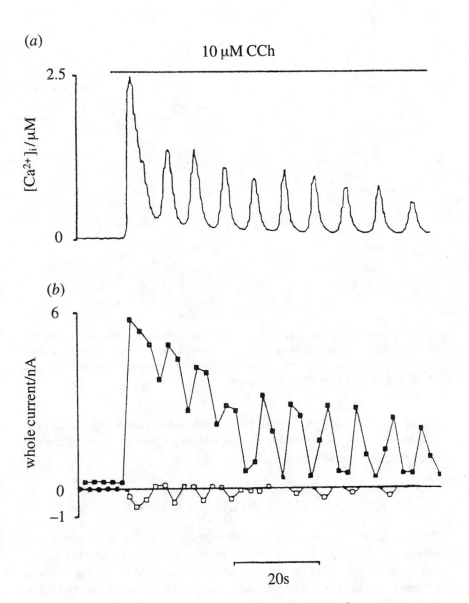

Figure 7. Parotid acinar cell: simultaneous recording of (a), $[Ca^{2+}]_i$, and (b), whole-cell current showing the synchrony between the increases in $[Ca^{2+}]_i$, an outwardly-directed Ca^{2+}-activated K^+ current (■), and an inward Ca^{2+}-activated Cl^- current (□) across the cell membrane. (From Gray, 1989.)

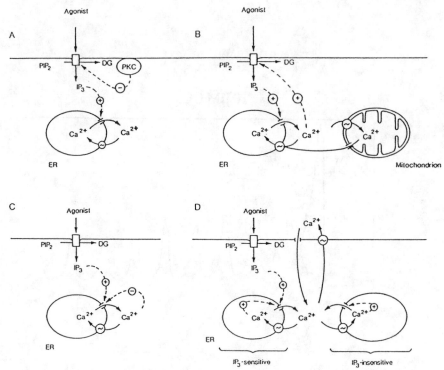

Figure 8. Models of cytosolic calcium oscillators involving receptor controlled oscillations in IP₃ (**A** and **B**) or second-messenger controlled oscillations (**C** and **D**) responsible for the periodic release and reuptake of calcium by the endoplasmic reticulum (ER). (From Berridge and Galione, 1988.).

(Figure 8*D*). This may be particularly important in partially depolarized acinar cells or where IP₃, or its derivative IP₄ together with IP₃, affects the entry of calcium across the plasma membrane.

ANTERIOR PITUITARY GLAND

The powerful control exerted by the anterior pituitary gland upon endocrine cells in the periphery is in turn governed by the action of the CNS on the anterior pituitary gland itself as well as by the feedback effects of increased circulating levels of hormones. The anterior pituitary is composed of at least five different cell types, and to study each of them, methods have had to be developed to effectively separate the different cells. This is a difficult task and a more convenient alternative has been to utilize cultured tumor cell lines derived from neoplastic pituitary cells. One such cell line, designated GH₃, is electrically excitable and secretes prolactin. By

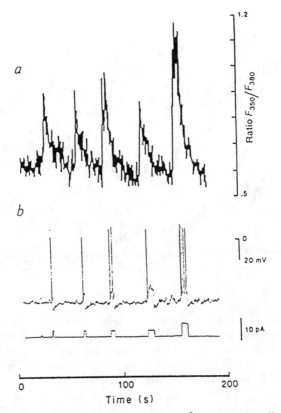

Figure 9. Action potential–induced alterations in $[Ca^{2+}]_i$ in a single cell of the pituitary GH_3B_6 cell line. Action potentials (**b**) were elicited by the injection of depolarizing current and were rapidly followed (**a**) by a corresponding increase in $[Ca^{2+}]_i$. (From Schlegel et al., 1987.)

monitoring changes in a fluorescent probe for calcium, it has been demonstrated that electrical depolarization induces action potentials closely correlated with an increase in intracellular calcium (Figure 9). In some cells action potentials occur spontaneously and are associated with Ca^{2+} influx via voltage-dependent calcium channels as indicated by a corresponding oscillation in $[Ca^{2+}]_i$ that is sensitive to block by calcium-entry blockers (e.g., nifedipine) (Schlegel et al., 1987). Normal somatotrophe cells also show a similar pattern of events. In the male rat approximately 50% of the pituitary cells are somatotrophes and, when isolated, two thirds of these cells secrete growth hormone (GH) spontaneously and the remainder only after stimulation with growth hormone–releasing hormone (GHRH). Most of the isolated cells secreting GH spontaneously demonstrated rhythmic high-amplitude transients of intracellular Ca^{2+} (Figure 10) varying between 50 and 500

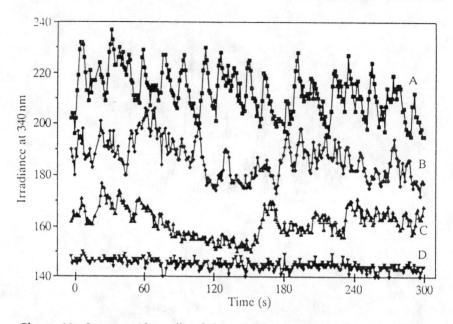

Figure 10. Somatotrophe cells of the anterior pituitary. Calcium oscillations measured simultaneously in three spontaneously active cells (**A, B, C**) compared to a non-active cell (**D**). There was a direct relationship between both frequency and amplitude of calcium oscillations and relative amount of GH secreted by each cell (i.e., **A>>B>C**). (From Thorner et al., 1988.)

nmol L^{-1} at a frequency of 2 to 13 pulses min^{-1}. The intracellular calcium transients are sensitive to $[Ca^{2+}]_o$ and to inorganic or organic calcium channel blockers, which also inhibit GH secretion. Likewise somatostatin (SRIF), which decreases $[Ca^{2+}]_i$ per se, as well as blocking the increase in $[Ca^{2+}]_i$ due to GHRH (Figure 11), also inhibits GH secretion. From these and the known effects of GHRH and SRIF on somatotrophes the following picture emerges for the cellular control of GH secretion. GHRH acts to promote the formation of cAMP via the G protein transduction pathway. cAMP activates protein kinase A, which phosphorylates the voltage-dependent Ca^{2+} channels and, by increasing their open state probability, increases $[Ca^{2+}]_i$ above that occurring by spontaneous channel opening and evokes exocytosis of GH. SRIF decreases both spontaneous and GHRH-induced release of GH via an inhibitory G protein, which not only decreases cAMP formation but also increases the outward membrane K^+ flux, hyperpolarizing the cell and diminishing the activation of voltage-operated Ca^{2+} channels (Figure 12). A more direct effect on the Ca^{2+} channels themselves is also possible, and it is likely too that DAG/protein kinase C plays some modulatory role in the operation of this system. The spontaneous oscillations of calcium seen in the somatotrophe may

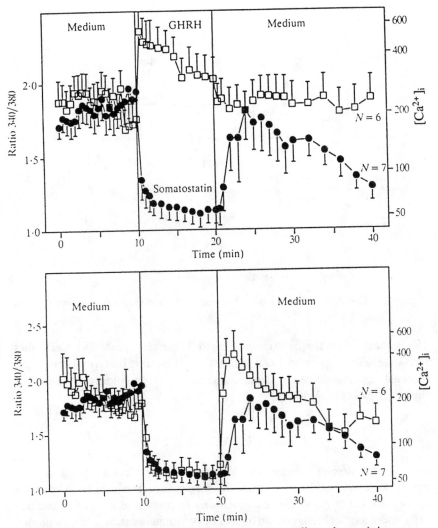

Figure 11. Somatotrophe cells of the anterior pituitary. Effect of growth hormone–releasing hormone (GHRH) or somatostatin on cytosolic free-calcium concentration, $[Ca^{2+}]_i$ (upper panel), or of GHRH and somatostatin together (□) or of somatostatin alone (●) on $[Ca^{2+}]_i$ (lower panel). (From Thorner et al., 1988.)

therefore represent cross-talk between several intracellular regulators interacting at the membrane level (Thorner et al., 1988). In view of this, it is important to note that *in vivo* GH is secreted into the circulation in a pulsatile manner under hypothalamic control.

Oscillations of $[Ca^{2+}]_i$ also occur in pituitary ACTH-secreting cells in response to CRH (Leong, 1988), and in gonadotrophes in response to GnRH (Limor et al.,

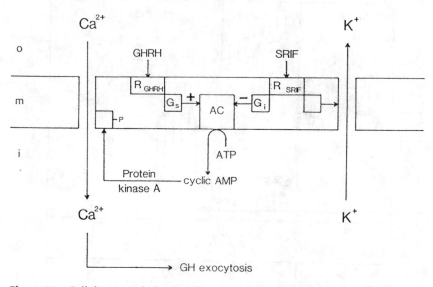

Figure 12. Cellular control of growth hormone secretion from the somatotrophe cell.

1987). Mechanisms similar to those of Figure 11 therefore probably underlie these responses. Furthermore, in isolated lactotrophes electrical activity occurs spontaneously and is inhibited by dopamine acting via D2 receptors to hyperpolarize the cell (Israel et al., 1987).

ADRENAL CORTEX

The output of steroids from the adrenal cortex, whether of corticosterone from the zona fasciculata or aldosterone from the zona glomerulosa, can be increased by specific stimuli (e.g., ACTH or angiotensin). In contrast to most other secretory cells, adrenocortical cells contain little or no preformed store of secretory product, rather the extra steroid for release is produced "on demand" by synthesis from precursors in response to an appropriate stimulus. The secretion of steroid does, however, require calcium, and under certain conditions, oscillations are seen in the adrenocortical cell membrane potential with large, slow "action potentials" occurring both in the fasciculata and glomerulosa cells (Matthews and Saffran, 1973). For example, when fasciculata cells are exposed to their normal stimulant, ACTH, the cells depolarize and action potentials are produced with a periodicity ranging from 2 to 12 s. These are not blocked by tetrodotoxin but are inhibited by agents that block membrane Ca^{2+} channels (Lymangrover et al., 1982).

In addition to cyclic oscillations in ionic conductance (Natke and Kabela, 1979), dose-dependent oscillations in $[Ca^{2+}]_i$ occur in adrenal glomerulosa cells stimu-

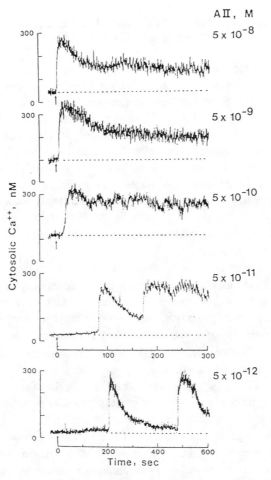

Figure 13. Adrenocortical glomerulosa cells showing oscillations in $[Ca^{2+}]_i$ in response to angiotensin II added, in the indicated concentrations, at the arrow; the dashed line represents the basal $[Ca^{2+}]_i$ concentration. (From Quinn et al., 1988.)

lated by angiotensin, an effect attributed to IP_3 activation of Ca^{2+} release (Quinn et al., 1988). Higher concentrations of angiotensin produce, with a decreased latency, a more persistent increase in $[Ca^{2+}]_i$ (Figure 13). Aldosterone output may be modulated through the frequency of these changes in $[Ca^{2+}]_i$ by activating the effector proteins for aldosterone production. Any Ca^{2+} influx across the plasma membrane will produce steep calcium gradients. Furthermore, where Ca^{2+} is released in close proximity to the intracellular stores, the effector proteins may experience large oscillations in calcium concentration; these may be damped out at more distant effector sites by Ca^{2+} binding and diffusion (Quinn et al., 1988).

FUNCTIONAL ASPECTS OF OSCILLATIONS IN SECRETORY CELLS

Although it is evident that other second messengers play important modulatory roles in cellular function, especially of secretory cells, the spatial and temporal characteristics of ionized Ca^{2+} distribution are fundamental to the control of secretion. As we have seen from the examples considered in detail in previous sections of this chapter, intracellular levels of ionized calcium can be determined by two major mechanisms: (1) the entry of calcium ions across the cell membrane from the extracellular environment and/or (2) the release of calcium from sequestered stores. Both processes display oscillatory behavior or periodicity. In the first of these processes, interacting changes in plasma membrane ionic conductance of a stimulated cell give rise to the periodic entry of calcium, whereas a receptor controlled/second messenger operated system initiates oscillations of $[Ca^{2+}]_i$ in the second of the two processes.

The question then arises, why should the frequency in the change of $[Ca^{2+}]_i$ generated by these mechanisms confer any greater benefits upon cellular activity than a simple and sustained change in amplitude? In fact, there are several sound advantages, economic, protective, and discriminatory, inherent in a frequency-modulated signaling system operated by Ca^{2+}. For example, the ionized $[Ca^{2+}]_o$: $[Ca^{2+}]_i$ ratio of > 10,000:1 creates a large ion gradient across the plasma membrane, and cellular homeostasis requires energy, especially for the outwardly directed Ca^{2+}-transport ATPase, which is costly metabolically. Brief transients of calcium sufficient to trigger calcium-dependent processes such as exocytosis would utilize the gradient to ensure rapid Ca^{2+} influx but, at the same time, avoid the high energy expenditure that a prolonged elevation of $[Ca^{2+}]_i$ would provoke by stimulating outward calcium pumping. Potential cytotoxic effects arising from a sustained increase in $[Ca^{2+}]_i$ (i.e., activation of phospholipases, proteases, and closure of gap junctions) are also avoided (Rink and Jacob, 1989). It is well recognized that a frequency-modulated system not only possesses a better signal-to-noise ratio than a simple amplitude-modulated system but also confers the benefit of multiple coding or discrimination based on three variables (i.e., spike frequency, amplitude, and shape), each of which might signal a different class of information (Rapp, 1987). Thus effector systems can be selectively tuned to respond to particular spike frequencies (i.e., FM transmission and detection) (Meyer and Stryer, 1988).

A frequency encoded biochemical regulation of secretion is therefore likely to provide both a more accurate and comprehensive signal transfer and control system than one depending upon amplitude alone. It is important to note, however, that the capacity to evoke a sustained response in certain intracellular pathways is not necessarily excluded by a rapidly switching initial signal but depends rather upon the actual time constants of the sequence involved. Thus if a comparatively slowly responding system (e.g., that for protein phosphorylation) is coupled sequentially

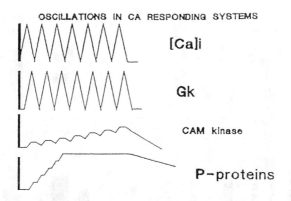

Figure 14. Cellular responses to oscillations in $[Ca^{2+}]_i$. Changes in $[Ca^{2+}]_i$ evoke rapid changes in membrane K^+ conductance via a Ca^{2+}-activated channel (Gk) as well as changes in the more slowly responding calmodulin-dependent kinase (CAM kinase) and in the phosphorylation of proteins (P-proteins). Note the progressive damping of the latter pathways. (From Sachs and Muallem, 1989.)

to a high frequency oscillator, then the ultimate output response is critically damped, resulting in a fixed, amplitude-like effect rather than the rapidly varying response of the initial signal (Figure 14).

So far oscillatory behavior has been considered almost exclusively from the standpoint of events in a single cell, but, in contrast to neurons and striatal muscle fibers where by design cross-talk is effectively eliminated, secretory cells are often coupled extensively through low resistance pathways via tight or gap junctions. This may facilitate secretory synchronization, oscillatory phase-locking, and entrainment of individual cells with economic benefits in functional or "pacemaker" terms following signal recognition (Matthews, 1974, 1982). A recent model for pancreatic β-cells illustrates just how a progressive recruitment of individual neighboring cells into strategically located functional clusters can effect a transition from stochastic spontaneous fluctuations of voltage and of $[Ca^{2+}]_i$ in a single cell to a "pacemaker" group of interconnected cells that exhibit a regular firing pattern and an enhanced $[Ca^{2+}]_i$ (Figure 15). As more cells become incorporated within the cluster, several important events can be identified: (1) the voltage becomes less noisy, (2) the bursts become more regular, (3) the duration of the burst period is extended, and (4) the plateau phase becomes longer. As a consequence of the lengthened plateau phase, the peak $[Ca^{2+}]_i$ level is increased and so, presumably, is insulin release. These properties are obviously important in ensuring a coordinated and concerted response of the endocrine pancreas to a wide variety of stimuli seeking the β-cell target. The integrated mechanisms involved form an excellent example of the functional significance of periodicity and the way in which secretory control may be optimized.

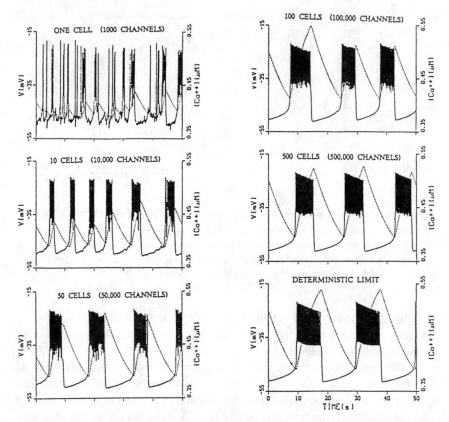

Figure 15. Model of progressive synchronization of pancreatic β-cells with increased cluster size. Each cell in the uncoupled state exhibits stochastic K⁺ channel behavior (i.e., of its 1000 voltage-gated K⁺ channels), but as the cells are gathered together in functional clusters communicating via low-resistance pathways, bursts of action potentials become tightly synchronized, the plateau phase from which the action potentials arise are extended in duration, and the mean $[Ca^{2+}]_i$ rises. (From Chay and Kang, 1988.)

REFERENCES

Berridge, M.J., & Galione, A. (1988). Cytosolic calcium oscillators. FASEB J. 2, 3074–3082.

Chay, T.R., & Kang, H.S. (1988). Role of single-channel stochastic noise on bursting clusters of pancreatic β-cells. Biophys. J. 54, 427–435.

Dean, P.M., & Matthews, E.K. (1968). Electrical activity in pancreatic islet cells. Nature 219, 389–390.

Grapengiesser, E., Gylfe, E., & Hellman, B. (1989). Three types of cytoplasmic Ca²⁺ oscillations in stimulated pancreatic β-cells. Arch. Biochim. Biophys. 268, 404–407.

Gray, P.T.A. (1988). Oscillations of free cytosolic calcium evoked by cholinergic and catecholaminergic agonists in rat parotid acinar cells. J. Physiol. 406, 35–53.

Gray, P.T.A. (1989). The relation of elevation of cytosolic free calcium to activation of membrane conductance in rat parotid acinar cells. Proc. R. Soc. Lond. B. 237, 99–107.

Israel, J.M., Kirk, C., & Vincent, J.D. (1987). Electrophysiological responses to dopamine of rat hypophyseal cells in lactotroph-enriched primary cultures. J. Physiol. 390, 1–22.

Leong, D.A. (1988). A complex mechanism of facilitation in pituitary ACTH cells: recent single cell studies. J. Exp. Biol. 139, 151–168.

Limor, R., Ayalon, D., Capponi, A.M., Childs, G.V., & Naor, Z. (1987). Cytosolic free calcium levels in cultured pituitary cells separated by centrifugal elutriation: effect of gonadotropin-releasing hormone. Endocrinology 120, 497–503.

Lymangrover, J.R., Matthews, E.K., & Saffran, M. (1982). Membrane potential changes of mouse adrenal zona fasciculata cells in response to adrenocorticotropin and adenosine 3', 5'-monophosphate. Endocrinology 110, 462–468.

Matthews, E.K. (1974). Bioelectrical properties of secretory cells. In: Secretory Mechanisms of Exocrine Glands. (Thorn, N.A. & Petersen, O.H., eds.), pp. 185–198. Munksgaard, Copenhagen.

Matthews, E.K. (1982). Pacemakers in secretory cells: mechanisms and possible functions. In: Cellular Pacemakers (Carpenter, D., ed.), Vol. I, pp. 303–323. Wiley, New York.

Matthews, E.K. (1985). Electrophysiology of pancreatic islet β-cells. In: The Electrophysiology of the Secretory Cell (Poisner, A.M. & Trifaró, J.M., eds.), pp. 93–112. Elsevier, Amsterdam.

Matthews, E.K. (1986). Calcium and membrane permeability. Brit. Med. Bull. 42, 391–397.

Matthews, E.K., Dean, P.M., & Sakamoto, Y. (1973). Biophysical effects of sulphonylureas on islet cells. In: Pharmacology and the Future of Man. Proc. 5th Int. Congr. Pharmacology, San Francisco 1972, Vol 3, pp. 221–229. Karger, Basle.

Matthews, E.K., & Saffran, M. (1973). Ionic dependence of adrenal steroidogenesis and ACTH-induced changes in the membrane potential of adrenocortical cells. J. Physiol. 234, 43–64.

Meyer, T., & Stryer, L. (1988). Molecular model for receptor-stimulated calcium spiking. Proc. Natl. Acad. Sci. USA 85, 5051–5055.

Natke, E., & Kabela, E. (1979). Electrical responses in cat adrenal cortex: possible relation to aldosterone secretion. Am. J. Physiol. 237, E158–E162.

Prentki, M., Glennon, M.C., Thomas, A.P., Morris, R.L., Matschinsky, F.M., & Corkey, B.E. (1988). Cell-specific patterns of oscillating free Ca^{2+} in carbamylcholine-stimulated insulinoma cells. J. Biol. Chem. 263, 11044–11047.

Quinn, S.J., Williams, G.H., & Tillotson, D.L. (1988). Calcium oscillations in single adrenal glomerulosa cells stimulated by angiotensin II. Proc. Natl. Acad. Sci. USA 85, 5754–5758.

Rapp, P.E. (1987). Why are so many biological systems periodic? Progress in Neurobiology 29, 261–273.

Rink, T.J., & Jacob, R. (1989). Calcium oscillations in non-excitable cells. TINS 12, 43–46.

Sachs, G., & Muallem, S. (1989). Sites and mechanisms of Ca^{2+} movement in non-excitable cells. Cell Calcium 10, 265–273.

Schlegel, W., Winiger, B.P., Mollard, P., Vacher, P., Wuarin, F., Zahnd, G.R., Wollheim, C.B., & Dufy, B. (1987). Oscillations of cytosolic Ca^{2+} in pituitary cells due to action potentials. Nature 329, 719–721.

Thorner, M.O., Holl, R.W., & Leong, D.A. (1988). The somatotrophe: an endocrine cell with functional calcium transients. J. Exp. Biol. 139, 169–179.

Yule, D.I., & Gallacher, D.V. (1988). Oscillations of cytosolic calcium in single pancreatic acinar cells stimulated by acetylcholine. FEBS Lett. 239, 358–362.

RECOMMENDED READINGS

Carpenter, D. (ed.) (1982). Cellular Pacemakers. Wiley, New York.

Poisner, A.M., & Trifaró, J.M. (eds.) (1985). The Electrophysiology of the Secretory Cell. Elsevier, Amsterdam.

Chapter 11

The Plasma Membrane as a Transducer and Amplifier

DAVID L. SEVERSON and
MORLEY D. HOLLENBERG

Fundamentals of Medical Cell Biology, Volume 5A
Membrane Dynamics and Signaling, pages 223–254
Copyright © 1992 by JAI Press Inc.
All rights of reproduction in any form reserved.
ISBN: 1-55938-309-7

INTRODUCTION

The plasma membrane plays a pivotal role in communicating information between the external and internal cellular milieu (and vice versa). This bidirectional communication is an essential feature of normal homeostasis in multicellular organisms. As summarized in Figure 1, there are a number of highly specialized membrane constituents that play key roles in the information transfer process. On

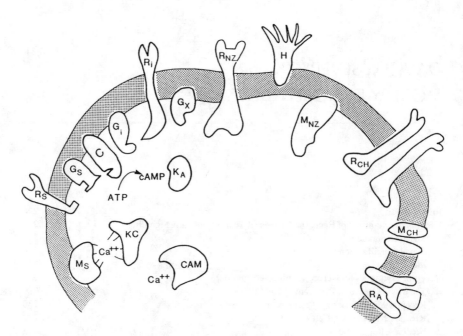

Figure 1. Transmembrane signaling mechanisms. Hypothetical molecular models are shown for the various transmembrane signaling reactions discussed in the text. Receptors that either stimulate (R_S) or inhibit (R_i) adenylate cyclase (C) via the stimulatory (G_S) or inhibitory (G_i) guanine nucleotide regulatory proteins are shown on the left, as is a G protein (G_x) that via a third type of receptor, R_x (not shown), regulates membrane phospholipase. Receptors that possess intrinsic enzymatic activity (tyrosine kinase, guanylate cyclase) are represented by R_{NZ}; M_{NZ} represents other membrane-associated enzymes that can be subjected to receptor-mediated regulation, and H represents a surface determinant (e.g., histocompatibility complex) that can interact with other cells to generate a transmembrane signal. Ion channel receptors, like the nicotinic acetylcholine receptor (R_{CH}), are shown on the right, along with other ion (M_{CH}) or metabolite/acceptor constituents (R_A) that may participate in transmembrane signaling. Potential messengers of the transmembrane signals like cAMP and calcium (Ca^{++}) are also shown along with their protein targets (K_A, cAMP-dependent protein kinase; KC, kinase C; CAM, calmodulin; M_S, membrane-docking protein).

the one hand, specific metabolites or ions may be transported directly via dedicated channels (M_{CH}; Figure 1) or acceptor proteins (R_A, Figure 1). In such cases, the transported substance (e.g., amino acid, cation, or vitamin) is the "signal" that is being sent to the intracellular space, and it is up to the intracellular milieu to respond to the increased (or decreased) amounts of the transported "messenger" substance. On the other hand, specific extracellular ligands may generate intracellular signals without crossing the plasma membrane. These ligands, involved in information transfer, include neurotransmitters, hormones, growth factors, and a variety of drugs. Such ligands must first be recognized with high affinity and specificity by their receptors (e.g., R_S, R_i, R_{CH}, R_{NZ}; Figure 1), and second, the process of ligand binding must be transduced into an amplified cellular signal that ultimately results in a cellular response. Often, the amplification process involves the generation of intracellular "second messenger" molecules. It is the dual function of (1) ligand recognition and (2) signal generation that distinguishes a pharmacological receptor from other membrane-localized information transfer constituents.

This chapter will deal solely with such membrane-localized receptors and will not discuss the mechanisms of action of agents like steroid hormones, which upon diffusing into cells and binding to their cytoplasmic/nuclear-localized receptors regulate cellular functions such as gene expression. The major focus of the sections to follow will be on two questions: (1) How does a receptor trigger a transmembrane signal? and (2) How is the receptor-triggered signal amplified? In large part, the chapter will deal with the signal transduction process that is mediated via the so-called guanine nucleotide binding regulatory proteins (also termed G or N proteins: see later).

GENERAL MECHANISMS OF RECEPTOR-TRIGGERED TRANSMEMBRANE SIGNALING

Signal Generation

Three basic mechanisms may account in general for the triggering of an intracellular signal by a variety of membrane-localized receptors:

1. The receptor may be a ligand-regulated ion channel (R_{CH}; Figure 1). The nicotinic receptor for acetylcholine (Changeux et al., 1987; Colquhoun et al., 1987) and the GABA/benzodiazepine receptor (Schofield et al., 1987; Levitan et al., 1988) are representative of this type of receptors.
2. The receptor may be a transmembrane enzyme (R_{NZ}; Figure 1), whereby the binding of a ligand to the extracellular domain activates a catalytic receptor domain that is situated at the inner aspect of the plasma membrane. It is now clear that the receptor for insulin, along with the epidermal growth factor receptor, the platelet-derived growth factor receptor, and other

tyrosine kinase receptors fit into this category (Yarden and Ullrich, 1988; Schlessinger, 1988).

3. The receptor, in a ligand-directed and GTP-dependent manner, may interact with so-called guanine nucleotide regulatory G (or N) protein oligomers (receptors of the type R_s, R_i, or R_x, interacting with G_s, G_i, and G_x; Figure 1: see G Protein-Mediated Transmembrane Signaling).

The key property of receptor function, irrespective of which of the three mechanisms is used, lies in the ability of the ligand (or its surrogate) to act as an allosteric regulator of the receptor's activity. Thus, for receptors, the process of ligand binding and receptor triggering can be seen as two distinct but related steps, such that it is possible for an agent to bind to a receptor (e.g., an antagonist) without triggering a cellular response. Moreover, both intramolecular (i.e., ligand-modu-lated allosteric effects) and intermolecular receptor dynamics (i.e., receptor mobility, aggregation, and internalization: see later) play important roles in the signal transduction process.

Signal Amplification

Once a ligand has triggered any of the three receptor mechanisms outlined, the initial signal must be greatly amplified so as to affect the entire cell. As with the triggering mechanisms, it is possible to single out only a small number of general cellular reactions that provide the key elements of signal amplification: (1) a change in membrane potential, resulting either from a direct ligand-induced change in ion flux via a receptor channel (R_{CH}; Figure 1) or from an indirect effect of receptor activation (e.g., a G protein-linked receptor; see later) on the function of membrane-localized ion channels (for example, M_{CH}; Figure 1), and (2) the initiation of a phosphorylation–dephosphorylation cascade, akin to the one that regulates the breakdown of glycogen to glucose.

Changes in Membrane Polarization

As has been repeatedly pointed out by Zierler and colleagues (Zierler and Rogus, 1981; Zierler, 1985) small changes in the cellular transmembrane potential can have an immediate and profound effect on the orientation (and presumably, therefore, on the function) of many membrane proteins. Thus, a small change in membrane polarization caused by a receptor-triggered process could immediately affect a number of voltage-regulated ion channels, like the ones for sodium and calcium. Changes in cellular calcium ion concentration and changes in trans-membrane calcium flux represent particularly important messenger/amplifiers of transmembrane signals, via interactions with proteins like calmodulin (CAM; Figure 1) and protein kinase C (KC; Figure 1). Additionally, a change in membrane potential can regulate the activity of an enzyme like phospholipase A_2 (Thuren et al., 1987), so as to generate a variety of arachidonic acid metabolites (prostanoids, leukotrienes) in response to a change in membrane potential. Thus, small receptor-triggered changes in membrane potential can be greatly amplified not only via voltage-sensitive ion channels but also via membrane-localized enzymes that are sensitive to changes in cell polarization.

Phosphorylation–Dephosphorylation Cascades

The triggering of a phosphorylation–dephosphorylation cascade represents perhaps one of the most versatile mechanisms of signal amplification. Classically, this mechanism is best understood in terms of the regulation of glycogenolysis whereby a few hundred molecules of cAMP, acting via a kinase cascade, can trigger a very large breakdown of glycogen. The mechanism of receptor-triggered cAMP formation will be discussed in further detail later in the sections describing G protein function. In many instances the receptor itself is a tyrosine kinase that when triggered by its specific ligand (e.g., insulin) phosphorylates a variety of cellular substrates. In terms of insulin action, the triggering of both phosphorylation (e.g., on ribosomal S-6 protein) and dephosphorylation (e.g., pyruvate dehydrogenase) reactions point to a key role for phosphorylation–dephosphorylation cascade reactions in amplifying the signal generated by the membrane-localized insulin receptor tyrosine kinase. As pointed out earlier, calcium flux can be rapidly altered by a receptor-triggered process. In turn, via its interaction with calmodulin, protein kinase C, and other molecules that can regulate phosphorylation–dephosphorylation reactions, calcium ions can participate in cascade reactions that amplify receptor-triggered signals. Thus, a variety of receptor-triggered signals can be amplified by a number of distinct phosphorylation–dephosphorylation cascade reactions.

The Role of G Proteins

The interactions of receptors with the G or N proteins (so-called because of the key role of guanine nucleotides in the function of these proteins: Rodbell, 1980; Gilman, 1987, 1989) are of utmost importance in triggering and amplifying a receptor-mediated signal. Those receptors that regulate the production of the second messenger cAMP represent the best understood examples of G protein activity. It is now known, as will be discussed in further detail later, that the enzyme adenylate cyclase (C; Figure 1) is controlled not by a direct interaction with the receptor itself, but rather by an indirect process, whereby the receptor triggers the release of a cryptic membrane-localized cyclase-regulatory polypeptide (αs) from the oligomeric G protein complex; intricate equilibria between the various constituents (receptor, G protein, and cyclase) are possible (e.g., Northup, 1985; Gilman, 1987; Levitzki, 1987; see G Protein Mediated Transmembrane Signaling and Figure 4). Much remains to be learned about the possible functions of the various α subunits present in the several distinct receptor-regulated G protein complexes (G_s, G_i, and G_x; Figure 1). To date, the activity of G proteins has been considered in terms of the regulation of adenylate cyclase, phospholipase C, and ion channel activity (see later).

Messengers

For some time, cAMP has been thought of as a prototype second messenger for the actions of agents like epinephrine or glucagon. Now, however, one might argue that the receptor itself may be thought of as one of the messenger molecules, and that the α subunits of the G protein oligomers may also be considered as messengers in the course of epinephrine or glucagon action. Furthermore, it has become evident (as will be outlined in detail under Classification of Receptors According to Their Mechanism of Transmembrane Signaling) that a single ligand–receptor interaction may trigger the release of not just one, but several messengers. Thus, it would appear to be inappropriate to focus on any one particular diffusible low molecular weight primary messenger for the multiple effects of any agent. Rather, it may be more fruitful to think in terms of a matrix of diffusible messengers, generated simultaneously by the combination of an individual ligand like insulin with its receptor. Substances that have been singled out for particular attention in terms of transmembrane signaling processes are (1) sodium ion (nicotinic cholinergic receptor); (2) potassium ion (muscarinic cholinergic and somatostatin receptors acting via a G_i protein); (3) chloride ion (γ-aminobutyric acid/benzodiazepine and glycine receptors); (4) cAMP (adenylate cyclase-coupled receptors like those for epinephrine, glucagon, and ACTH); (5) cGMP (atrial natriuretic factor receptor); (6) hydrogen ion (the triggering of growth factor receptors can increase intracellular pH); (7) calcium ion (for muscarinic cholinergic and α_1-adrenergic receptors); (8) diacylglycerol (many neurotransmitter receptors); along with the concomitantly released messenger (9) inositol trisphosphate and other inositol polyphosphates (many neurotransmitter and peptide receptors); and (10) the glycan-containing messengers that have been described in connection with insulin action. It is important to point out that in the case of many agents, not one, but three of these messengers, namely calcium, diacylglycerol, and inositol 1,4,5-trisphosphate, may be involved in a complex bifurcating pathway triggered by the hydrolysis of membrane phosphoinositides (Berridge, 1987). Furthermore, the action of phospholipases to liberate arachidonate (either directly from membrane phospholipid or indirectly via the subsequent action of diacylglycerol lipase on phosphoinositide-derived diacylglycerol) can result in the formation of prostanoids and/or leukotrienes that in turn can serve in a cascade manner as messengers triggering distinct receptor systems. Thus, the potential simultaneous release of multiple messengers during the course of the action of a single agonist like insulin renders difficult, if not impossible, the interpretation of overall cell response in terms of the generation of a *single* primordial messenger.

RECEPTOR DYNAMICS AND TRANSMEMBRANE SIGNALING

The Mobile Receptor Paradigm

The concept of a receptor as a "mobile" or "floating" membrane constituent has evolved with the development of understanding of the general properties of cell surface proteins. Along with studies of immunoglobulin receptors (e.g., for IgE; Metzger and Ishizaka, 1982), studies of the insulin receptor have contributed in a major way to the concept that receptor mobility and cross-linking are key factors in generating a transmembrane signal. The "mobile" or "floating" receptor model, described in more detail elsewhere (Cuatrecasas and Hollenberg, 1976; Hollenberg, 1985), permits the receptor to interact with multiple effector moieties within the plane of the membrane. Some of the key observations that stimulated the development of this model relate to the ability of multiple ligands (e.g., ACTH, epinephrine, glucagon, etc.) to activate adenylate cyclase in a single target tissue like the rat adipocytes; the data pointed to the stimulation of the same enzyme (i.e., adenylate cyclase) by each hormone acting via its own independent receptor. The key tenet of the model developed independently by several research groups lies in the putative ability of the ligand, upon binding to the receptor, to change the interaction of the receptor with other membrane components. Thus, the entity LR, resulting from the combination of a ligand, L, with its receptor, R,

$$L + R \rightleftharpoons LR \rightleftharpoons LR^*$$

results in a conformationally active form of the receptor R^* that can go on to form effector complexes of the kind,

$$LR^* + E \rightleftharpoons LR^*E$$

wherein E represents an effector molecule involved in the process of cell activation. As will be discussed later, the G protein oligomer represents the "effector" moiety for many hormone receptor systems. A number of variations of this model have been developed (e.g., Levitzki, 1974; DeHaën, 1976; Boeynaems and Dumont, 1977, 1980). For instance, although the equations illustrate an association model, wherein ligand binding promotes receptor–effector coupling, an alternative possibility is a "dissociation" model: a precoupled inactive effector–receptor complex is dissociated to yield an active effector E^* when the ligand binds to the receptor,

$$L + RE \rightleftharpoons LR + E^*$$

In principle, the mobile receptor model does not restrict the number of distinct effector moieties with which the ligand–receptor complex might interact. This property could readily permit a single ligand–receptor complex to trigger concurrently a variety of transmembrane signals.

Receptor Microclustering, Patching, and Internalization

Observations with a number of ligands, including insulin, epidermal growth factor–urogastrone (EGF-URO), low density lipoprotein (LDL), immunoglobulins, and transferrin, have revealed that, subsequent to ligand binding, many receptors (or acceptors, like the ones for LDL and transferrin) follow a common sequence of mobile reactions as outlined in Figure 2 and summarized elsewhere (King and Cuatrecasas, 1981). In the absence of their specific ligands, receptors can be diffusely distributed over the cell surface. However, as illustrated in Figure 2, at physiological temperatures, the binding of a ligand can lead to a rapid microclustering (receptor microclusters, containing perhaps 2 to 10 receptors) and a reduction in receptor mobility, accompanied by the progressive aggregation of ligand–receptor complexes into immobile patches (aggregates containing tens to hundreds of receptors) that can be visualized by fluorescence photomicrography. (Schlessinger, et al., 1978). In cultured fibroblasts, the microclustering event is thought to precede the formation of patches that can be seen in the fluorescence microscope. Subsequent to the formation of the comparatively large receptor aggregates, the ligand–receptor complexes can be either shed into the medium or taken into the cell (internalized). Receptor internalization (Pastan and Willingham, 1981) appears to be an ongoing process that is accelerated when a ligand such as insulin binds to its receptor. It is not clear whether or not receptor occupation is a prerequisite for forming small receptor clusters in all cell types. For instance, in adipocytes there are data to indicate that insulin receptors exist as small clusters prior to the addition of insulin (Jarett and Smith, 1987). The mechanism(s) that lead to microclustering, aggregation, and internalization of receptors (or acceptors) are poorly understood. In many cells, such as fibroblasts, internalization appears to occur at specific sites on the cell surface—the so-called bristle-coated pit. In some cell types (e.g., adipocytes or hepatocytes) receptors may be localized and internalized at sites other than the coated pit regions. Subsequent to aggregation, the receptor can be internalized via an endocytotic process into a cellular compartment that appears to be distinct from the lysosome (Figure 2). The intracellular receptor-bearing vesicles, which in contrast with lysosomes are not phase-dense in the electron microscope and are acid phosphatase negative, have been termed "endosomes" or "receptosomes." The latter term emphasizes the role of these specialized endocytotic vesicles in the process of receptor-mediated endocytosis (Pastan and Willingham, 1981). One possible fate of such receptor-bearing endosomes is fusion with lysosomes, followed by the lysosomal degradation of the receptor (so-called receptor processing). An alternative route that the receptosome may follow leads back to the cell surface via a recycling process that reintegrates the receptor into the plasma membrane. A possible fusion of the receptosome with other intracellular organelles (e.g., nuclear envelope) cannot be ruled out but has yet to be documented. At present, little is known about the factors that control either the internalization process or the trafficking process that may lead, on the one hand,

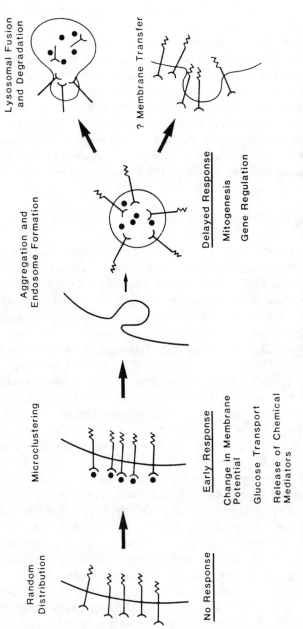

Figure 2. Receptor dynamics and cell activation. As discussed in the text, cell responses that are rapidly regulated probably involve the initial microclustering event. Delayed effects may be caused by a receptor that is internalized in the endosomal organelle. The topography of the endosome would permit the intracellular portion of the receptor (zig-zag line) to interact with a variety of intracellular constituents located at considerable distances from the plasma membrane. In the course of its intracellular migration, the receptor-bearing endosome could ultimately fuse either with the lysosomes or with other membrane structures (e.g., Golgi elements or nuclear membranes), resulting in a further relocation of the receptor.

231

to lysosomal receptor processing or, on the other hand, to a recycling of the receptor back to the cell surface. The intracellular receptor domains very likely play an important role in this trafficking process. There is also little known about the possible role(s) for the degradation products (ligand or receptor fragments) that may be released into the cytoplasm as a result of the endosomal and lysosomal degradation (processing) events. For quite a number of receptors, including those that interact with G proteins and those that possess intrinsic tyrosine kinase activity, the factors that govern receptor internalization and trafficking are presently being studied with great interest. In view of the peripatetic nature of the hormone–receptor complex, migrating from the cell surface to the cytoplasmic space, a key question to answer is what role (if any) do these receptor migratory pathways play in the process of transmembrane signaling? The following section will deal with this question.

Receptor Mobility and Cell Activation

It has been appreciated for some time that cell surface antigens can be triggered to patch and cap. Now, it is apparent that the cross-linking of cell surface receptors for immunoglobulins (like IgE; Metzger and Ishizaka, 1982; Kanner and Metzger, 1983) or for polypeptides like insulin is a key event for cell activation. In terms of polypeptide hormone action, essential observations underlining the importance of receptor microclustering came from work with anti-insulin receptor antibodies obtained either from insulin-resistant patients (Kahn et al., 1978) or from rabbits immunized with purified insulin receptor (Jacobs et al., 1978). In summary, intact antibodies or bivalent (Fab')$_2$ antibody preparations were able to mimic most of the actions of insulin (e.g., stimulation of glucose and amino acid transport) in a variety of target cells (hepatocytes, adipocytes, muscle, fibroblasts) except for the stimulation of DNA synthesis (Kahn et al., 1981). The intact antibodies, like insulin, were also able to cap insulin receptors on intact cells (Schlessinger et al., 1980). However, monovalent Fab' antireceptor antibody preparations that could compete effectively for insulin binding were not only unable to mimic insulin action but were also effective competitive antagonists of insulin in an adipocyte glucose oxidation assay; the biological activity of the monovalent Fab' fragments could be restored by the addition of a second anti-Fab' antibody (Kahn et al., 1981). Results supporting the importance of receptor microaggregation for cell activation have also come from work with antibodies directed against luteinizing hormone–releasing hormone (LHRH) antagonists, wherein the antagonists could be turned into agonists in the presence of cross-linking antibodies (Hopkins et al., 1981; Conn et al., 1982; Gregory et al., 1982). The antibody-induced aggregation of the LHRH receptors could be observed at the electron microscopic level (Hopkins et al., 1981). Although receptor microaggregation is associated with cell activation, as mentioned earlier, the microclustering of adipocyte insulin receptors can be observed in the absence of insulin (Jarett and Smith, 1977). Thus, receptor clustering may prove to be necessary but not necessarily sufficient to initiate a cell response for insulin and other agents. Furthermore, the role of receptor microaggregation in terms of activating cells via a G protein mechanism has yet to be adequately defined, although it is now clear that in such systems, the ligand-occupied receptors must interact in some way with the G protein oligomer.

As outlined in Figure 2, subsequent to receptor microclustering, both the ligand and the receptor can be internalized. An unresolved question is does the internalized ligand–receptor complex play a role in cell activation? In terms of the rapid actions of hormones like insulin (e.g., stimulation of glucose transport), internalization per se would take place at too slow a rate to play a role in cell triggering. The "down regulation" process might, nonetheless, function in terms of modulating overall cell sensitivity. However, in relation to some of the delayed effects of agents like nerve growth factor (e.g., neurite outgrowth) or insulin (mitogenesis, gene regulation), a role for the internalized receptor–ligand complex has been hypothesized. The detection of the retrograde transport and nuclear binding of nerve growth factor in nerve cells (Johnson et al., 1978; Yankner and Shooter, 1979) and observations of the nuclear binding of EGF-URO (Johnson et al., 1980; Savion et al., 1981) and insulin (Goldfine et al., 1977; Podlecki et al., 1987; Jarett and Smith, 1987) are in keeping with this hypothesis. The demonstrated action of intracellularly administered insulin in frog oocytes also argues strongly in favor of this possibility (Miller, 1988). Overall, the process of cell activation related to receptor dynamics is pictured according to the outline in Figure 2. Early responses (changes in membrane potential, metabolite transport) are thought to be triggered in concert with the microclustering of receptors. Delayed responses (mitogenesis, gene regulation) are pictured as possibly involving internalized receptor that is relocated to a specific cellular compartment. Thus, the temporally distinct actions of certain ligands like insulin may relate directly to the topographically distinct dynamic events (microclustering, followed by internalization) that occur over quite different time frames subsequent to ligand binding. In this context, the continued internalization of a receptor may be required to sustain a delayed cellular response to an agent like insulin or nerve growth factor. Thus, transmembrane signaling would have two cytoanatomic tiers and two associated time frames.

Feedback Regulation

Irrespective of the pathway that triggers a cell response (either microclustering or internalization), it is now evident that the entire activation process can be subject to feedback regulation. Thus, the same receptor-triggered reactions that initiate a cell response (e.g., activation of a phosphorylation cascade) can feed back on the receptor itself to turn off a receptor-driven process. For instance, phosphorylation of the β-adrenergic receptor, when it is occupied by an agonist, is believed to play a role in the desensitization of cells to adrenergic agents (Lefkowitz and Caron, 1988), and protein kinase C-mediated phosphorylation is believed to play a role in receptor internalization and recycling (Hunter et al., 1984; Lin et al., 1986). Furthermore, β-adrenergic stimulation, presumably via a cAMP-regulated kinase, causes a down regulation of insulin receptors (Pessin et al., 1983), and elevation of cellular cAMP causes a reduction in insulin receptor kinase activity (Stadtmauer and Rosen, 1986). Thus, in principle, each step of the mechanisms of signal generation and amplification outlined in this chapter can be a target for feedback regulation.

G PROTEIN-MEDIATED TRANSMEMBRANE SIGNALING

As outlined earlier, the mechanisms whereby a G protein mediates receptor-triggered signals represent perhaps the most versatile (and, at present, best understood) processes related to signal transduction and amplification. The key observations that set the stage for understanding the role of G proteins were (1) the discovery that cAMP is a second messenger for the action of many hormones like epinephrine (Sutherland et al., 1968) and (2) the discovery that GTP plays a key role in the ability of glucagon to trigger adenylate cyclase in cell membranes (Rodbell et al., 1971). Furthermore, the isolation and analysis of variants of a cell line (S-49 lymphoma) that were resistant to the ability of isoproterenol to activate adenylate cyclase (so-called cyc⁻ S49 cells) provided essential insight for understanding the G protein system (Gilman, 1987, 1989). The sections that follow describe in some detail both observations leading to an understanding of the G protein system and the details of a number of the G protein systems that have been characterized to date.

Gs and the Stimulation of Adenylate Cyclase

Regulation of cAMP Formation: The Role of GTP

cAMP (adenosine 3′,-5′-monophosphate) was the first intracellular second messenger that was shown to be capable of mediating the effects of many hormones (for examples, catecholamines interacting with the β-adrenergic receptor, glucagon, ACTH, vasopressin and the V_2 receptor). Of particular importance was the experimental observation by Sutherland and co-workers (Sutherland et al., 1968) that this hormonal response (stimulation of the formation of cAMP from ATP, catalyzed by the enzyme adenylate cyclase) could be observed in broken-cell preparations consisting of plasma membranes. The coupling mechanism between the plasma membrane receptor and the adenylate cyclase catalytic unit was unknown until it was reported that GTP was required for activation of adenylate cyclase by glucagon in hepatic plasma membranes (Rodbell et al., 1971; Rodbell, 1980; Limbird, 1981). This critical requirement for GTP at micromolar concentrations had not been observed in previous investigations due to the presence of contaminating guanine nucleotides in crude membrane preparations and, most importantly, in the ATP substrate for the adenylate cyclase reaction, which requires high (millimolar) concentrations of ATP because of ATPase activities in the plasma membrane preparations. By utilizing low concentrations of a non-hydrolyzable ATP analogue (APP[NH]p) as the adenylate cyclase substrate, Rodbell and co-workers were able to demonstrate that GTP enhanced the stimulation of cyclase activity in hepatic plasma membranes by glucagon. Furthermore, non-hydrolyzable analogues of GTP (Gpp[NH]p) were able to increase cyclase activity in the *absence* of glucagon (Salomon et al., 1975). Therefore, the function of the glucagon–receptor complex is to facilitate activation of the adenylate cyclase catalytic unit by guanine nucleotides.

In addition to enhancing the hormone activation of adenylate cyclase activity determined *in vitro*, GTP decreased the affinity of agonists (but not antagonists) for receptors in ligand-binding experiments (Rodbell, 1980; Limbird, 1981). The majority of binding sites in frog erythrocyte membranes for isoproterenol were characterized as having a K_d of 24 nM in the absence of GTP, whereas in the presence of GTP, the K_d for isoproterenol was increased to 1.4 µM (Lefkowitz et al., 1984). When the binding experiment was performed with alprenolol (an antagonist of β-adrenergic receptors), a K_d of 12 nM was observed with or without GTP.

Effects of GTP: Requirement of the Presence of a GTP-Binding Protein and Isolation of G_S

Chromatography on GTP-Sepharose columns was able to demonstrate that the effects of GTP on cyclase activity required the presence of a protein fraction that could be physically resolved from the receptor and the adenylate cyclase catalytic unit (Limbird, 1981; Gilman, 1987). This protein was called G_s, because it was a GTP-binding protein (G protein) that stimulated adenylate cyclase activity. Therefore, G_s is the transducing element that couples receptor occupation to stimulation of adenylate cyclase activity (Figure 3).

As alluded to earlier, a major technical advance was made possible by experiments using a mutant cell line of S49 lymphoma cells termed cyc⁻ (Casperson and Bourne, 1987). Membranes from cyc⁻ cells could not respond, in terms of adenylate cyclase activation, because they had no G_s protein and therefore had very low basal adenylate cyclase activity; neither GTP analogues nor hormones triggered the enzyme in these membranes. However, hormone-sensitive adenylate cyclase activity in cyc⁻ membranes could be restored by the addition of detergent extracts of wild-type membranes containing G_s wherein cyclase activity had first been inactivated by *N*-ethylmaleimide (Gilman, 1989). The availability of cyc⁻ membranes resulted in the development of a reconstitution assay for G_s (increase in adenylate cyclase activity) that permitted efforts to be directed to the purification and characterization of the G_s protein oligomer (Northup, 1985).

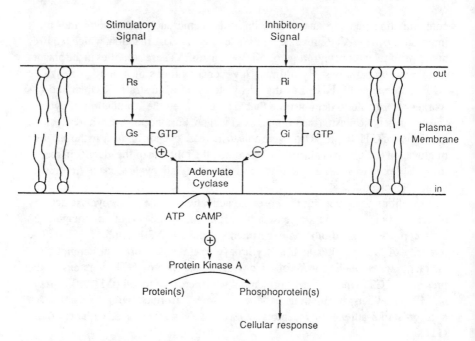

Figure 3. Regulation of adenylate cyclase activity in the plasma membrane by stimulatory and inhibitory signals, each interacting with specific receptors (R_s, R_i) coupled to specific G proteins (G_s and G_i). The conversion of ATP to cAMP results in the activation of the cAMP-dependent protein kinase A; the resulting phosphorylation of protein(s) that are specific to a particular cell then initiates the appropriate response to the signal molecules.

G_s: An Oligomer with Complex Dissociation/Reassociation Equilibria Involving GTP

The G_s protein is a heterotrimer (Gilman, 1987), consisting of αs, β, and γ subunits (Table 1). The αs subunit contains the binding site for guanine nucleotides. The mechanism proposed for the transducing function of G_s in agonist activation of adenylate cyclase is shown in Figure 4 (Gilman, 1987). The G_s protein in an unstimulated plasma membrane is proposed to exist as an oligomer with bound GDP (G_s-GDP). The binding of an agonist or signal molecule (A) to a specific receptor (R) results in the formation of a ternary complex, which is followed by the dissociation of GDP. The slow rate of spontaneous dissociation of GDP from G_s is the rate-limiting step in this transmembrane signaling system; interaction of G_s-GDP with an agonist–receptor complex accelerates the rate of dissociation, so this step constitutes the "on-switch." The rapid association of GTP then produces a conformational change, resulting in dissociation of the ternary complex. The

Table 1. Properties of G Protein Subunits

Subunit	kDa[a]	Toxin Sensitivity	Function
αs (4)	44.5–46	Cholera toxin	Stimulation of adenylate cyclase Regulation of Ca^{2+} channels.
αi (3)	40–41	Pertussis toxin	Inhibition of adenylate cyclase (weak) Regulation of cardiac K^+ channels
αt	40	Cholera and pertussis toxin	Stimulation of retinal rod cGMP-phosphodiesterase
αo	39	Pertussis toxin	Unknown
β (2)	35, 36	—	Anchor α subunit to membrane Interaction of G protein oligomer with receptor
γ (3)	8–11	—	Inhibition of α subunit effects Direct effector regulation?

[a]Molecular masses derived either from the individual cDNAs or from SDS-polyacrylamide gel electrophoresis.

affinity of the receptor for agonist is lowered, and the released agonist and receptor can then interact with additional G_s-GDP in the membrane, so that the recycling A:R complex can result in a *catalytic* effect on activation of G_s and *amplification* of the original signal. The activated G_s-GTP undergoes dissociation to yield the βγ subunit and αs-GTP, which activates the effector system (adenylate cyclase catalytic unit) so that ATP is converted to cAMP (Figure 4). The catalytic activity

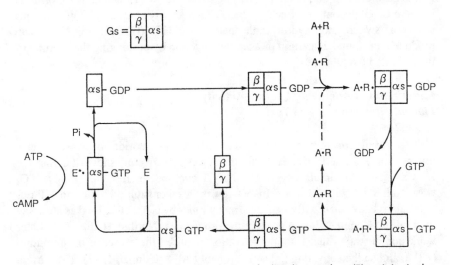

Figure 4. Proposed scheme for the stimulation of adenylate cyclase (E) activity in the plasma membrane by a stimulatory signal (A) interacting with its receptor (R) and the G_s oligomeric protein (heterotrimer consisting of αs, β, and γ subunits). Amplification of the original signal is the result of two catalytic events: activation of G_s by the re-cycling A.R. complex, and the turnover number for the adenylate cyclase enzyme. Inactivation of the cyclase catalytic unit is the result of the GTPase activity (GTP → GDP and P_i) of the αs subunit of G_s.

of the cyclase-αs-GTP complex results in further *amplification* of the stimulatory signal. The turnover number for adenylate cyclase is 1000 min^{-1}, therefore several hundred molecules of cAMP can be synthesized during the lifetime of single αs-GTP (Gilman, 1987). Therefore, *amplification* of the cAMP second messenger at the level of the plasma membrane involves both the catalytic activation of G_s and the activity (turnover number) of the adenylate cyclase catalytic unit. The subsequent effect of cAMP on the cAMP-dependent protein kinase (kinase A) will further amplify the signal, ultimately resulting in the appropriate cell response.

The G protein subunit dissociation model shown in Figure 4 also accounts for the observation that GTP reduces affinity of receptors for agonists (see earlier). The high affinity state is represented by the binding of agonists to receptor complexed with G_s (GDP-dissociated); in the presence of GTP, the ternary complex dissociates and produces free receptor, which has a lower affinity for binding of the agonist.

The αs subunit, in addition to having a GTP binding site, also exhibits an intrinsic GTPase activity, and so inactivation of the adenylate cyclase effector unit is due to the hydrolysis of GTP to yield GDP (Figure 4). The turnover number of the αs-GTPase activity is relatively slow (10 min^{-1}), so the lifetime of αs-GTP is normally sufficient (several seconds) to allow for a significant stimulation of cyclase activity (Freissmuth et al., 1989). The inactive adenylate cyclase catalytic unit dissociates and the αs-GDP subunit reassociates with βγsubunits to form the inactive G_s-GDP complex. Because the GTPase activity of the α subunit functions as an "off-switch," non-hydrolyzable analogues of GTP (GTPγS; Gpp[NH]p) can produce a profound activation of adenylate cyclase by prolonging the lifetime of the activated αs-subunit.

Stimulation of Adenylate Cyclase Activity by Fluoride Ion and Cholera Toxin

The stimulation of adenylate cyclase activity by fluoride ion was an early observation of Sutherland and co-workers. It was subsequently shown that fluoride ion was unable to stimulate cyclase activity in cyc$^-$ membranes, indicating that G_s was required for fluoride action. In fact, it was the reconstitution of fluoride-stimulated adenylate cyclase activity in cyc$^-$ membranes by G_s that was used as the assay to monitor G_s content during purification procedures (Northup et al., 1980). Later work demonstrated that fluoride activation required the presence of aluminum (Al^{3+}), and so the activating ligand is probably AlF_4^- (Gilman, 1987). The current proposal for fluoride activation of adenylate cyclase is that the AlF_4^- mimics the γ-phosphate in GTP, and so forms a GTP-like complex with GDP bound to G_s, thus initiating the activation cycle (Figure 4).

The αs subunit of G_s is a substrate for cholera toxin, with the result that an arginine residue is ADP-ribosylated (Figure 5). The covalent modification of αs by ADP-ribosylation results in a decrease in GTPase activity; therefore the αs-GTP

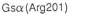

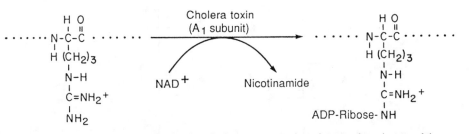

Figure 5. Reaction sequence for the cholera toxin-catalyzed ADP-ribosylation of the α subunit of G_s. Utilization of $[^{32}P]NAD$ as a reactant results in the incorporation of radioactivity into the αs subunit of G_s (arginine amino acid residue). ADP-ribosylation of αs by cholera toxin results in a decrease in GTPase activity, and so adenylate cyclase activation, according to the scheme shown in Figure 4, is prolonged.

subunit is maintained in its active form for continued adenylate cyclase stimulation. In addition to being an experimental tool to produce a sustained activation of adenylate cyclase, the ADP-ribosylation of G_s can be utilized as a probe to identify the presence of αs subunits in cell extracts by autoradiography of SDS-polyacrylamide electrophoresis gels, for example, if the NAD^+ utilized in the reaction sequence (Figure 5) is radioactively labeled with ^{32}P. However, cholera toxin also stimulates the ADP-ribosylation of other G proteins (Table 1; see later) and so the results from these types of experiments must be interpreted carefully.

G_i and the Inhibition of cAMP Formation

A number of different signal molecules (adenosine interacting with A_1 receptors; acetylcholine with muscarinic receptors; catecholamines with $α_2$ receptors) result in inhibition of adenylate cyclase activity and a decrease in the cellular content of cAMP (Figure 3). This inhibitory effect requires GTP and a GTP-binding protein termed G_i that is separate and distinct from G_s (Limbird, 1981; Gilman, 1987). For example, somatostatin produces a GTP-dependent inhibition of adenylate cyclase activity in cyc⁻ membranes. This G_i protein was subsequently purified and characterized as a heterotrimer with βγ subunits that were identical with the βγ subunits in G_s, and a unique αi subunit (Table 1). A mechanism for adenylate cyclase inhibition by G_i has been proposed to involve subunit exchange (Gilman, 1987) and is shown in Figure 6. As outlined earlier, adenylate cyclase activity is stimulated by αs-GTP, formed as a result of interaction of the stimulatory signal (A) with its receptor (R_s) and G_s. An inhibitory signal (B) can interact with its receptor (R_i) and with G_i-GDP. This is followed by dissociation of GDP, association with GTP, release of B·R, and dissociation of G_i-GTP to yield βγ and αi-GTP subunits in a manner that is exactly analogous to the stimulatory scheme. When the isolated subunits were tested for their ability to inhibit adenylate cyclase, the

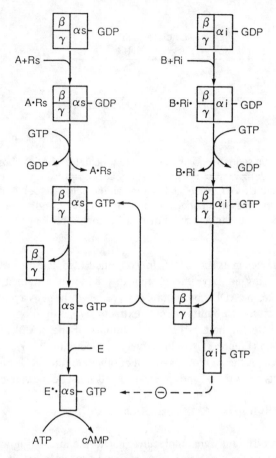

Figure 6. Proposed scheme for the regulation of adenylate cyclase (E) activity by a stimulatory signal (A) interacting with a receptor (R_s) coupled to G_s (cf. Figure 4) and by an inhibitory signal (B) interacting with a receptor (R_i) coupled to G_i. Inhibition of adenylate cyclase activity may be a direct effect of the α_i-GTP subunit or due to the binding of the stimulatory α_s-GTP subunit by $\beta\gamma$ subunits (derived from G_i) to reconstitute a G_s oligomer.

greatest inhibition was observed with $\beta\gamma$ subunits, presumably by complexing with the α_s-GTP to re-form the oligomer. Therefore, this inhibitory mechanism requires the presence of G_s. However, as noted earlier, somatostatin can inhibit adenylate cyclase activity in cyc⁻ membranes that are devoid of G_s, indicating that α_i-GTP must be capable of the "direct" inhibition of adenylate cyclase. The relative quantities of G_i and G_s in the plasma membrane will be a factor in cyclase inhibition produced by this subunit dissociation model. The content of G_i is typically greater than G_s (Gilman, 1987); therefore the $\beta\gamma$ subunits in G_i could act as a reservoir to regulate the content and activity of αs. One of the other implications of this model is that $\beta\gamma$ subunits derived from any of the G proteins (Table 1) could potentially inhibit cyclase activity.

The G_i protein is also subject to toxin modification. In this case, the αi subunit is ADP-ribosylated by pertussis toxin, but unlike the cholera toxin-modification of αs, the ADP-ribosylation of αi occurs on a cysteine residue. The ADP-ribosylated G_i protein is uncoupled from inhibitory receptors; therefore, pertussis toxin pretreatment prevents the inhibition of adenylate cyclase activity by signal molecules acting through G_i (Figure 3). Modification of G_i by pertussis toxin in the presence of radioactively labeled NAD^+ can be used to identify pertussis toxin substrates, but ADP-ribosylation is not restricted to G_i proteins (Table 1).

Phototransduction

Phototransduction represents a special case of signal transduction where the signal is not a chemical but rather a photon of light (Stryer, 1986). Retinal photoreceptor cells, consisting of rods and cones, convert light into nerve impulses. In rod outer segments, cation-specific channels in the plasma membrane are open in the dark because of high levels of intracellular cGMP that directly stimulate the channel (Figure 7). The "light receptor" in rods is a protein, rhodopsin, that contains an 11-*cis*-retinal prosthetic group that functions as a chromophore. Rhodopsin exhibits substantial homology with receptors (e.g., β-adrenergic and muscarinic receptors) that interact with G proteins (Dohlman et al., 1987).

Transducin (Gt) and the Activation of cGMP Phosphodiesterase

A specific G protein, transducin (Gt), is involved in the *phototransduction* process. Transducin consists of a unique αt subunit, plus β and γ subunits (Table 1). As described previously for receptors coupled to stimulation of adenylate cyclase (Figure 4), photoexcited rhodopsin (R*) forms a complex with Gt-GDP, which is followed by dissociation of GDP and association of GTP (Figure 7). Photoexcited rhodopsin is then released and can activate another transducin molecule, resulting in a signal *amplification*. A single photoexcited rhodopsin catalyzes the activation of 500 transducin molecules (Stryer, 1986). Gt-GTP then dissociates into $\beta\gamma$ and αt-GTP subunits, and αt-GTP activates the cGMP-dependent phosphodiesterase so that cGMP is hydrolyzed to GMP (Figure 7). As a result, the cation channel is closed and the light-induced hyperpolarization is transmitted to the synaptic body of the rod cells, which eventually results in a nerve impulse that is processed by the retina before transmission to the visual cortex of the brain. cGMP-dependent phosphodiesterase in rod outer segments consists of three subunits (α, β, γ); phosphodiesterase activity in the dark is inhibited by constraint imposed by the γ subunit. Activation of cGMP-dependent phosphodiesterase by αt-GTP is the consequence of the binding of inhibitory γ subunits by αt-GTP. The

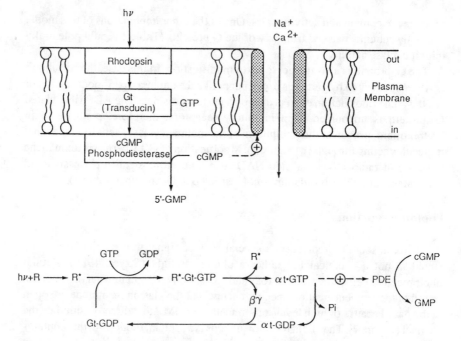

Figure 7. Regulation of phototransduction in rods by the light (hν)-induced activation of rhodopsin (R), which is coupled to the stimulation of cGMP phosphodiesterase (PDE) activity by transducin (Gt). The resulting fall in cGMP content results in closure of a cation-specific channel and hyperpolarization of the rod photoreceptor cell.

catalytic activity of the phosphodiesterase (turnover number of 3700 s^{-1}; Stryer, 1986) results in additional signal *amplification*, and as a result, phototransduction is a remarkably sensitive process. The overall amplification provided by transducin activation and cGMP-phosphodiesterase activity is about 10^5 (Stryer, 1986).

The intrinsic GTPase activity of the αt subunit reverses the activation of phosphodiesterase (Figure 7). Interestingly, transducin is a substrate for both cholera and pertussis toxins (Table 1). Recently, a transducin-like protein has been localized in cones (Casey and Gilman, 1988; Freissmuth et al., 1989) where it is assumed to activate a phosphodiesterase and regulate the phototransduction process in relation to color vision.

Heterogeneity of G Proteins

Alpha Subunits

G proteins are heterotrimers (α, β, γ subunits) and are classified according to the identity of their distinct α subunits (Table 1), which have specific functions such as stimulation of adenylate cyclase (αs) and cGMP-dependent phosphodiesterase (αt) activities. All of the α subunits have a high affinity guanine nucleotide binding site and exhibit intrinsic GTPase activity, but their susceptibility to ADP-ribosylation by toxins is variable (Casey and Gilman, 1988; Freissmuth et al., 1989). A G_o protein (where o stands for "other") was discovered as the major G protein in brain, where it constitutes approximately 1% of membrane protein. The function of G_o and its αo subunit is not known, but presumably G_o will regulate yet to be determined pertussis toxin-sensitive processes. Recently, two new G proteins have been characterized from cDNA-deduced sequences. (Freissmuth et al., 1989; Gilman, 1989). G_s-olf (α-olf, 44.7 kDa) is presumed to activate olfactory adenylate cyclase and thus participate in the transduction process of olfaction. The α subunit of G_z (40.9 kDa) has the interesting property of not being a substrate for either cholera or pertussis toxins. This further underscores the limitations in relying on toxin sensitivity to characterize G protein regulatory events.

Additional heterogeneity exists within some of the classes of G proteins. (Freissmuth et al., 1989). As noted earlier, distinct transducin molecules are present in retinal rods and cones. Four forms of the αs subunit originate by alternative splicing of mRNA (Table 1). In contrast, αi exists in three forms that are the product of distinct genes. At present, this heterogeneity has not been associated with any functional significance.

Beta-Gamma Subunits

The β subunits associated with most G proteins can be resolved into two forms (β_{35} and β_{36}) that are products of two genes (Table 1); transducin is an exception with only one form of the β subunit (β_{36}). At least three forms of γ subunits can be distinguished. Consequently, heterogeneity of the individual α, β, and γ subunit proteins can result in a remarkable degree of heterogeneity for the G protein oligomers (Gilman, 1989). Originally, it was thought that a single $\beta\gamma$ subunit might be present in all G proteins with α subunits having the exclusive role in functional selectivity of action. Although the $\beta\gamma$ complex from G_s, G_i, and G_o is functionally interchangeable with the appropriate α subunit, the αs subunit can distinguish between transducin $\beta\gamma$ and $\beta\gamma$ from non-retinal G proteins; the γ subunit may confer this specificity (Freissmuth et al., 1989). Given the potential for subunit exchange, as proposed for the mechanism for cAMP inhibition (Figure 6), then heterogeneity among G protein subunits could result in a great deal of complexity and fine regulation of signal transduction processes. Certainly, free $\beta\gamma$ subunits have the potential to inhibit processes stimulated by α subunits. In addition to a structural role for the $\beta\gamma$ subunit in the attachment of the G protein oligomers to membranes and interaction with receptors, a great deal of current effort is attempting to determine if $\beta\gamma$ subunits can directly regulate effector systems (for example, inhibition of adenylate cyclase and cAMP-phosphodiesterase activities, and activation of cardiac K^+ channels, see later).

Phosphoinositide Turnover

Generation of Phosphoinositide-Derived Second Messengers

A great deal of interest has focused in recent years on the generation of two second messengers, inositol 1,4,5-trisphosphate (IP_3) and diacylglycerol (DAG), by the phospholipase C-catalyzed degradation of membrane phosphatidylinositol 4,5-bisphosphate (PIP_2) (Berridge, 1987). The hydrolysis of PIP_2 (Figure 8) can be stimulated by a variety of signal molecules, such as norepinephrine (interacting with α_1-receptors), vasopressin (V_1 receptors), angiotensin II, platelet-derived growth factor, thrombin, and many others. The second messenger roles for IP_3 and DAG are due to release of calcium from intracellular stores and activation of the calcium- and phospholipid-dependent protein kinase C, respectively (Figure 8).

Evidence for the Involvement of the G Proteins in Phospholipase C Activation

Considerable evidence has accumulated in favor of the hypothesis that a G protein (G_p) is involved in the transduction process (see Figure 8) between the receptor and stimulation of a phosphoinositide-specific phospholipase C (Berridge, 1987; Gilman, 1987; Fain et al., 1988). The affinity of vasopressin for binding to hepatic plasma membranes is decreased by GTP and vasopressin stimulated GTPase activity in membranes from cultured hepatocytes, but these effects of vasopressin were not altered by cholera or pertussis toxin treatments. Pertussis toxin inhibits agonist-stimulated phosphoinositide breakdown in some, but not all, cell systems (Fain et al., 1988). This suggests heterogeneity of the putative G_p transducer. Fluoride ions (presumably AlF_4^-) can also stimulate phosphoinositide breakdown in some intact and broken cell preparations. More direct evidence for the involvement of a G_p protein in phosphoinositide breakdown has come from investigations with permeabilized cells where the addition of guanine nucleotides enhances the accumulation of the diacylglycerol second messenger. Phospholipase C activity can be measured in membranes where the endogenous PIP_2 pool has been prelabeled by preincubation of intact cells with [^{3}H] inositol or $^{32}P_i$; the subsequent addition of GTPγS to these membranes results in stimulation of phospholipase C activity. A stimulatory effect of GTPγS on cytosolic phospholipase C activity measured with an exogenous [^{3}H]PIP_2 substrate has also been observed (Majerus et al., 1986).

The involvement of G_p in the stimulation of phospholipase C (Figure 8) activity must be considered as probable, but not proven, because although GTPγS stimulates phospholipase C activity in crude preparations, reconstitution experiments with addition of various G protein oligomers and subunits to purified phosphoinositide-specific phospholipase C preparations have not resulted in any stimulation of enzyme activity (Rhee et al., 1989). One of the technical problems

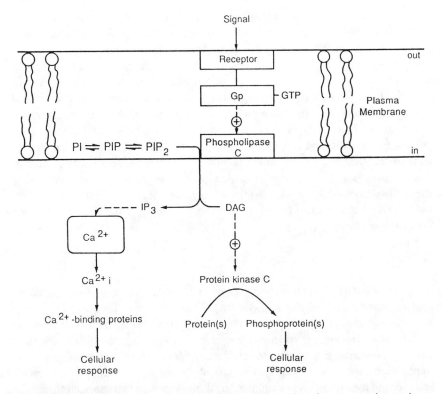

Figure 8. Regulation of phosphoinositide turnover in the plasma membrane by a signal molecule interacting with a specific receptor coupled to the stimulation of phospholipase C activity by a GTP-binding protein (G_p). As a consequence, phosphatidylinositol 4,5-bisphosphate (PIP_2) is hydrolyzed to produce two second messengers: inositol 1,4,5-trisphosphate (IP_3), which releases Ca^{2+} from the internal membrane stores so that the interaction of Ca^{2+}_i with Ca^{2+}– binding proteins like calmodulin can produce a cellular response, and diacylglycerol (DAG), which activates the Ca^{2+}- and phospholipid-dependent protein kinase C so that serine-phosphorylation of cellular proteins may produce a specific cellular response or modulate the cellular response to calcium.

in working with phospholipase C is that the presentation of an exogenous PIP_2 substrate for *in vitro* determinations of membrane-bound enzyme activity is much more difficult than the addition of ATP to adenylate cyclase assays, for example, due to requirements for emulsification of an exogenous radiolabeled lipid substrate and the presence of endogenous phospholipids in the plasma membrane. An increase in phosphoinositide-specific phospholipase C activity in plasma membrane preparations incubated with the appropriate stimulatory signal has never been observed. It is also possible that an additional inhibitory factor(s) may be involved in the regulation of phosphoinositide-specific phospholipase C. In contrast with adenylate cyclase, purified phosphoinositide-specific phospholipase C preparations are highly active in the absence of guanine nucleotides or G proteins. Therefore, an additional protein that inhibits phosphoinositide-specific phospholipase C activity may be present in the plasma membrane. Activation of phospholipase C by the putative G_p protein could then be the result of the removal of this inhibitory protein, much like the activation of cGMP-dependent phosphodiesterase by transducin involves removal of the inhibitory constraint of the γ subunit of the phosphodiesterase. Isolation of this potential inhibitory factor will be required in order to reconstitute a G protein-sensitive phospholipase C effector system from the purified components. Analysis of the primary sequence of one of the purified isoenzymes of phospholipase C reveals domains that may interact with negative regulatory factors (Rhee et al., 1989), by analogy to the sequence of some oncogene products.

G Protein Gating of Ion Channels

Acetylcholine (ACh) increases potassium (K^+) conductance in atria by opening specific potassium channels (I_{K-ACh}) in the plasma membrane. The current evidence that this ligand-regulated potassium channel is gated directly by a G protein has been reviewed recently (Brown and Birnbaumer, 1988; Hartzell, 1988). The time course for the change in membrane conductance induced by ACh is relatively slow, implicating a transducing step(s). The possible involvement of diffusible intracellular second messengers was eliminated by the observation that application of ACh outside cell-attached patches had no effect on the I_{K-ACh} channel activity measured within the patches, and that ACh activated I_{K-ACh} channels in excised patches of membranes.

The activation of the potassium channel by ACh requires GTP and is blocked by pertussis toxin, suggesting that G_i may gate this potassium channel (Figure 9). In excised membrane patches, I_{K-ACh} channels are activated by the G_i oligomer preactivated by GTPγS and isolated αi subunits (Yatani et al., 1988), confirming that the gating action of the G_i protein on the channel is direct. Some studies have shown that βγ subunits can also activate the I_{K-ACh} channel (Neer and Clapham, 1988), but the physiological significance of this observation remains controversial (Hartzell, 1988).

Calcium channels may also be regulated directly and indirectly by G proteins, specifically G_s (Table 1). The involvement of G_s in the β-adrenergic stimulation of adenylate cyclase results in protein kinase A–catalyzed phosphorylation and stimulation of the dihydropyridine-sensitive calcium channel; the addition of GTPγS, or activated G_s and αs-GTP, to phosphorylated calcium channels in excised patches further stimulates calcium channel activity (Brown and Birnbaumer, 1988). Therefore, specific G proteins, such as G_i and G_s, can couple receptors to more than one effector system.

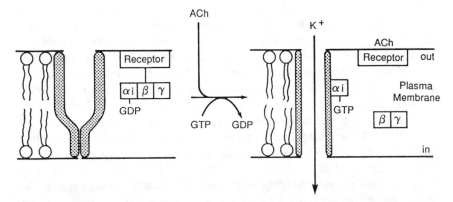

Figure 9. Acetylcholine (ACh) interacts with muscarinic receptors in cardiac atrial cells, with the result that K⁺channels are opened by a direct gating action of G_i. Thus, G_i can couple receptors to more than one effector system (adenylate cyclase, Figure 3, and ion channels).

Summary and Future Prospects for Research with G Proteins

G proteins are oligomeric proteins consisting of α, β, and γ subunits that perform an important role as *transducing elements* by coupling the occupancy of plasma membrane receptors by signal molecules (ligands) to the simulation of effector systems such as adenylate cyclase (G_s) or cGMP-dependent phosphodiesterase (G_t, transducin). As a consequence of this signal transduction sequence, the content of intracellular second messenger molecules is regulated in order to mediate the appropriate cellular response. The ligand–receptor activation of the G protein oligomer is a catalytic event that, together with the catalytic activity (turnover number) of the enzyme effector system, results in marked signal *amplification.*

In keeping with the predictions of the mobile receptor paradigm of hormone action, it is now clear that specific G proteins can interact with multiple receptors. For example, G_s couples receptors specific for catecholamines, ACTH, and glucagon to adenylate cyclase stimulation. Therefore, the precise molecular mechanism for the interaction of the G protein oligomer with these receptors will be an important area of future investigation. The deduced amino acid sequence of the cloned β-adrenergic receptor suggests that the receptor contains seven trans-membrane spanning regions (24–28 hydrophobic amino acid residues) connected by a series of extracellular and cytoplasmic loops (Dohlman et al., 1987; Lefkowitz and Caron, 1988). Application of the powerful techniques of molecular biology to produce mutant receptors with deletions in selected domains and receptors subjected to site-directed mutagenesis should delineate the domain(s) of the receptor that interact with the appropriate G protein (Lefkowitz et al., 1989). Other receptors that are coupled to G proteins such as muscarinic acetylcholine receptors and rhodopsin are also characterized by the presence of seven transmembrane spanning regions that exhibit substantial homology to the β-adrenergic receptor (Dohlman et al., 1987). The expression of chimeric receptor genes where structural domains can be interchanged between receptors will also be a powerful tool to test for functional changes in the interaction of chimeric receptors with specific G proteins (Lefkowitz et al., 1989).

Recently, the cloned structure of adenylate cyclase has been reported (Krupinski et al., 1989). Therefore, application of the molecular biological approaches outlined previously for receptors should delineate the domains of the adenylate cyclase molecule involved in the interaction with αs-GTP, αi-GTP, and perhaps βγ subunits. Also, as predicted by the mobile receptor model of hormone action, it has now been demonstrated that receptor-activated G proteins can couple with more than one effector system, as illustrated by the inhibition of adenylate cyclase activity and the opening of cardiac potassium channels by G_i. It is perhaps not surprising, therefore, that the proposed structure for adenylate cyclase contains transmembrane sequences that resemble plasma membrane ion channels. Another important area of investigation in the future will be to assess the functional significance of the potential heterogeneity (e.g., three isoforms of αi) within the oligomeric G proteins. Further research is also required to establish the signaling systems regulated by G_o and G_z.

The G protein oligomers that are involved in signal transduction belong to a larger family of GTP-binding proteins that are characterized as having high-affinity binding sites for guanine nucleotides, variable GTPase activity, but a much smaller size (approximately 21 kDa). Some examples of these latter proteins are ADP-ribosylation factor (ARF), *ras* gene products, and yeast YPT1 and SEC4 gene products (Casey and Gilman, 1988; Freissmuth et al., 1989). In most cases, the physiological function for the low molecular weight GTP-binding proteins is not known. ARF containing bound GTP (or GTPγs) is required for the ADP-ribosylation of $G_s\alpha$ by cholera toxin (Figure 5), but it not clear if this represents the only function of ARF. The product of normal mammalian *ras* genes are 21-kDa GTP-binding proteins that exhibit an intrinsic GTPase activity that can be increased by a GTPase-activating protein (GAP) of approximately 120 kDa (Santos and Nebreda, 1989). Interestingly, transforming *ras* proteins have greatly reduced GTPase activity that is not increased by GAP. It has been proposed, therefore, that *ras* proteins can exist in an equilibrium between an inactive conformation containing bound GDP and an activated conformation containing GTP that can then interact with as yet unknown targets. Thus, oncogenic mutations, by reducing GTPase activity, would result in an accumulation of *ras* proteins in the activated conformation and thus produce a sustained activation of some signal transduction pathway(s) that could eventually result in cellular transformation (Santos and Nebreda 1989). More research is required to identify the effector systems regulated by *ras* proteins. The GTP-binding protein products of yeast YPT1 and SEC4 genes

(23 and 23.5 kDa, respectively) may regulate secretory processes (Bourne, 1988). As outlined previously for the *ras* gene products, GTP binding and hydrolysis by the YPT1 and SEC4 gene products may control "switching" between different GTP-binding protein conformations that can selectively recognize a protein component in one membrane and thus mediate vectorial transport to another membrane compartment. An analogous mechanism has been proposed for the involvement of GTP-binding proteins (for example, elongation factor tu) in protein synthesis in prokaryotes (Allende, 1988).

The oligomeric G proteins share some key features with the low molecular weight GTP-binding proteins. In both instances, the binding of GTP produces a conformational change in the respective proteins. For the low molecular weight GTP-binding proteins, this conformational change directly regulates the effector system (e.g., a translating ribosome or a membrane protein). In the case of oligomeric G proteins, the GTP-induced conformational change results in dissociation of the ligand–receptor complex and dissociation of the oligomeric structure to yield the free α subunit that typically regulates the effector system (an enzyme or ion channel). The dissociation of GDP and association of GTP in the oligomeric G protein is triggered by the ligand–receptor complex, whereas the guanine nucleotide exchange reaction for low molecular weight GTP-binding proteins often requires a specific exchange factor (Allende, 1988). In both cases, GTPase activity results in termination of the effect. In summary, oligomeric G proteins are involved in signal transduction and amplification, and the low molecular weight GTP-binding proteins are conformational switches that confer directionality to a process (Bourne, 1988). Further research is required to determine if there are biological processes that are regulated by both the oligomeric and low molecular weight G proteins.

LOOKING TO THE FUTURE: CLASSIFICATION OF RECEPTORS ACCORDING TO THEIR MECHANISM OF TRANSMEMBRANE SIGNALING

Throughout this chapter, reference has been made to a variety of agonists in terms of the mechanisms whereby these agents trigger a cellular signal. What has become evident over the past few years, as a result of the detailed sequencing of a variety of receptors, is that there are perhaps more similarities between groups of the receptors themselves than there are similarities between the various ligands that activate the receptors. For instance, the receptors for insulin, insulin-like growth factor I, EGF-urogastrone, platelet-derived growth factor, and colony-stimulating factor I all have in their structures highly homologous domains that confer upon these receptors their tyrosine kinase activity (Schlessinger, 1988; Yarden and Ullrich, 1988). Likewise, the receptors for β-adrenergic agents, muscarinic agents,

Table 2. Receptors and Their Transmembrane Signaling Mechanisms[a]

		Receptor	
Signaling Mechanism[b]		*Pharmacological Classification*[c]	*Subtype*[c]
R_{CH}:	ion channel regulation		
	Sodium/potassium	Nicotinic	Nicotinic, N1 to N5
	Chloride	GABA/benzodiazepine	$GABA_{A_1}$ to $GABA_{A_3}$
		Glycine	
R_{NZ}:	ligand-regulated enzyme		
	Tyrosine kinase	Colony-stimulating factor 1	
		EGF-urogastrone	
		Insulin	
		Insulin-like growth factor 1	
		Platelet-derived growth factor	PDGF-A, PDGF-B
	Guanylate cyclase	Atrial natriuretic factor	
R_S, R_i, etc:	G protein coupled	ACTH	
		Adrenergic	α_1, α_2, β_1, β_2
		Angiotensin	A-1, A-2
		Cholinergic	Muscarinic, M_1–M_5
		Dopaminergic	D_1, D_2
		Glucagon	
		Histaminergic	H_1, H_2
		LHRH	
		Serotonergic	5HT-1a, 5HT-1c
		Substance K (tachykinins)	
		Vasopressin	V_1, V_2

[a]The list of receptors is intended to be representative and not exhaustive.
[b]The symbols R_{CH}, R_{NZ}, R_S, and R_i are shown in Figure 1; the mechanisms are discussed in the text.
[c]Receptors are listed for which complete sequence data are available and/or for which a mechanism of action has been clearly established. The receptor subtypes when available, as determined by structure–activity studies or by molecular cloning, are also listed.

and serotonergic agents all have common elements that are presumably related to their abilities to interact with G proteins (Caron and Lefkowitz, 1988). An emerging theme is that the common thread linking groups of receptors, in terms of their structures, is their transmembrane signaling mechanism. Thus, in the future, it should prove possible to classify receptors not only according to their triggering ligands but also according to their signaling mechanisms. A representative example of such a classification is shown in Table 2, wherein the receptors mentioned in this chapter are listed. The value of viewing receptors in this way relates to the ability of predicting structures for newly discovered receptors for which only preliminary data may be available in terms of their activity (e.g., dependence of action on GTP would predict a β-adrenergic–like structure; the regulation of Na^+/K^+ flux would point to domains present in the nicotinic receptor). In the future, one can look forward to the detailed study, in a variety of receptors, of the specific protein domains that are involved intimately in the generation of transmembrane signals (for a brief synopsis, see Hollenberg, 1990). Thus, one can anticipate in the near future, a detailed description at the molecular level of many of the steps involved in transmembrane signaling, starting with the triggering of a receptor and ending with a response generated by the amplified receptor signal.

REFERENCES

Allende, J.E. (1988). GTP-mediated macromolecular interactions: the common feature of different systems. FASEB J. 2, 2356–2367.

Berridge, M.J. (1987). Inositol trisphosphate and diacylglycerol: two interacting second messengers. Ann. Rev. Biochem. 56, 159–193.

Boeynaems, J.M., & Dumont, J.E. (1977). The two-step model of ligand-receptor interaction. Mol. Cell. Endocrinol. 7, 33–47.

Boeynaems, J.M., & Dumont, J.E. (1980). Outlines of Receptor Theory. Elsevier/North Holland Biomedical, Amsterdam.

Brown, A.M., & Birnbaumer L. (1988). Direct G protein gating of ion channels. Am. J. Physiol. 254, H401–H410.

Bourne, H.R. (1988). Do GTPases direct membrane traffic in secretion? Cell 53, 669–671.

Casperson, G.F., & Bourne H.R. (1987). Biochemical and molecular genetic analysis of hormone-sensitive adenylyl cyclase. Ann. Rev. Pharmacol. Toxicol. 27, 371–384.

Casey, P.J., & Gilman, A.G. (1988). G-protein involvement in receptor-effector coupling. J. Biol. Chem. 263, 2577–2580.

Changeux, J-P., Giraudat, J., & Dennis, M. (1987). The nicotinic acetylcholine receptor: molecular architecture of a ligand-regulated ion channel. TIPS 8, 459-466.

Colquhoun, D., Ogden, D.C., & Mathie, A. (1987). Nicotinic acetylcholine receptors of nerve and muscle: functional aspects. TIPS 8, 465–472.

Conn, P.M., Rogers, D.C., Stewart, J.M., Neidel, J., & Sheffield, T. (1982). Conversion of a gonadotropin-releasing hormone antagonist to an agonist. Nature 96, 653–655.

Cuatrecasas, P., & Hollenberg, M.D. (1976). Membrane receptors and hormone action. Adv. Protein Chem. 30, 251–451.

De Haen, C. (1976). The non-stoichiometric floating receptor model for hormone-sensitive adenylate cyclase. J. Theor. Biochem. 58, 383–400.

Dohlman, H.G., Caron, M.G., & Lefkowitz, R.J. (1987). A family of receptors coupled to guanine nucleotide regulatory proteins. Biochemistry, 26, 2657–2664.

Fain, J.N., Wallace, A., & Wocjcikiewicz, R.J.H. (1988). Evidence for involvement of guanine nucleotide-binding regulatory proteins in the activation of phospholipases by hormones. FASEB J. 2, 2569–2574.

Freissmuth, M., Casey, J., & Gilman, A.G. (1989). G proteins control diverse pathways of trans-membrane signaling. FASEB J. 3, 2125–2131.

Gilman, A.G. (1987). G proteins: transducers of receptor-generated signals. Ann. Rev. Biochem. 56, 615–649.

Gilman, A.G. (1989). G proteins and regulation of adenylyl cyclase. J. Amer. Med. Assoc. 262, 1819–1825.

Goldfine, I.D., Vigneri, R., Cohen, D., Pliam, N.B., & Kahn, C.R. (1977). Intracellular binding sites for insulin are immunologically distinct from those on the plasma membrane. Nature 269, 698–700.

Gregory, H., Taylor, C.L., & Hopkins, C.R. (1982). Luteinizing hormone release from dissociated pituitary cells by dimerization of occupied LHRH receptors. Nature 300, 269–271.

Hartzell, H.C. (1988). Regulations of cardiac channels by catecholamines, acetylcholine and second messenger systems. Prog. Biophys. Molec. Biol. 52, 165–247.

Hollenberg, M.D. (1985). Receptor models and the action of neurotransmitters and hormones: some new perspectives. In: Neurotransmitter Receptor Binding, 2nd ed. (Yamamura HI et al., eds.) pp. 1–39. Raven, New York.

Hollenberg, M.D. (1990). Receptor triggering and receptor regulation: Structure-activity relationships from the receptors point of view. J. Medicinal Chem. 33, 1275–1281.

Hopkins, C.R., Semoff, S., & Gregory, H. (1981). Regulation of gonadotropin secretion of the anterior pituitary. Philos. Trans. R. Soc. Lond. [Biol] 296, 73–81.

Hunter, T., Ling, N., & Cooper, J.A. (1984). Protein kinase C phosphorylation of the EGF receptor at a threonine residue close to the cytoplasmic face of the plasma membrane. Nature 311, 480–483.

Jacobs, S., Chang, K-J., & Cuatrecasas, P. (1978). Antibodies to purified insulin receptor have insulin-like activity. Science 200, 1283–1284.

Jarett, L., & Smith, R.M. (1977). The natural occurrence of insulin receptors in groups on adipocyte plasma membranes as demonstrated with monomeric ferritin-insulin. J. Supramol. Struct. 6, 45–59.

Johnson, E.M. Jr., Andres, R.Y., & Bradshaw, R.A. (1978). Characterization of the retrograde transport of nerve growth factor (NGF) using high specific activity $[^{125}I]$ NGF. Brain. Res. 150, 319–331.

Johnson, L.K., Vlodavsky, I., Baxter, J.D., & Gospodarowicz, D. (1980). Nuclear accumulation of epidermal growth factor in cultured rat pituitary cells. Nature 287, 340–343.

Kahn, C.R., Baird, K.L., Jarett, D.B., & Flier, J.S. (1978). Direct demonstration that receptor crosslinking or aggregation is important in insulin action. Proc. Natl. Acad. Sci. USA 75, 4209–4213.

Kahn, C.R., Baird, K.L., Flier, J.S., Grunfeld, C., Harmon, J.T., Harrison, L.C., Karlsson, F.A., Kasuga, M., King, G.L., Lang, U.C., Podskalny, J.M., & Van Obberghen, E. (1981). Insulin receptors, receptor antibodies, and the mechanism of insulin action. Recent Prog. Horm. Res. 37, 472–538.

Kanner, B.I., & Metzger, H. (1983). Crosslinking of the receptors for immunoglobulin E depolarizes the plasma membrane of rat basophilic leukemia cells. Proc. Natl. Acad. Sci. USA 80, 5744–5748.

King, A.C., & Cuatrecasas, P. (1981). Peptide hormone-induced receptor mobility, aggregation, and internalization. N. Engl. J. Med. 305, 77–88.

Krupinski, J., Coussen, F., Bakalyar, H.A., Tang, W.J., Feinstein, P.G., Orth, K., Slaughter, C., Reed, R.R., & Gilman, A.G. (1989). Adenylyl cyclase amino acid sequence: possible channel- or transporter-like structure. Science 244, 1558–1564.

Lefkowitz, R.J., & Caron, M.G. (1988). Adrenergic receptors. Models for the study of receptors coupled to guanine nucleotide regulatory proteins. J. Biol. Chem. 263, 4993–4996.

Lefkowitz, R.J., Caron, M.G., & Stiles, G.L. (1984). Mechanisms of membrane-receptor regulation. Biochemical, physiological, and clinical insights derived from studies of the adrenergic receptors. N. Eng. J. Med. 310, 1570–1579.

Lefkowitz, R.J., Kobilka, B.K. & Caron, M.G. (1989). The new biology of drug receptors. Biochem. Pharmacol. 18, 2941–2948.

Levitan, E.S., Schofield, P.R., Burt, D.R., Rhee, L.M., Wisden, W., Köhler, M., Fujita, N., Rodriguez, H.F., Stephenson, A., Darlison, M.G., Barnard, E.A., & Seeburg, P.H. (1988). Structural and functional basis for GABA$_A$ receptor heterogeneity. Nature 335, 76–79.

Levitzki, A. (1974). Negative cooperativity in clustered receptors as a possible basis for membrane action. J. Theor. Biol. 44, 367–372.

Levitzki, A. (1987). Regulation of adenylate cyclase by hormones and G-proteins. FEBS Lett. 211, 113–118.

Limbird, L.E. (1981). Activation and attenuation of adenylate cyclase. The role of GTP-binding proteins as macromolecular messengers in receptor-cyclase coupling. Biochem. J. 195, 1–13.

Lin, C.R., Chen, W.S., Lazar, C.S., Carpenter, C.D., Gill, G.N., Evans, R.M., & Rosenfeld, M.G. (1986). Protein kinase C phosphorylation at Thr 654 of the unoccupied EGF receptor and EGF binding regulate functional receptor loss by independent mechanisms. Cell 44, 839–848.

Majerus, P.W., Connolly, T.M., Deckmyn, H., Ross, T.S., Bross, T.E., Ishii, H., Bansal, V.S., & Wilson, D.B. (1986). The metabolism of phosphoinositide-derived messenger molecules. Science 234, 1519–1526.

Metzger, H., & Ishizaka, T. (1982). Transmembrane signalling by receptor aggregation: the mast cell receptor for IgE as a case study. Fed. Proc. Fed. Am. Soc. Exp. Biol. 41, 7–34.

Miller, D.S. (1988). Stimulation of RNA and protein synthesis by intracellular insulin. Science 240, 506–509.

Neer, E.J. & Clapham, D.E. (1988). Roles of G protein subunits in transmembrane signalling. Nature 333, 129–134.

Northup, J.K. (1985). Overview of the guanine nucleotide regulatory protein system, N_S and N_i, which regulate adenylate cyclase activity in plasma membranes. In: Molecular Mechanisms of Transmembrane Signalling (Cohen, P., & Houslay, M.D. eds.), pp. 91–116. Elsevier Science Publishers, Amsterdam.

Northup, J.K., Sternweis, P.C., Smigel, M.D., Schleifer, L.S., Ross, E.M., & Gilman, A.G. (1980). Purification of the regulatory component of adenylate cyclase. Proc. Natl. Acad. Sci. USA 77, 6516–6520.

Pastan, I.H., & Willingham, M.C. (1981). Receptor-mediated endocytosis of hormones in cultured cells. Ann. Rev. Physiol. 43, 239–250.

Pessin, J.E., Gitomer, W., Oka, Y., Oppenheimer, C.L., & Czech, M.P. (1983). β-adrenergic regulation of insulin and epidermal growth factor receptors in rat adipocytes. J. Biol. Chem. 258, 7386–7394.

Podlecki, D.A., Smith, R.M., Kao, M., Tsai, P., Huecksteadt, T., Branderberg, D., Lasher, R.S., Jarett, L., & Olefsky, J.M. (1987). Nuclear translocation of the insulin receptor: a possible mediator of insulin's long term effects. J. Biol. Chem. 262, 3362–3368.

Rhee, S.G., Suh, P.G., Ryu, S.H., & Lee, S.Y. (1989). Studies of inositol phospholipid-specific phospholipase C. Science 244, 546–550.

Rodbell, M. (1980). The role of hormone receptors and GTP-regulatory proteins in membrane transduction. Nature 284, 17–22.

Rodbell, M., Birnbaumer, L., Pohl, S.L., & Krans, H.M.J. (1971). The glucagon-sensitive adenyl cyclase system in plasma membranes of rat liver. J. Biol. Chem. 246, 1877–1882.

Salomon, Y., Lin, M.C., Londos, C., Rendell, M., & Rodbell, M. (1975). The hepatic adenylate cyclase system. I. Evidence for transition states and structural requirements for guanine nucleotide activation. J. Biol. Chem. 250, 4239–4245.

Santos, E., & Nebreda, A.R. (1989). Structural and functional properties of ras proteins. FASEB J. 3, 2151–2163.

Savion, N., Vlodavsky, I., & Gospodarowicz, D. (1981). Nuclear accumulation of epidermal growth factor in cultured bovine corneal endothelial and granulosa cells. J. Biol. Chem. 256, 1149–1154.

Schlessinger, J. (1988). The epidermal growth factor receptor as a multifunctional allosteric protein. Biochemistry 27, 3119–3123.

Schlessinger, J., Schechter, Y., Willingham, M.C., & Pastan, I. (1978). Direct visualization of binding, aggregation, and internalization of insulin and epidermal growth factor on living fibroblast cells. Proc. Natl. Acad. Sci. USA 75, 2659–2663.

Schlessinger, J., Van Oberghen, E., & Kahn, C.R. (1980). Insulin and antibodies against insulin receptor cap on the membrane of cultured human lymphocytes. Nature 286, 729–731.

Schofield, P.R., Darlison, M.G., Fujita, N., Burt, D.R., Stephenson, F.A., Rodriguez, H., Rhee, L.M., Ramachandran, J., Reale, V., Glenrose, T.A., Seeburg, P.H., & Barnard, E.A. (1987). Sequence and functional expression of the $GABA_A$ receptor shows a ligand-gated receptor super-family. Nature 328, 221–227.

Stadtmauer, L., & Rosen, O.M. (1986). Increasing the cAMP content of IM-9 cells alters the phosphorylation state and protein kinase activity of the insulin receptor. J. Biol. Chem. 261, 3402–3407.

Stryer, L., (1986). Cyclic GMP cascade of vision. Ann. Rev. Neurosci. 9, 87–119.

Sutherland, E.W., Robison, G.A., & Butcher, R.W. (1968). Some aspects of the biological role of adenosine 3', 5'-monophosphate (cyclic AMP) . Circulation 37, 279– 306.

Thuren, T. , Tulkki, A.-P., Virtanen, J.A. & Kinnunen, P.K.J. (1987). Triggering of the activity of phospholipase A2 by an electric field. Biochemistry 26, 4907–4910.

Yankner, B.A., & Shooter, E.M. (1979). Nerve growth factor in the nucleus: interaction with receptors in the nuclear membrane. Proc. Natl. Acad. Sci. USA 76, 1269–1273.

Yarden, Y., & Ullrich, A. (1988). Growth factor receptor tyrosine kinases. Ann. Rev. Biochem. 57, 443–478.

Yatani, A., Mattera, R., Codina, J., Graf, R., Okabe, K., Padrell, E., Iyengar, R., Brown, A.M., & Birnbaumer, L. (1988). The G protein-gated K^+ channel is stimulated by three district Gi α-subunits. Nature 336, 680–682.

Zierler, K. (1985). Membrane polarization and insulin action. In: Insulin: Its Receptor and Diabetes. (Hollenberg, M.D., ed.), pp. 141–179. Marcel Dekker, New York.

Zierler, K., & Rogus, E.M. (1981). Effects of peptide hormones and adrenergic agents on membrane potentials of target cells. Fed. Proc. 40, 121–124.

Chapter 12

The Phosphoinositide Cascade

LOWELL E. HOKIN

Fundamentals of Medical Cell Biology, Volume 5A
Membrane Dynamics and Signaling, pages 255–303
Copyright © 1992 by JAI Press Inc.
All rights of reproduction in any form reserved.
ISBN: 1-55938-309-7

INTRODUCTION

In recent years, there has been widespread interest in the agonist-stimulated breakdown and turnover of myoinositol-containing phospholipids. Although these lipids comprise less than 10% of the total cellular phospholipid, their stimulated turnover accounts for a considerably higher percentage of total phospholipid turnover. The phosphoinositide field is currently the most cited field in biochemistry (excluding molecular biology). Based on one abstract service, the number of papers published in this area is around 150 per month or five papers per day. It now seems clear that the phosphoinositide effect is a multi-regulatory mechanism involving the release of various moieties of the parent molecules, which serve as second messengers subserving myriad changes in the cell elicited by receptor activation induced by a wide variety of agonists. In addition to the well established second messengers, inositol 1,4,5-trisphosphate [I(1,4,5)P$_3$] and 1,2-diacylglycerol (DAG), evidence is mounting pointing to additional second messengers, such as inositol 1,3,4,5-tetrakisphosphate [I(1,3,4,5)P$_4$] and arachidonic acid.

There are numerous reviews on this subject (see e.g., Berridge and Irvine, 1984; Downes and Michell, 1985; Hokin, 1985; Abdel-Latif, 1986; Majerus et al., 1986; Berridge, 1987b; Rana and Hokin, 1990). Recently reviews have tended to focus on one aspect of the phosphoinositide response or one cell type. This review covers most aspects of the field.

It is impossible to do justice here to all of the workers in this field. Whenever possible, reviews will be quoted. Where no reviews exist, examples of original papers are quoted. A more comprehensive bibliography is presented in a recent review by Rana and Hokin (1990).

HISTORICAL BACKGROUND

Discovery of the "Phosphoinositide Effect"

The discovery of the "phosphoinositide effect" arose out of an accidental observation over 35 years ago. The details of this discovery have been recently reviewed (Hokin, 1987; Rana and Hokin, 1990). Essentially, the incorporation of ^{32}Pi into the "RNA" fraction isolated by the then current method of nucleic acid isolation was increased considerably on cholinergic stimulation of enzyme secretion in pigeon pancreas slices. It eventually turned out that an alkaline hydrolytic product(s) of phospholipids contaminating the "RNA" fraction was responsible for the stimulation. When the phospholipids were looked at directly, they were found to show a remarkable increase in turnover (increased specific activity of ^{32}P-labeled phospholipids on stimulation with acetylcholine [ACh] or carbachol) (Hokin and Hokin, 1953, 1954). This increased turnover was confined to phosphatidylinositol

(PI) and phosphatidic acid (PA) (Hokin and Hokin, 1955, 1958a). It turned out that the phosphoinositide effect was a broadly based mechanism thrown into play on activation of a variety of receptors (Hokin, 1968).

Discovery of the Phosphatidylinositol Cycle

On the basis of kinetic studies utilizing [^{32}P]orthophosphate and [^{3}H]myoinositol labeling in the avian salt gland, a scheme was proposed, called the "phosphatidylinositol-phosphatidic acid" (PI-PA) cycle (Hokin and Hokin, 1964b; Rana and Hokin, 1990) in which, on stimulation with ACh, PI breaks down to DAG and inositol 1-phosphate, catalyzed by PI phosphodiesterase (phospholipase C), and DAG is phosphorylated by ATP to form PA. On removal of ACh, PA is converted back to PI by the sequential actions of CTP-PA cytidyl transferase and PI synthase. If albatross salt gland slices were incubated without and with ACh, the average PI-phosphorus in the stimulated tissue was 40% lower than that in the control (Hokin-Neaverson, 1974). Later (Hokin-Neaverson, 1977), a similar decrease in mass of PI and a roughly equivalent increase in mass of PA were observed on stimulation of mouse pancreas with ACh or CCK/PZ and on stimulation of parotid with ACh or epinephrine (Jones and Michell, 1974). This confirmed and extended the studies on the kinetics of ^{32}P$_i$ and [^{3}H]myoinositol labeling as well as the chemical measurements in the avian salt gland made more than a decade earlier. If brain cortex slices were incubated with and without ACh in the presence of [^{14}C]glycerol, there was no increased incorporation of radioactivity into PI or PA (Hokin and Hokin, 1958b). This suggested that, at least in brain, the DAG moiety was conserved in the PI cycle. This made it highly likely that DAG kinase was the enzyme responsible for the stimulated incorporation of ^{32}P$_i$ into PA and PI, rather than sn-glycero-3-phosphate transacylase.

In this initial version of the PI cycle, only the breakdown of PI was considered (work on the isolation and characterization of PIP and PIP$_2$ and their metabolism did not begin until a decade after the discovery of the PI effect). With the inclusion of PIP and PIP$_2$ in the PI cycle, many investigators suggested that the "breakdown" of PI is, in fact, phosphorylation to PIP and that there is no direct hydrolysis of PI. Support for direct hydrolysis of PI has recently been obtained. Dixon and Hokin (1989) have carried out a kinetic analysis of the cyclic inositol phosphate pathway (see later) in carbachol-stimulated pancreatic minilobules. Kinetic analysis of this pathway is more straightforward than that of the noncyclic inositol phosphate pathway, which has numerous branch points. With the cyclic inositol phosphates, metabolism proceeds as follows: I(c1:2,4,5)P$_3$ → I(c1:2,4)P$_2$ → I(c1:2)P → I(1)P. Only I(c1:2)P can be decyclized (Connolly, et al., 1986b). Also, this pathway proceeds much more slowly than the noncyclic inositol phosphate pathway, making kinetic analysis much easier. This analysis showed that about half of the I(c1:2)P formed is by the sequential dephosphorylation of I(c1:2,4,5)P$_3$, leaving the other half of I(1:2)P formed by direct breakdown of PI. Ackermann, Gish, Honchar, and

Sherman (1987) have also provided evidence for direct breakdown. They found that there was a 10-fold greater content of I(1)P (isolated from a chiral column) than I(4)P in both pilocarpine-stimulated and -unstimulated lithium-treated rat brains *in vivo,* and because I(4)P is the major hydrolytic product of the I(1,4)P$_2$-phosphatase reaction in cell-free preparations (Inhorn et al., 1987), they concluded that a major source of I(1)P must be hydrolysis of PI. This conclusion is, of course, based on enzymatic specificities determined in cell-free systems, and it can always be argued that the situation in agonist-stimulated cells is different.

Early Studies Suggesting an Involvement of Inositol Lipids in Ca^{2+}-Dependent Responses

It had been known since the time of Ringer over a century ago (Ringer, 1883) that Ca^{2+} is essential for contraction of the heart. Beginning with the classical work of Katz on ACh release from nerve terminals and of Douglas on stimulus–secretion coupling (see Katz, 1969; Douglas, 1974; Rasmussen and Barrett, 1984), it was becoming apparent that Ca^{2+} was an essential link in many agonist-evoked responses. Namely, the physiological response required extracellular Ca^{2+}, and increases in intracellular Ca^{2+} derived from intracellular stores or from extracellular Ca^{2+} were seen.

By presoaking pigeon pancreas slices in EDTA, followed by incubation without and with Ca^{2+}, it was found that in the absence of Ca^{2+}, amylase secretion in response to ACh was inhibited 98%, but the stimulated ^{32}P incorporation into PI and PA was inhibited only 38% and 41%, respectively (Hokin, 1966). These results showed loose coupling between the phospholipid effect and enzyme secretion and suggested that the phospholipid effect was concerned with some step in the "overall process of excitation and secretion other than exocytotic enzyme secretion *per se*" (Hokin, 1966). These studies also showed that the phospholipid effect was partially independent of Ca^{2+}. These observations were part of the underpinnings of the Ca^{2+} gating hypothesis proposed by Michell in his seminal review a decade later (Michell, 1975). That is to say, when Michell surveyed a large number of tissues showing the PI response, he noted: (1) the stimulus-response mechanism required extracellular Ca^{2+}, and rises in intracellular Ca^{2+} occurred on agonist stimulation; (2) the PI effect was independent or only partially dependent on Ca^{2+} in many tissues; and (3) in some tissues, Ca^{2+} ionophores stimulated the physiological response but not PI breakdown. Taken together, these observations favored PI breakdown as antecedent to elevated cell Ca^{2+}, and it was suggested that PI breakdown opened "Ca^{2+} gates" in the plasma membrane. More direct support for the "Ca^{2+} gating" hypothesis was obtained by Berridge and Fain in the blowfly salivary gland (Fain and Berridge, 1979a,b), where stimulation with 5-HT caused PI loss and increased exchange of Ca^{2+} in the epithelial cells. Supramaximal stimulation with 5-HT caused a loss of the small pool of responsive PI and a fall in Ca^{2+} transport (desensitization). Incubation of washed glands with myoinositol restored both PI sensitivity to 5-HT and Ca^{2+} transport. Although there was considerable controversy in the early years over the Michell hypothesis, the concept proved to be correct, albeit not in its details.

Early Studies on Polyphosphoinositide Turnover

As early as 1962, the polyphosphoinositides were shown to turn over rapidly in brain, and it was suggested that they may play an important physiological role (Brockerhoff and Ballou, 1962). Later, it was shown that there was substantial radioactivity in polyphosphoinositides in a variety of tissues incubated with $^{32}P_i$ (Santiago-Calvo et al., 1964). If sea gull salt gland slices were prelabeled with $^{32}P_i$, followed by addition of ACh, there was a drop in radioactivity in PIP and PIP_2 to 68% and 77% of that of control slices, respectively (Santiago-Calvo et al., 1964). PI-kinase and PIP-kinase were also discovered at that time by incubating erythrocyte membranes with $[\gamma\text{-}^{32}P]ATP$, which led to the labeling of PIP and PIP_2 in their monoesterified phosphates (Hokin and Hokin, 1964a). If exogenous PI was added to erythrocyte membranes, the incorporation of ^{32}P from $[\gamma\text{-}^{32}P]ATP$ into PIP was stimulated, providing further evidence for PI kinase. PI-kinase was demonstrated shortly afterwards in brain (Colodzin and Kennedy, 1965). Somewhat later, PI-kinase was reported to be present in plasma membranes other than that of the erythrocyte (Michell and Hawthorne, 1965; Harwood and Hawthorne, 1969).

Formation of Inositol Phosphates

In the late 1960s, Durell, Garland, and Friedel (1969) reported the formation of inositol mono- and bisphosphates in synaptosomes in response to ACh, and they suggested that the primary reaction in response to ACh might be a phosphodiesteric cleavage of PIP and PIP_2. In the late 1970s, Abdel-Latif, Akhtar, and Hawthorne (1977) showed a rapid breakdown of PIP_2 in iris smooth muscle after muscarinic or α-adrenergic stimulation. Subsequently, they showed that the breakdown of PIP_2 was accompanied by the formation of I(1)P, inositol bisphosphate (IP_2), and inositol trisphosphate $[I(1,4,5)P_3,]$ (Akhtar and Abdel-Latif, 1980). However, they did not believe that the release of inositol phosphates was involved in Ca^{2+} mobilization, because this response was partially dependent on Ca^{2+} and could be mimicked by a Ca^{2+} ionophore. Stimulation of PIP_2 breakdown by Ca^{2+} ionophore in this tissue has now been shown to be a secondary effect attributed to the release of Ca^{2+}-mobilizing neurotransmitters (Akhtar and Abdel-Latif, 1984).

Recent developments that have led to the recognition that phosphoinositides play a central role in transduction of signals from Ca^{2+} mobilizing receptors are discussed later.

MODERN ERA

Stimulation of Polyphosphoinositide Metabolism

Even though polyphosphoinositide breakdown in response to agonists had been described earlier (see previously), little attention was paid to it until the early 1980s, when reports describing the rapid breakdown of polyphosphoinositides were published (Weiss et al., 1982; Creba et al., 1983; Thomas et al., 1983). This led to a revival of interest in polyphosphoinositides, because, unlike the effect in iris smooth muscle reported earlier, the vasopressin-stimulated breakdown of polyphosphoinositides in liver was independent of Ca^{2+} (Creba et al., 1983).

Recent studies of the effects of receptor activation on phosphoinositide metabolism in various tissues have focused on three aspects: (1) effects of agonist stimulation on hydrolysis of polyphosphoinositides; (2) characterization of products formed as a result of this hydrolysis; and (3) determination of physiological functions of these products. We now review these various aspects.

Breakdown of Polyphosphoinositides

PIP2 hydrolysis is rapid and precedes that of PI. The rapid breakdown of polyphosphoinositides following receptor activation has now been reported in numerous tissues in response to stimuli, such as neurotransmitters, growth factors, hormones, and light. Several reviewers have tabulated agonists and tissues where a polyphosphoinositide response has been demonstrated (Downes and Michell, 1985; Abdel-Latif, 1986; Berridge, 1986; Sekar and Hokin, 1986). A key feature of the studies summarized in these reviews is that breakdown of PIP_2 and PIP is rapid, usually observed within seconds, and precedes that of PI.

PIP2 hydrolysis is independent of calcium mobilization. Based on the Ca^{2+} requirements for the phosphoinositide effect in various tissues, it is now generally accepted that under physiological conditions of ionic strength and pH, the Ca^{2+} concentration required for the breakdown of at least PIP_2 is at or below the resting cytosolic Ca^{2+} concentration (0.1–1 μM). In a recent evaluation of the Ca^{2+} requirement for phospholipase C activity in liver cells (Renard et al., 1987), where the $[Ca^{2+}]_i$ was clamped at 29 nM, which is below resting $[Ca^{2+}]$, vasopressin caused a two-fold stimulation of $I(1,4,5)P_3$ formation at the clamped $[Ca^{2+}]_i$, indicating that $I(1,4,5)P_3$ formation was not dependent on a rise in $[Ca^{2+}]_i$. Furthermore, $I(1,4,5)P_3$ formation was little affected by clamping $[Ca^{2+}]_i$ at the resting level of 193 nM and up to 1130 nM. The phospholipase C reaction at the low $[Ca^{2+}]_i$ concentrations can occur in the presence of G proteins and GTP. Therefore, cytosolic Ca^{2+} is not rate-limiting for PIP_2 cleavage (see Rasmussen and Barrett,

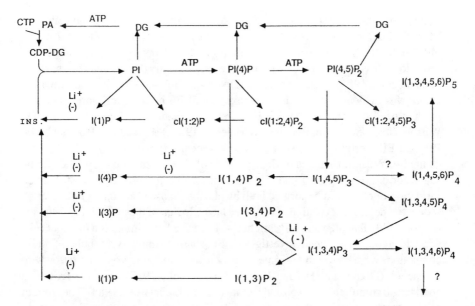

Figure 1. The phosphoinositide cascade. The cascade is initiated by agonist-stimulated phosphodiesteratic cleavage of PIP_2. All inositol phosphates are D-isomers. (−) indicates sensitivity to lithium.

1984), although in some cases it may be rate-limiting for PI or PIP cleavage (Majerus et al., 1984; Hokin, 1985).

Products of Polyphosphoinositide Hydrolysis

Formation of I(1,4,5)P3. The phosphodiesteratic cleavage of PIP_2 by phospholipase C leads to the formation of DAG and $I(1,4,5)P_3$ (Kemp et al., 1961a,b; Thompson and Dawson, 1964). The reaction can be measured by determining the loss of PIP_2, the formation of DAG, or the formation of $I(1,4,5)P_3$.

The inositol trisphosphate produced upon hydrolysis of PIP_2 can be either inositol 1,4,5-trisphosphate [$I(1,4,5)P_3$] or inositol 1,2-cyclic 4,5-trisphosphate (cIP_3), depending on the source of the hydroxyl group utilized in the reaction. The entry of the hydroxyl group from water produces $I(1,4,5)P_3$, whereas the use of the resident hydroxyl at the 2-position of the inositol ring leads to production of cIP_3.

Of the various inositol phosphates, $I(1,4,5)P_3$ appears first upon agonist stimulation, and this is then followed by the appearance of $I(1,3,4,5)P_4$, $I(1,3,4)P_3$, and $I(1,4)P_2$, in that order (Wollheim and Biden, 1986). Figure 1 depicts the phosphoinositide cascade. Rapid agonist-induced formation of $I(1,4,5)P_3$; (usually reaching a peak within a few seconds) has been reported in numerous tissues (see earlier for reviews).

Formation of cIP3. In the late 1950s, Dawson (1959) showed that incubation of PI with PLC released, in addition to inositol phosphate, inositol 1,2-cyclic phosphate (cIP). However, the formation of cIP on agonist-stimulated PI breakdown in cells was not shown until relatively recently (Dixon and Hokin, 1985; Graham et al., 1987). It was suggested (Dixon and Hokin, 1985) that the formation of cIP was evidence for the direct breakdown of PI, because this compound is a product of PLC action on PI. This interpretation became more complicated when Wilson, Bross, Sherman, Berger, and Majerus (1985a) showed that PLC action on PIP and PIP_2 produced inositol 1,2-cyclic 4-bisphosphate (cIP_2) and cIP_3, and Connolly, Wilson, Bross, and Majerus (1986b) showed that no enzyme could be found for decyclizing cIP_2 and cIP_3, in contrast to cIP. Thus, agonist-stimulated formation of cIP could be formed both by phospholipase C action on PI and/or by sequential dephosphorylation of cIP_3 (see later). In lieu of these enzymatic studies, the possibility that cIP_3 and cIP_2 might also be formed on stimulation of intact cells with an agonist has been investigated. Substantial amounts of both of these compounds were shown to be formed in pancreatic minilobules on stimulation with carbachol (Dixon and Hokin, 1987b; Sekar et al., 1987) and in parotid on cholinergic stimulation (Dixon and Hokin, 1987a; Hughes et al., 1988). The amount of cIP_3 formed equalled the amount of $I(1,4,5)P_3$ formed at 30 min (Sekar, et al., 1987). Ishii, Connolly, Bross, and Majerus (1986) showed the formation of a small amount of cIP_3 on thrombin stimulation of platelets. cIP_3 forms very slowly but accumulates over time in both pancreas (Dixon and Hokin, 1987b) and parotid (Dixon and Hokin, 1987a; Hughes et al., 1988). The current data are compatible with the concept that cIP_3 might be playing a second messenger role at later times but is unlikely to do so at short times. The relative rates of formation of $I(1,4,5)P_3$ and cIP_3 on incubation with the purified seminal vesicle enzyme is reported to be 10:1 (Wilson et al., 1985a). Assuming that this ratio does not change during the course of agonist stimulation, the relative abundance of these compounds at any given time would depend on these rates of formation and further metabolism.

Two enzymes are involved in the removal of $I(1,4,5)P_3$ (i.e., 3'-kinase and 5'-phosphatase [see later]. cIP_3 is not phosphorylated by the 3'-kinase, and the phosphatase is only about one tenth as active toward cIP_3 as $I(1,4,5)P_3$ (Connolly et al., 1987). This can explain the different kinetic behavior of these compounds in agonist-stimulated pancreas and parotid.

I(1,3,4,5)P4 formation. The $I(1,4,5)P_3$ formed on agonist stimulation is subject to further metabolism by either a kinase that is specific for the 3 position of inositol or a phosphomonoesterase specific for the 5 position. The former produces $I(1,3,4,5)P_4$, and the latter produces $I(1,4)P_2$. The $I(1,4,5)P_3$-3-kinase was first shown in brain (Irvine et al., 1986a) and shortly after in liver (Hansen et al., 1986), parotid (Hawkins et al., 1986), and other tissues. The enzyme activity was first shown to be stimulated by Ca^{2+} in RINm5F cells (Biden and Wollheim, 1986) and somewhat later in other tissues and cells. It appears likely that $I(1,4,5)P_3$-induced intracellular Ca^{2+} mobilization enhances the formation of I(1,3,4,5)P4, thus functioning as a negative regulator of levels of I(1,4,5)P3 and cytosolic calcium levels.

Formation of I(1,3,4)P3. The I(1,3,4)P3 isomer was first identified in parotid gland (Irvine et al., 1984) and has subsequently been seen in many types of cells . The formation of I(1,3,4)P3 involves phosphorylation of I(1,4,5)P3 to form I(1,3,4,5)P4, which is then dephosphorylated to I(1,3,4)P3 (Batty et al., 1985). Compared to I(1,4,5)P3, which has a half-life of about 4 s, I(1,3,4)P3 appears to turn over much more slowly. Consequently, when cells are stimulated by agonists, the proportion of I(1,3,4)P3 relative to I(1,4,5)P3 continues to rise after I(1,4,5)P3 reaches a peak or steady-state level. Li^+ selectively enhances accumulation of I(1,3,4)P3 but has no effect on the accumulation of I(1,4,5)P3 (Burgess et al., 1985). Li^+ has been a useful tool because it amplifies the inositol phosphate response, but because it causes selective accumulation of some but not all inositol phosphates, it distorts their quantitative proportions.

Inositol Phosphates as Second Messengers

I(1,4,5)P3. Cytosolic Ca^{2+} can be mobilized either from intracellular stores or from the extracellular compartment or both. Initially, the Ca^{2+} is mobilized from intracellular stores (Schulz and Stolze, 1980; Rasmussen and Barrett, 1984), and this is followed by an influx of Ca^{2+} from the extracellular compartment. Much of the cytosolic Ca^{2+} derived from intracellular stores is subsequently released from the cell, although there is evidence to indicate that some of the I(1,4,5)P3-releasable Ca^{2+} is taken up by mitochondria (Biden et al., 1986). Elevated intracellular Ca^{2+} is sustained at later times by an influx of extracellular Ca^{2+} (Kojima et al., 1985; Reynolds and Dubyak, 1985), which maintains the response and also tends to replenish the depleted intracellular store. Based on the hypothesis first enunciated by Michell and on studies in the insect salivary gland by Fain and Berridge, it was anticipated that a second messenger generated during the breakdown of polyphosphoinositides in the plasma membrane may be involved in the mobilization of intracellular Ca^{2+}.

Within the last few years, overwhelming evidence has accumulated that implicates I(1,4,5)P3 as the molecule that links receptor-activated phosphoinositide breakdown to Ca^{2+} mobilization from intracellular stores. These studies were initiated in 1983 with the demonstration by Streb, Irvine, Berridge, and Schulz (1933) that the addition of I(1,4,5)P3 to permeabilized pancreatic acinar cells resulted in the release of Ca^{2+} from these cells. Subsequently, at least 40 reports have appeared confirming this property of I(1 ,4,5)P3 in permeabilized cells or microsomal fractions representing a wide variety of tissues (for a recent list of tissues and references, see Berridge, 1987b). The release of Ca^{2+} in permeabilized cells is rapid and occurs at less than micromolar concentrations of I(1,4,5)P3. This release by I(1,4,5)P₃ is transient, due to its rapid metabolism, and is accompanied by Ca^{2+} reuptake into the ER or a related structure.

Although the demonstration that I(1,4,5)P₃ releases Ca^{2+} from permeabilized cells was crucial in establishing the link between I(1,4,5)P₃ formation and Ca^{2+} mobilization, experiments involving intracellular injection of I(1,4,5)P₃ in intact cells were important in confirming this function of I(1,4,5)P₃. Thus, intracellular injection of I(1,4,5)P₃ was shown to mimic stimulation by light of *Limulus* photoreceptor cells (Fein et al., 1984). The formation of I(1,4,5)P₃ and mobilization of intracellular Ca^{2+} are both very rapid events, and because I(1,4,5)P₃ is rapidly metabolized to either I(1,4)P₂ or I(1,3,4,5)P₄ (Irvine et al., 1986b), it is not easy to see a clear separation of these two events. Using a rapid mixing device. Tashjian, Heslop, and Berridge (1987) attempted to determine subsecond changes in inositol polyphosphates in GH₄Cl cells. They demonstrated that changes in inositol polyphosphates and cytosolic Ca^{2+} could be detected as early as 400 to 500 ms.

On the basis of cell fractionation experiments and the use of metabolic inhibitors, I(1,4,5)P₃ appears to release Ca^{2+} from a membrane fraction that is a component of the ER (Streb et al., 1984) An electron microprobe X-ray analysis of smooth muscle (Bond et al., 1984) and intracellular injection of I(1,4,5)P₃ at various depths in *Xenopus* oocytes (Busa et al., 1985) have suggested localization of the I(1,4,5)P₃-sensitive pool of Ca^{2+} in a region of ER that is close to the cell surface. I(1,4,5)P₃ binding sites appear to be relatively abundant in a liver fraction enriched in plasma membrane, and the release of Ca^{2+} from this fraction in response to (1,4,5)P₃ appears to be four-fold higher than the ER.

Specificity of inositol phosphates. The specificity of the 1,4,5-isomer of inositol trisphosphate for Ca^{2+} release has been demonstrated in pancreas (Streb et al., 1983) and Swiss mouse 3T3 cells (Irvine et al., 1986b). I(1)P and I(1,4)P₂, which are products of I(1,4,5)P₃ hydrolysis and which can also be derived from direct breakdown of PI and PIP by phospholipase C, were unable to release Ca^{2+}. cIP₃ appears to be slightly active (see later). It appears that the phosphates on the 4 and 5 positions of the inositol ring are necessary for Ca^{2+} release, because I(4,5)P₂ and I(2,4,5)P₃ are somewhat active, although their potencies are much lower than that of I(1,4,5)P₃.

Identification of the I(1,4,5)P₃ receptor. Binding studies with I(1,4,5)P₃ (Baukal et al., 1985; Spat et al., 1986) suggest that I(1,4,5)P₃ stimulates Ca^{2+} release from the ER via interaction with a specific receptor. Mitochondrial membranes appear to lack the I(1,4,5)P₃ binding sites. Further support for the existence of a receptor derives from specific photoaffinity labeling of the receptor (Hirata et al., 1985), as well as the recent purification of the I(1,4,5)P₃ receptor from rat cerebellum by Supattapone, Worley, Baraban, and Snyder (1988). The binding studies with the I(1,4,5)P₃ receptor show a high degree of specificity toward $I(1,4,5)P_3$. The molecular mechanism by which $I(1,4,5)P_3$ releases Ca^{2+} from the ER remains to be elucidated. Indirect evidence based on the temperature insensitivity of $I(1,4,5)P_3$-induced Ca^{2+} release suggests that binding of $I(1,4,5)P_3$ to the ER may lead to opening of a Ca^{2+} channel (Muallem et al., 1985; Joseph and Williamson, 1986). K^+ is required and acts as a counter ion in $I(1,4,5)P_3$-induced Ca^{2+} release. This is supported by the ability of potassium channel blockers to inhibit $I(1,4,5)P_3$-induced Ca^{2+} release from rat brain microsomes (Shah and Pant, 1988). In addition to its well-established role in Ca^{2+} release from intracellular stores, $I(1,4,5)P_3$ may also have a potential role in stimulation of extracellular Ca^{2+} influx (Parker and Miledi, 1987; Snyder et al., 1988).

cIP3. Wilson and associates (1985b) found that microinjection of cIP_3 into *Limulus* photoreceptor cells caused an increase in conductivity, and its potency in this regard was at least five times that of $I(1,4,5)P_3$. The release of Ca^{2+} was also observed, but cIP_3 was not as potent as $I(1,4,5)P_3$ in this respect. They also found that cIP_3 released Ca^{2+} from permeabilized platelets, although it was inactive below the EC_{30} for $I(1,4,5)P_3$-induced Ca^{2+} release; at the EC_{50}, both compounds were equipotent. Irvine, Letcher, Lander, and Berridge (1986b) reported that in permeabilized 3T3 fibroblasts, cIP_3 was equipotent to $I(1,4,5)P_3$ in releasing Ca^{2+}. More recent studies indicate that cIP_3 is only about one twentieth to one tenth as potent (Crossley et al., 1988; Meyer et al., 1988; Lee and Hokin, 1989; Willcocks et al., 1989). Because it is difficult to remove all of $I(1,4,5)P_3$ from cIP_3 preparations, it is possible that in some of the previous studies where cIP_3 was found to be quite active in releasing Ca^{2+} there was sufficient contamination of cIP_3 with $I(1,4,5)P_3$ to cause considerable Ca^{2+} release by "cIP_3.". Although it now appears that cIP_3 is not a second messenger for Ca^{2+} release, it seems unlikely that the accumulation of large amounts of cIP_3 on agonist stimulation in some tissues is of no biological significance. At present, cIP_3 remains a viable candidate for an undiscovered second messenger function.

I(1,3,4,5)P4. There is recent suggestive evidence that $I(1,3,4,5)P_4$ may be a second messenger in Ca^{2+} influx. In sea urchin eggs, intracellular injection of $I(1,3,4,5)P_4$ plus $I(2,4,5)P_3$ [the latter active in releasing Ca^{2+} but not phosphorylated to $I(1,3,4,5)P_4$] elevated the fertilization envelope, an event that is dependent on the presence of extracellular Ca^{2+} (Irvine and Moor, 1986). Ca^{2+} influx was not measured in this study. More recently, Morris, Gallacher, Irvine, and Petersen (1987) reported that $I(1,3,4,5)P_4$ and $I(2,4,5)P_3$ [an active congener of $I(1,4,5)P_3$ in mobilizing intracellular calcium, but not phosphorylated] interact synergistically to activate Ca^{2+}-dependent potassium currents in the plasma membranes of lacrimal acinar cells. Also, $I(1,3,4,5)P_4$ potentiates $I(1,4,5)P_3$-induced Ca^{2+} transients (Joseph et al., 1987). If $I(1,3,4,5)P_4$ is a second messenger, it should act via a receptor. Highly specific binding sites for this molecule have been detected in adrenal cortex (Enyedi and Williams, 1988). However, there is controversy regarding the role of I(1,3,4,5)P4 in Ca^{2+} influx. Snyder, Krause, and Welsh (1988) have recently shown that intracellular injections of some I(1,4,5)P3 isomers [but not I(1,3,4,5)P4] could activate Ca^{2+} influx in *Xenopus* oocytes.

Hill, Dean, and Boyenton (1988) have presented evidence that indicates a potential role for I(1,3,4,5)P4 in the resequestration of calcium released by I(1,4,5)P3. They report that, unlike I(1,4,5)P3, the Ca^{2+} released by I(2,4,5)P3 from calcium-loaded permeabilized liver cells is not resequestered. Because liver cells can metabolize I(1,4,5)P3, but not I(2,4,5)P3, they surmised that a I(1,4,5)P3 metabolite may be required for sequestration of calcium. This metabolite appears to be I(1,3,4,5)P4 because only it and no other inositol phosphate was able to cause resequestration when added 30 s after the addition of I(2,4,5)P3.

IP5 and IP6. Although IP5 and IP6 were recognized as natural metabolites of inositol as early as 1969 (Cosgrove, 1969), their detection in mammalian cell types has been quite recent (Heslop et al., 1985; Tilly et al., 1987). Their relative distribution in various regions of the brain following [^{3}H]myoinositol labeling was recently reported (Vallejo et al., 1987). IP5 was tentatively determined to be inositol 1,3,4,5,6-pentakisphosphate [I(1,3,4,5,6)P5]. Both labeled IP5 and IP6 were found in approximately equal proportions in the midbrain and the hypothalamus. IP5 was more abundant in the medulla oblongata, and IP6 in the hippocampus, whereas the corpus striatum lacked IP5. The biosynthetic pathways for the synthesis of IP5 and IP6 have not been adequately characterized. According to a recent report, I(1,3,4,5,6)P5 is synthesized by the phosphorylation of I(1,4,5,6)P4 by a kinase that shows specificity for the 3-OH position (Stephens et al., 1988). Presumably, IP6 is produced by the phosphorylation of I(1,3,4,5,6)P5.

So far, there is no evidence implicating IP5 and IP6 in intracellular signaling, as their levels do not show the transient alterations characteristic of inositol phosphate second messengers. However, the report by Jackson, Hallam, Downes, and Hanley (1987) supports the possibility that IP5 and IP6 may act as extracellular signals. They have reported that ionophoresis of IP5 and IP6 into the nucleus tractus solitarius, which controls cardiovascular homeostasis, produces rapid and transient decreases in both mean arterial pressure and heart rate. Undoubtedly, further work is required to clarify the biosynthesis, degradation, and function of these two inositol polyphosphates.

Enzymatic Pathways

Reviews on the enzymatic pathways for the metabolism of inositol phosphates have recently appeared (Majerus et al., 1988; Shears, 1989). Figure 1 shows the complex metabolism of inositol phosphates on activation of phospholipase C. The

interested reader is encouraged to consult these reviews for detailed information and references.

I(1,4,5)P3. Advances in HPLC techniques for separation of various inositol phosphates have made it possible to determine the dephosphorylation and phosphorylation routes for I(1,4,5)P3 with some degree of certainty. Two alternative routes for the complete dephosphorylation of I(1,4,5)P3 have been described. One of these is initiated by a phosphomonoesterase that specifically removes the 5'-phosphate from I(1,4,5)P3, leading to the production of I(1,4)P2. The enzyme catalyzing this reaction was initially detected in human red blood cell membranes (Downes et al., 1982) and has been purified from platelets (Connolly et al., 1985) and liver (Storey et al., 1984).

The I(1,4,5)P3-3-kinase has been detected in many tissues and cells on agonist stimulation and involves phosphorylation of I(1,4,5)P3 to form I(1,3,4,5)P4. Dephosphorylation of the latter forms I(1,3,4)P3. The first enzyme in this pathway [i.e., I(1,4,5)P3-3-kinase] is maximally activated by Ca^{2+}/calmodulin with an enzyme:calmodulin ratio of 1:1. The second step [i.e., the conversion of I(1,3,4,5)P4 to I(1,3,4)P3] is catalyzed by the 5'-phosphatase that dephosphorylates I(1,3,4,5)P4 to I(1,3,4)P2 [see formation of I(1,3,4)P3 earlier]. This enzyme has a much higher affinity for I(1,3,4,5)P4 than for I(1,4,5)P3. The higher affinity for I(1,3,4,5)P4 may have physiological significance because, in the presence of substantial amounts of I(1,3,4,5)P4, dephosphorylation of I(1,4,5)P3 to I(1,4)P2 would be inhibited.

I(1,3,4)P3. Investigations on dephosphorylation of I(1,3,4)P3 have employed both tissue homogenates and intact cells. In the initial studies, radiolabeled I(1,3,4)P3 was added to homogenates of a variety of cells, and it was reported to be converted to a single species of IP2, which was identified as I(3,4)P2, indicating a 1-phosphatase. Formation of I(3,4)P2 was also detected on incubation of a rat brain homogenate with $[4,5-^{32}P]I(1,3,4,5)P4$. The 1-phosphatase, which can dephosphorylate both I(1,3,4)P3 and I(1,4)P2, has also been purified. The lithium sensitivity of this enzyme is consistent with the reported ability of Li^+ to enhance accumulation of I(1,3,4)P3 and I(1,4)P2 in several tissues or cells.

Some reports have shown that besides I(3,4)P2, I(1,3)P2 is also produced upon dephosphorylation of I(1,3,4)P3. This has been shown in several cell types. These observations indicate a rather complex dephosphorylation pathway for I(1,3,4)P3. At least in some tissues, I(3,4)P2 is first converted to I(3)P and then to free inositol, whereas I(1,3)P2 is first metabolized to I(1)P and subsequently to free inositol. Unlike the conversion of I(1,3,4)P3 to I(3,4)P2, the conversion of I(1,3,4)P3 to I(1,3)P2 is Li^+ insensitive. Studies also suggest that I(1,3,4)P3 may represent yet another branch-point in the metabolism of inositol phosphates. There are reports that I(1,3,4)P3 can be phosphorylated to form I(1,3,4,6)P4 by an I(1,3,4)P3-6-kinase. The further metabolic fate of I(1,3,4,6)P4 is unknown.

I(1,4)P2. It was initially thought that I(1,4)P2 is first dephosphorylated to either I(1)P or I(4)P, which is then dephosphorylated to free inositol. However, investigations by Inhorn, Bansal, and Majerus (1987) suggest that I(1,4)P2 is only dephosphorylated to I(4)P, which is further degraded to inositol and inorganic phosphate by an inositol monophosphate phosphatase. These investigators incubated [^{3}H-inositol][4-^{32}P]I(1,4)P2 with brain homogenates in the presence of excess unlabeled I(l)P to suppress the degradation of any I(1)P that might be formed. They failed to detect any [^{3}H]I(1)P. The only product detected was [^{3}H-inositol][4-^{32}P]I(4)P. They were also able to purify an enzyme that removed the 1-phosphate from both I(1,4)P2 and I(1,3,4)P3, but not from I(1)P. The enzyme was Li$^+$ sensitive and was termed inositol polyphosphate-1-phosphatase. Results consistent with the conversion of I(1,4)P2 to I(4)P and subsequently to inositol have been reported for several cell types. On the basis of the data, it appears very likely that the hydrolysis of I(1,4,5)P3 via I(1,4)P2 leads to the formation of I(4)P and not I(1)P. However, the most abundant inositol monophosphate measured after agonist stimulation appears to be I(l)P (Siess, 1985; Ackermann et al., 1987). This is considered to be derived from direct breakdown of PI and from the hydrolysis of cIP (Ross & Majerus, 1986) (see also earlier), which in turn is derived either from PI by phospholipase C action (Dawson, 1959; Dixon and Hokin, 1989) or from dephosphorylation of cIP3 by a 5- and 4-phosphatase (Connolly et al., 1986b). Studies in rat pituitary cells (Imai & Gershengorn, 1986) also support the possibility that most of I(l)P is derived from the direct action of phospholipase C on PI. However, there are other sources of I(1)P from the non-cyclic dephosphorylation pathways, for example, from I(1,3,4)P3 via I(1,3)P2 (Bansal et al., 1987).

Role of GTP-Binding Proteins in Receptor Coupling to Phospholipase C

Evidence Supporting the Involvement of G Proteins

The subject of G proteins and their involvement in transduction of agonist signals has been recently reviewed (Litosch and Fain, 1986; Fain, 1987; Gilman, 1987). Studies using permeabilized cells or isolated membranes have proven invaluable in establishing the involvement of GTP-binding proteins in receptor-mediated activation of phospholipase C. In such preparations, agonist-induced responses, although somewhat attenuated, can be mimicked by the addition of Ca^{2+} and potentiated by the addition of GTP or GTP analogues. Using such an approach, Gomperts (1983) demonstrated that guanine nucleotides stimulated a Ca^{2+}-dependent secretion of histamine from permeabilized mast cells. He proposed that a guanine nucleotide binding protein was involved in Ca^{2+}-dependent secretory events, and it was suggested that the stimulation of secretory responses by GTP or its analogues was due to the stimulation of phosphoinositide breakdown. The successful demonstration of the potentiating effects of GTP analogues in per-meabilized mast cells led to similar studies in numerous cell types. Potentiation of

agonist-induced phosphoinositide breakdown by GTP, GTP-γ-S, or Gpp(NH)p has been widely reported. In addition to the potentiation of agonist-induced responses, the ability of GTP-γ-S or Gpp(NH)p alone to stimulate the breakdown of endogenously labeled phosphoinositides has been demonstrated in many cell types. In many of these studies, the stimulatory effect of GTP analogues was blocked in the presence of GDP-β-S, indicating the involvement of a G protein, because GDP-β-S arrests G proteins in their inactive state.

Identity of G Proteins Involved in the Activation of Phospholipase C

The investigations on the identity of the putative G proteins that may be involved in the receptor-mediated activation of phospholipase C have so far relied heavily on the use of ADP-ribosylation as a tool (for review, see Ui, 1986). Two bacterial toxins, namely, cholera toxin and pertussis toxin, catalyze an NAD$^+$-dependent ribosylation that leads to inactivation of the protein. Nakamura and Ui (1983) were the first to show the involvement of a pertussis toxin substrate in mediating the stimulatory effect of a Ca^{2+}-mobilizing agonist (i.e., pretreatment of mast cells with pertussis toxin inhibited histamine [H1] secretion in response to compound 48/80). They subsequently showed that the inhibition of histamine secretion was due to the inhibition of phosphoinositide breakdown (Nakamura and Ui, 1985). Certain observations suggest that the protein ribosylated was either G_i or another protein very similar to G_i. Because A23187-induced histamine release was not affected by pertussis toxin treatment, the pertussis toxin-sensitive protein was postulated to act at a step between the receptor occupation and phospholipase C activation. The inhibition by pertussis toxin of phosphoinositide breakdown in response to Ca^{2+}-mobilizing agonists has been reported in many cell types in the past few years. In some cases, phosphoinositide breakdown also appears to be sensitive to cholera toxin (Imboden et al., 1986). The failure of some agonist-induced phosphoinositide responses to be inhibited by either pertussis toxin or cholera toxin has led to the suggestion that the coupling of phospholipase C to receptors may be mediated by a unique G protein, which has been named G_p or N_p. However, the existence of G_p or what determines its sensitivity or insensitivity to pertussis or cholera toxin remains to be defined pending its isolation.

G Proteins Enhance Ca^{2+} Sensitivity of Phospholipase C

It has been widely reported that membranes derived from [^{32}P]- or [^{3}H]myoinositol-labeled cells undergo phosphoinositide hydrolysis when incubated in the presence of Ca^{2+}. However, the concentration of Ca^{2+} required for this stimulation is very high. In the presence of GTP or nonhydrolyzable analogues of GTP, the Ca^{2+} requirement is reduced to the physiological range for resting cells (Bradford and Rubin, 1986). The Ca^{2+} requirement for platelet cytosolic phospholipase C was reduced 100-fold in the presence of GTP-γ-S (Deckmyn et al., 1986). A similar reduction of Ca^{2+} requirement was reported for thrombin-induced secretion in permeabilized platelets (Haslam and Davidson, 1984). These observations suggest that GTP-binding proteins increase the sensitivity of phospholipase C to Ca^{2+}, although the mechanism for this enhancement is not as yet defined.

Protein Kinase C (C-kinase)

The other product of phospholipase C-mediated breakdown of phosphoinositides is DAG. DAG acts as a second messenger via activation of protein kinase C (C-kinase), which, in addition to influencing many cellular processes, is also involved in the attenuation of the phosphoinositide response itself. Some key features of this enzyme will be reviewed here. More detailed information is available in several recent reviews (Berridge, 1984; Niedel and Blackshear, 1986; Nishizuka, 1986).

Discovery and Mechanism of Activation

C-kinase was discovered by Nishizuka and associates in 1977 (see Nishizuka et al., 1984) and subsequently shown to require Ca^{2+} and phospholipid for activity (Nishizuka et al., 1984). Two key observations heightened interest in C-kinase: (1) DAG was found to stimulate the enzyme by lowering its Ca^{2+} requirement into a physiological range (0.1–1.0 μM) (Takai et al., 1979); and (2) the potent tumor promoter, phorbol myristate acetate (PMA), stimulated the enzyme by substituting for DAG (Castanga et al., 1982). This latter property quickly caught the attention of many workers in the field of chemical carcinogenesis.

Fully active C-kinase is believed to be a quaternary complex consisting of phospholipid, Ca^{2+}, DAG, and the enzyme. The ternary complex lacking DAG displays activity but only at a 100-fold higher concentration of Ca^{2+}. Phosphatidylserine was found to be the most effective phospholipid in reconstituting enzyme activity.

Acceptance of DAG as a Second Messenger

Although the activation of C-kinase by DAG clearly indicated the possibility of a second messenger function for DAG, it was important to show that the cellular mass of DAG increased after the application of a cell surface stimulus and that the exogenous addition of cell-permeable DAGs induced responses identical to those of growth factors, hormones, and neurotransmitters. Using a variety of methods, a rapid net increase in the mass of cellular DAG upon stimulation has been reported in a number of tissues and cells (Banschbach et al., 1974, 1981; Rittenhouse-Simmons, 1979; Bocckino et al., 1985; Preiss et al., 1987). An enrichment of stearate and arachidonate in the incremental fraction of DAG on stimulation with agonists has suggested that the DAG must be derived from inositol phospholipids (Rittenhouse-Simmons, 1979; Cockcroft and Allan, 1984). More recently, Kennerly (1987) has shown that in stimulated mast cells, half of this incremental fraction in DAG is derived from non-phosphoinositide lipids. Non-phosphoinositide sources of DAG have also been reported in several other systems.

Over the course of the past few years, several cell-permeable DAGs have been synthesized and evaluated for their ability to mimic the responses induced by cell surface stimuli. Cell-permeable DAGs have proven to be valuable tools for the study of C-kinase–mediated responses, and studies with these agents have provided additional support to the second-messenger status of DAG. Early studies with synthetic DAGs were largely unsuccessful because they were carried out with long-chain DAGs, which did not interact with cells due to phase separation in an aqueous environment. Nishizuka and his colleagues were able to overcome this problem by using 1-oleoyl-2-acetylglycerol (OAG). Initial studies with this compound in platelets, neutrophils, and mast cells indicated that OAG could function as a second messenger for secretion (Nishizuka, 1984). The initial success with OAG encouraged Bell and his associates to synthesize and test DAGs with varying chain lengths as activators of C-kinase. Optimal biological activity in many intact cell systems was seen with DAGs containing saturated acyl chain lengths of six to nine carbon atoms (Lapentina et al., 1985). Two important issues relating to the use of cell-permeable synthetic DAGs are (1) Do they really mimic the action of endogenous DAGs? and (2) Are their effects always mediated via activation of C-kinase? The first question has not been adequately addressed. With regard to the second question, some investigators have attempted to establish the involvement of C-kinase by blocking the effect of synthetic DAGs, which can be done by down-regulating C-kinase (see later) or by use of chemicals that inhibit C-kinase activity. Unfortunately, most inhibitors used in the latter studies lack the desired specificity or have additional effects of their own. A case in point is H7, which appears to inhibit several protein kinases (Garland et al., 1987). More recently, staurosporine has been reported to be a more potent and possibly a more selective inhibitor of C-kinase (King and Rittenhouse, 1989).

Identification of C-kinase as the Phorbol Ester Receptor

Phorbol esters have been used extensively as a substitute for DAG in studying C-kinase. It has been assumed that they act by similar mechanisms. However, the resemblance between DAG and phorbol esters is relevant only in the context of C-kinase activation and may not be applicable to those effects of phorbol esters or DAGs that are not mediated by C-kinase activation. Nevertheless, in understanding how DAGs activate C-kinase, it may be helpful to review some key advances that have led to the current concepts regarding the mechanism of action of phorbol esters.

The report by Castanga and co-workers (1982) that PMA could substitute for DAG and activate C-kinase *in vitro* was a major breakthrough in understanding the mechanism of phorbol ester action. The dose-response curve for PMA activation of C-kinase was similar to the saturation curve for PMA binding to the receptor, as were the structure–activity relationships for enzyme activation when compared with receptor binding, suggesting that C-kinase may act as a phorbol receptor. Furthermore, when the enzyme was maximally stimulated by either PMA or DAG alone, the addition of the other activator did not lead to additional stimulation, indicating that both of these agents acted by a common mechanism. Soon after this report, PMA was shown to immediately activate C-kinase in intact cells (Sano et al., 1983). Additional support for the idea that C-kinase was a phorbol ester receptor came from studies on receptor purification. It was invariably observed that both the binding activity and the Ca^{2+}/phospholipid-dependent enzyme activity copurified in the same fraction. Several laboratories have reported a cytosolic form of the receptor that requires PS and copurifies with C-kinase and, in general, after reconstitution, exhibits the same order of affinities for phorbol ester analogues as seen with the receptor in cells or subcellular fractions, or by biological assays (Ashendel et al., 1983; Kikkawa et al., 1983; Niedel et al., 1983).

At present, there is general agreement that C-kinase is the major cellular receptor for phorbol esters, but there is insufficient evidence to conclude either that it is the only receptor or that all phorbol ester effects are mediated through activation of C-kinase. On the basis of solubility characteristics of the receptor, most cell types appear to have several operationally distinct pools of the receptor, but the receptor from each compartment requires phosphatidylserine and Ca^{2+} for reconstitution and always copurifies with C-kinase activity.

Translocation of C-kinase

The term translocation, in the context of C-kinase, refers to the change in the intracellular site occupied by the enzyme upon exposure of certain cell types to phorbol esters, and in some cases, natural agonists. The phenomenon of C-kinase translocation was first observed by Kraft and Anderson (1983) in a study of the subcellular distribution of C-kinase activity in EL4 mouse thymoma cells following phorbol ester exposure. Since then, numerous reports have appeared that demonstrate that phorbol esters can induce a rapid association of C-kinase with the cellular particulate fraction in a variety of cell types. A recent review on C-kinase has cited 12 papers between 1984 and 1986, covering 11 cell types, where phorbol esters have been shown to cause translocation of C-kinase activity from the cytosolic fraction to a particulate fraction (Niedel and Blackshear, 1986).

Although the translocation of C-kinase is fairly consistently observed in response to active phorbol esters, natural agonists show more variability. For example, in 3T3 fibroblasts, no shift of C-kinase from cytosol to particulate fraction was observed on treatment of cells with platelet-derived growth factor, fibroblast growth factor, and bombesin, even though the shift occurred readily in response to DAG or PMA (Niedel and Blackshear, 1986). There is no unifying hypothesis, as yet, to account for the variable responses outlined, nor is the molecular basis for the association of C-kinase with membranes understood. With respect to phorbol esters, the generally held view is that these agents can intercalate into various membrane structures, where they can replace the DAG as a component of the quaternary complex. The translocation apparently results from the high affinity of C-kinase for phorbol ester. Because there is some evidence to suggest that Ca^{2+} increases the binding affinity of C-kinase for DAG (Dougherty and Niedel, 1986), $I(1,4,5)P_3$ could play an important role in translocation of C-kinase because of its calcium-elevating effects. One caveat regarding the translocation hypothesis is that artifacts can be created by homogenization and cell fractionation. For example, homogenization may release loosely bound membranous C-kinase into the soluble fraction.

C-kinase Substrates

Although there is general agreement that C-kinase plays a role in the initiation and/or modulation of receptor-linked cellular responses (see later), the precise nature or molecular details of this involvement remain elusive. Identification of cellular protein targets for C-kinase should provide clues to the mechanism(s) by which C-kinase modulates cellular responses. Consequently, a number of recent investigations have focused on possible C-kinase targets. Several reviews have dealt with various aspects of C-kinase substrates. Lists of some of the proteins believed to be phosphorylated by C-kinase either *in vitro* or in intact cells have been compiled (Kuo et al., 1984; Niedel and Blackshear, 1986; Witters and Blackshear, 1987). At that time, this list included eight receptors and 18 endogenous proteins or enzymes. However, great caution should be exercised in extrapolating results from cell-free preparations to intact cells.

Regulation of C-kinase Activity and Substrate Specificity

Prolonged exposure of intact cells to active phorbol esters is known to lead to a gradual decline of C-kinase activity. This phenomenon has been described as down regulation of C-kinase. Down regulation has proven a useful tool for investigating the importance of C-kinase–dependent pathways involved in the actions of a variety of hormones, neurotransmitters, and growth factors. In some cases, possible involvement of C-kinase in a particular response has been strengthened by showing that microinjection of purified C-kinase into cells restores the responses lost following C-kinase down regulation (Pasti et al., 1986).

In several early studies, it was noted that treatment of cells with active phorbol esters led to development of resistance to subsequent additions. Studies in several cell lines showed that the development of resistance to further effects was due to the disappearance of specific phorbol ester receptors. Solanki and Slaga (1984) showed that the loss of phorbol ester receptors was not due to internalization of receptors, supporting metabolism and degradation of the receptors as the mechanism of down regulation. Furthermore, down regulation was shown to be reversible (i.e., receptor number gradually returned to normal following the removal of the down regulating stimulus). The mechanism of down regulation remains to be determined.

Investigations within the past few years have revealed that C-kinase exists as a family of multiple subspecies with subtle individual characteristics (Coussens et al., 1986; Knopf et al., 1986; Ono et al., 1987). In brain tissue, structures of the subspecies have been deduced from the analysis of their cDNA sequences. The structures of the subspecies (α, βI, βII, and γ) are highly homologous and reveal four conserved and five variable regions. Two of the variable regions are within the regulatory domain of the structure. Therefore, subtle differences in the regulation of various subspecies may be expected. Type I C-kinase (encoded by the γ sequence) is less sensitive to DAG but significantly activated by relatively low concentrations of free arachidonate. Type II C-kinase (derived from splicing of βI and βII sequences) exhibits substantial activity at basal Ca^{2+} levels and responds well to DAG and, to some extent, arachdonate. Type III C-kinase (encoded by the α sequence) is most sensitive to 1-stearoyl-2-arachidonylglycerol (the major species of DAG derived from inositol phospholipids). This subspecies of C-kinase can be activated by high concentrations of arachidonate and Ca^{2+} in the absence of phospholipid. The physiological significance of these subspecies differences remains to be clarified.

In addition to the subtle regulatory differences between the C-kinase subspecies outlined, another important factor may be their differential expression in various tissues. On the basis of Northern blot analysis, expression of C-kinase subspecies appears to be tissue- or cell-type specific. Additionally, combined biochemical, immunological, and cytochemical approaches have revealed significant differences in the regional expression of various subspecies in rat brain (Kitano et al., 1987; Yoshida et al., 1988). The molecular basis of this differential gene expression is yet to be elucidated.

Modulatory Functions of C-kinase in Phosphoinositide-Mediated Signal Transduction

The phosphoinositide signal transduction mechanism is unique in the sense that receptor activation leads to production of two messengers, $I(1,4,5)P_3$ and DAG, each of which in turn either independently or together leads to production or mobilization of other messengers and modulators. Together, the two branches of this pathway represent a highly versatile mechanism to control a host of cellular responses.

In recent years, it has become increasingly clear that the two branches of the

phosphoinositide messenger system not only interact with each
interact in rather complex ways with other regulatory molecules. C-
to play a pivotal role in control of the signal transduction process at

Modulatory interactions related to Ca^{2+} signaling. These inte
to affect Ca^{2+} signaling, either directly via modulation of influx and efflux or
indirectly via modulation of polyphosphoinositide formation or degradation. These
interactions are

Feedback inhibition of polyphosphoinositide turnover. It appears that one
of the important functions of the DAG–C-kinase pathway may be to inhibit Ca^{2+}
signaling. This action of C-kinase may prevent the rise of cytosolic Ca^{2+} to
unphysiological levels (Berridge, 1986). In a number of cell types, elevation of
intracellular free Ca^{2+} by Ca^{2+}-mobilizing agonists known to act by receptor-
mediated stimulation of inositol phospholipid turnover has been shown to be
inhibited by phorbol esters (Orellana et al., 1985; Rittenhouse & Sasson, 1985).

The phorbol ester-induced inhibition of the agonist-stimulated rise in intracellular free Ca^{2+} is believed to be a consequence of the inhibition of inositol
phospholipid turnover because in many instances the inhibition of responsiveness
was correlated with the inhibition of formation of inositol phosphates. The synthetic cell-permeant diacylglycerol, OAG, also inhibits agonist-dependent inositol
phosphate generation (Watson and Lapetina, 1985).

Although it has been generally accepted that C-kinase activation, either by
endogenous DAGs or by phorbol esters, can modulate phosphoinositide responses
by feedback inhibition, the mechanisms underlying this inhibition appear to be
quite diverse. There are numerous instances where phorbol ester treatment of cells
modifies agonist responses via C-kinase–mediated phosphorylation of the receptor
(e.g., the cholinergic muscarinic receptor) (Safran et al., 1987). The generally held
view is that receptor phosphorylation reduces the affinity for agonists or interferes
with agonist binding by promoting internalization of the receptor.

In many cases, phorbol ester treatment inhibits phosphoinositide responses but
without affecting the affinity or the number of receptors examined (e.g., in
astrocytoma cells, activation of the muscarinic ACh receptor is linked to PIP$_2$
hydrolysis, and PIP$_2$ hydrolysis is inhibited by TPA, but the binding properties of
the receptor are not altered) (Orellana et al., 1985). In these cases, it is thought by
some that the coupling of receptors to phospholipase C is interrupted as a result of
C-kinase–induced phosphorylation of guanine nucleotide binding regulatory
protein(s), in analogy with adenylate cyclase (Katada et al., 1985). At present, a
role for C-kinase in the modulation of phosphoinositide breakdown via phosphorylation of G proteins remains plausible but not firmly established.

Removal of calcium from the cytoplasmic compartment. In addition to the
feedback inhibition of the phosphoinositide response outlined earlier, Ca^{2+}

signaling can also be suppressed by activation of pumps that remove Ca^{2+} from the cytosol. Evidence for the involvement of C-kinase in this process has been obtained in GH3 cells, where phorbol ester treatment was shown to reduce not only the agonist-induced Ca^{2+} rise but also the K^+ depolarization–induced Ca^{2+} increase (Drummond, 1985).

Acceleration of I(1,4,5)P3 degradation. The rate of I(1,4,5)P3 degradation may be an important factor in modulating I(1,4,5)P3-mediated intracellular Ca^{2+} mobilization. In this context, the observations in platelets (Connolly et al., 1986a; King and Rittenhouse, 1989) suggest that C-kinase, by phosphorylating inositol 5′-trisphosphatase, enhances the dephosphorylation of I(1,4,5)P3. However, this does not appear to be the case in permeabilized RINm5F cells (Biden et al., 1988), where phorbol esters failed to enhance the rate of degradation of exogenously added I(1,4,5)P3.

Interaction with voltage-dependent Ca^{2+} channels. In many types of cells, the rise in cytosolic Ca^{2+} after activation of Ca^{2+}-mobilizing receptors appears to be biphasic. As described earlier, first there is a transient rise attributed to I(1,4,5)P3-induced intracellular Ca^{2+} mobilization. This increase is then followed by a return to a new steady state, which is similar or somewhat higher than the prestimulatory state. Voltage-sensitive Ca^{2+} channels in the plasma membrane appear to play an important role in maintaining cytosolic Ca^{2+} in the second phase (Jy and Haynes, 1987). C-kinase may modulate the activity of these channels. PMA was found to inhibit the binding of Ca^{2+}-channel antagonists to PC-12 cells (a neural cell line), supporting the possible involvement of the C-kinase branch in the modulation of voltage-dependent Ca^{2+} channels (Messing et al., 1986).

Synergism between Ca^{2+} and C-kinase. One of the earliest questions concerning the roles of the two messengers Ca^{2+} and DAG was whether either one or both of these messengers was required for sustained cellular responses. In many tissues, it was observed that although activation of C-kinase alone by phorbol esters or elevation of cytosolic Ca^{2+} alone by ionophores induced small responses, activation of both branches was required for induction of sustained maximal responses equivalent to those induced by natural agonists. This synergism between the two branches has been reported in many tissues and cell types (see Rasmussen et al., 1986).

The biochemical mechanism for the synergistic interaction between the I(1,4,5)P3 and C-kinase branches of the phosphoinositide-signaling pathway is not clear. Rasmussen has attempted to explain this synergism by proposing that C-kinase may be the predominant determinant of sustained cellular responses, and that I(1,4,5)P3-induced Ca^{2+} release plays an essential role in the optimal stimulation of C-kinase. An interesting feature of C-kinase modulation is that although it affects in a positive manner the events that lie distal to the formation of $I(1,4,5)P_3$, it negatively modulates the formation of $I(1,4,5)P_3$.

Interaction with cAMP. cAMP and the DAG branch of the phosphoinositide cascade appear to interact with each other in complex ways involving positive as well as negative modulatory interactions. Phorbol esters have been shown to augment the production of cAMP in response to activation of β-adrenergic receptors (see, for example, Bell et al., 1985). The mechanism for this potentiation, although not clear, may involve either a facilitation of the interaction between adenylate cyclase and GTP-binding proteins or an inhibitory interaction with G_i. The latter mechanism is plausible because C-kinase has been shown to phosphorylate G_i in platelets (Katada et al., 1985).

Arachidonate and Its Metabolites

This subject has received detailed treatment in a recently published book (Waite, 1989). Arachidonate metabolites (i.e., prostaglandins, leukotrienes, and thromboxanes, which are collectively referred to as eicosanoids) are potent regulators of various physiological responses. Because mammalian cells do not store eicosanoids and because unstimulated cells contain very little, if any, unesterified arachidonate, the levels of these compounds are primarily regulated by the availability of free arachidonate, which must be liberated mainly from esterified lipids (Lands and Samuelsson, 1968). It appears that PI in all mammalian tissues is rich in arachidonate at the 2 position of glycerol, as first shown by Keenan and Hokin (1964) (see Majerus et al., 1988). The polyphosphoinositides have also been shown to contain a large proportion of arachidonate at the 2 position of glycerol (Baker and Thompson, 1972). Besides phosphoinositides, PC and PE are also esterified with arachidonate to varying degrees. Irvine (1982) has discussed in detail the various mechanisms by which arachidonate levels are modulated in mammalian cells.

Increased release of arachidonate and subsequent synthesis of eicosanoids appear to be related to the agonist-induced phosphoinositide responses in several tissues (see Rana and Hokin, 1990). This is supported by the observation that in many cell types and tissues, activation of the same receptors that control phosphoinositide breakdown also results in the liberation of arachidonate and/or eicosanoids. Agonists that have been shown to stimulate the release of arachidonate and/or eicosanoids in various tissues or cell types include thrombin, muscarinic cholinergic agonists, $α_2$-adrenergic agonists, serotonin, vasopressin, angiotensin II, bradykinin, substance-P, f-Met-Leu-Phe, caerulein, adrenocorticotropin hormone, EGF, GTP-γ-S, IgE-specific antigen, oxytocin, interleukin-1, and the platelet-activating factor. Although it is now well established that arachidonate is released from phospholipids in response to certain stimuli, it is still not known for most tissues what proportion of the total arachidonate released is derived from phosphoinositides and what is its mechanism of release.

Mechanisms of Arachidonate Liberation

Depending upon the agonist and the type of tissue involved, several mechanisms for arachidonate release appear to exist. Two mechanisms have been investigated in some detail. One of these involves sequential actions of phospholipase C and DAG- and monoacyl glycerol (MAG)-lipase, whereas the other involves direct action of phospholipase A_2 on phospholipids. We review here these and other possible mechanisms.

Phospholipase C–DAG-lipase pathway. In platelets (Rittenhouse, 1982; Majerus et al., 1984) and pancreatic minilobules (Dixon and Hokin, 1984), there is agonist-evoked liberation of free arachidonate, and at least in pancreas this appears to proceed exclusively via the sequential actions of DAG- and MAG-lipase on DAG liberated from the phosphoinositides. Because DAG-lipase preferentially cleaves the fatty acyl group at the *sn*-1 position, arachidonate liberation requires the action of MAG-lipase on the *sn*-2 position of MAG. In pancreatic minilobules, several observations support the PLC-DAG lipase pathway for release of arachidonate (Dixon and Hokin, 1989): (1) Stimulation of phosphoinositide breakdown with the secretogogue caerulein was associated with an increase in the steady-state level of 1-stearoyl, 2-arachidonoyl-*sn*-glycerol and a release of substantial amounts (up to 50% of the total PI breakdown [see later] of stearate, arachidonate, and glycerol. (2) The accumulation of PA, DAG, glycerol, and fatty acids accounted for the loss in PI, indicating that there was no substantial contribution of other pathways leading to arachidonate. (3) The DAG-lipase inhibitor RHC 80267 (Sutherland and Amin, 1982) reduced the secretogogue-stimulated liberation of free stearate, arachidonate, and glycerol and elevated the steady-state level of 1-stearoyl, 2-arachidonoyl-*sn*-glycerol. RHC 80267 had no effect on phospholipids, suggesting that the RHC 80267 effects were specific in this system. (4) Stimulation of pancreatic minilobules with either caerulein or carbachol did not give rise to increased amounts of lysophosphatidylinositol, glycerophosphorylinositol, or glycerophosphorylinositol mono- or bisphosphate (Dixon and Hokin, 1984; Sekar et al., 1987)—products of phospholipase A_2 action on phosphoinositides. These observations in the pancreas provide strong evidence that stimulation of phosphoinositide breakdown generates arachidonate via the sequential actions of phospholipase C and DAG and MAG lipases.

Phospholipase A_2 pathway. Diacylphospholipids. In neutrophils (Lapetina et al., 1980), the presence of phospholipase A_2 specific for PA has been described, and it has been implicated in the release of arachidonate from PA produced during

PI turnover. But it was subsequently shown that half of the arachidonate that was released on stimulation of platelets with thrombin occurred before any rise in PA (Neufeld & Majerus, 1983). However, this does not exclude phospholipase A_2 action on PA at a later time point. There appear to be multiple pathways for release of arachidonate in platelets (see later).

Another pathway that has been implicated in arachidonate release in platelets is a phospholipase A_2-mediated breakdown of PC, PE, and PI (Bills et al., 1977). Phospholipase A_2-mediated hydrolysis of PC is a major source of arachidonate in antigen-stimulated mast cells (Yamada et al., 1987), whereas both PC and PI appear to be hydrolyzed in PMN leukocytes (Meade et al., 1986). The independence of this pathway from phospholipase C-mediated turnover of phosphoinositides is indicated by a report (Nakashima et al., 1987b) that arachidonate release in platelets could be initiated by neomycin, which is an inhibitor of phospholipase C.

Alkylacyl phospholipids. Another source of arachidonate that so far has received little attention is ether phospholipids. In some tissues, alkylacyl-glycero-3-phosphorylcholine appears to be a significant source of metabolizable arachidonate. In a recent study with rat platelets (Colard et al., 1986), stimulation with thrombin was shown to cause loss of arachidonate from diacyl-*sn*-glycero-3-phosphoinositol (PI) and alkylacyl- and diacyl-*sn*-glycero-3-phosphorylcholine. Further work is required to establish metabolic pathways for ether phospholipids in other tissues.

Ca^{2+} Dependency of Phospholipase A_2 and Its Relationship to the Phosphoinositide Cascade

The affinity of phospholipase A_2 for Ca^{2+} is much lower (higher K_d) than that of phospholipase C (Billah et al., 1980), suggesting that phospholipase A_2 action on phospholipids is likely to be stimulated by agonist-evoked rises in Ca^{2+}. Thus, arachidonate release may respond to the transient rises in Ca^{2+} evoked by $I(1,4,5)P_3$. The possibility that a rise in intracellular Ca^{2+} may be important in activating phospholipase A_2 is supported by the observation that Ca^{2+} ionophores are potent activators of phospholipase A_2 (Laychock and Putney, 1982; Okano et al., 1985). However, the $I(1,4,5)P_3$-induced rise in cytosolic Ca^{2+} in most tissues appears to be far below the high micromolar or low millimolar concentrations required to activate phospholipase A_2 *in vitro*. Some recent reports suggest that G proteins are involved in phospholipase A_2 activation. For example, in platelets (Nakashima et al., 1987a), GTP-γ-S stimulates phospholipase A_2-mediated arachidonate release. It is possible that activation of G proteins may enhance the Ca^{2+} sensitivity of phospholipase A_2, as has been shown for phospholipase C (see G proteins). Also, in thyroid and endothelial cells, the G protein involved in arachidonate release appears to be distinct from the one involved in phospholipase C activation, because arachidonate release, but not inositol phosphate release, was sensitive to pertussis toxin (Burch et al., 1986a).

Involvement of C-kinase in Arachidonate Release

Several observations support a potential role for C-kinase in arachidonate release. In several cell types (see for example, Kolesnick and Paley, 1987; Parker et al., 1987), treatment with PMA releases arachidonate and/or eicosanoids. Like the agonist-induced release of arachidonate (see earlier), the PMA-induced release also appears to involve both phospholipase A_2 and phospholipase C acting on different phospholipids, depending on the tissue.

Role of Released Arachidonate

A detailed discussion of various functions of eicosanoids is beyond the scope of this review. Only some relatively recent developments are emphasized here.

Arachidonate metabolites as mitogens. Arachidonate or its metabolites have been suggested to play an undefined role in cell replication (Habenicht et al., 1985; Burch et al., 1986b). Hydroxyeicosatetraenoic acids (HETEs) have been shown to act as mitogens in several types of cells (see, for example, Chan et al., 1985).

Possible involvement of arachidonate and/or its metabolites in phospholipase C action. A number of investigators have reported that arachidonate and/or its metabolites are capable of activating phospholipase C in intact cells (Siess et al., 1983; Zeitler and Handwerger, 1985) and cell-free preparations (Laychock and Putney, 1982). In placental cells, stimulation of [^{3}H]myoinositol release by arachidonate from labeled cells was insensitive to lipoxygenase or cyclooxygenase inhibitors, indicating that the effect was mediated either directly by arachidonate or indirectly by metabolites that were insensitive to these inhibitors. Because arachidonate metabolites are usually released from the cell, they can have physiologically important effects on the same cell or neighboring cells that may possess the appropriate receptors. In this respect, the lipoxygenase product leukotriene B_4 appears to be a good candidate for the amplification of the phospholipase C signal because it is exported from cells and has been shown to act as an agonist, leading to receptor-mediated activation of phospholipase C (Andersson et al., 1986) and Ca^{2+} mobilization (Goldman et al., 1985).

Arachidonate and Ca^{2+} mobilization. Another possible role for arachidonate that so far has been seen in only a few systems is its effect on Ca^{2+} movements. Kolesnick, Musacchio, Thaw, and Gershengorn (1984) showed that 3 mM arachidonate added exogenously to cloned pituitary GH3 cells stimulated $^{45}Ca^{2+}$ efflux and prolactin secretion. Eicosatetraenoic acid and indomethacin, inhibitors of the lipoxygenase and cyclooxygenase pathways for arachidonate metabolism, respectively, did not inhibit either of these effects, suggesting a direct effect of arachidonate on $^{45}Ca^{2+}$ efflux and prolactin secretion.

Arachidonate itself has been shown to mobilize Ca^{2+}, from isolated organelles, such as mitochondria (Roman et al., 1979) and sarcoplasmic reticulum (Cheah, 1981). More recently, utilizing pancreatic islets (Wolf et al., 1986) and liver microsomes (Chan and Turk, 1987), arachidonate was shown to release Ca^{2+} from a pool that was distinct from the $I(1,4,5)P_3$-sensitive pool.

Activation of potassium channels. Very recently, two laboratories independently reported that arachidonic acid opened potassium channels (i.e., in neonatal rat atrial cells [Kim and Clapham, 1989] and smooth muscle cells [Ordway et al., 1989]). This may be the best evidence yet of a second messenger role for arachidonic acid.

Arachidonate and cGMP. Arachidonate, as well as a few other fatty acids, is capable of activating guanylate cyclase in some cell-free systems (Goldberg and Haddox, 1977), which suggests that cGMP levels may be regulated by arachidonate levels. An elevation in cGMP, but not cAMP, levels has been observed in several tissues that show enhanced phosphoinositide turnover (see Michell, 1975). However, attempts to stimulate cGMP formation by addition of arachidonate to intact cells have generally failed (Sekar and Hokin, 1987). This does not rule out the possibility that a distinct arachidonate pool inaccessible to exogenously added arachidonic acid may be involved in stimulation of cGMP formation.

Activation of C-kinase. Arachidonate metabolism may play a key role in C-kinase activation (Nishizuka, 1988; see also earlier). Arachidonate-derived oxygenation products, particularly lipoxin A (5,6,15 L-trihydroxy-7,9,11,13-eicosatetraenoic acid), are potent intracellular activators of C-kinase and may also be involved in the modulation of substrate specificity of this enzyme (Hansson et al., 1986).

Possible Compartmentalization of Arachidonate

In the exocrine pancreas, the elevation in PI-derived arachidonate on stimulation of enzyme secretion with several agonists is several orders of magnitude greater than the stimulated formation of PGE_2 and $PGF_{2\alpha}$ (Banschbach and Hokin-Neaverson, 1980; Bauduin et al., 1981; Dixon and Hokin, 1984). The difference between the amount of arachidonate released and prostaglandin formed could possibly be due to increased arachidonate in compartments that are inaccessible to prostaglandin-synthesizing enzymes. Such a possibility is supported by the following observations. First, in human platelets, arachidonate released from phospholipids is preferentially utilized by cyclooxygenase rather than lipooxygenase (Sautebin et al., 1983). Second, although zymosan and ionophore both liberate arachidonate in neutrophils, only the arachidonate released by ionophore was converted to HETES (Walsh et al., 1981).

Phosphoinositides, Cell Proliferation, and Oncogenes

Several reviews on this subject have recently appeared (Ian, 1985; Whitman et al., 1986; Berridge, 1987a; Weinstein, 1987). Evidence favoring a role for phosphoinositide turnover in cell proliferation and oncogenesis is presented here.

Mitogens and Phosphoinositide Turnover

Historically, the first studies suggesting a possible link between growth stimulation and phosphoinositide turnover were studies in T lymphocytes by Fisher and Mueller (1971). They observed that the mitogen phytohemagglutinin specifically stimulated incorporation of [^{32}P]orthophosphate into PI and PA at early time points prior to any changes in radioactivity in other phospholipids. More recently, it has been shown that many mitogens stimulate rapid PI turnover when added to quiescent 3T3 fibroblasts. In addition to increased PI turnover following exposure to mitogens, there is an increased production of I(1,4,5)P$_3$ (Berridge et al., 1984), increases in DAG and activation of C-kinase (Habenicht et al., 1981), and intracellular Ca^{2+} mobilization (McNeil et al., 1985).

That the turnover of phosphoinositides constitutes a crucial step in the signaling pathway for some mitogens is supported by an observation that the microinjection of an antibody to PIP$_2$ leads to complete abolition of nuclear labeling with [^{3}H]thymidine in NIH 3T3 cells in response to PDGF or bombesin (Matuoka et al., 1988). A similar conclusion was reached from an earlier study that showed that neomycin, an inhibitor of PIP$_2$ phosphodiesteratic cleavage, inhibited thrombin-induced proliferation in hamster fibroblasts (see earlier). This inhibition was observed only at doses of neomycin that were sufficient to block thrombin-induced phosphoinositide turnover completely. It should be borne in mind that neomycin inhibits PIP$_2$ breakdown by forming a salt complex with PIP$_2$, and this is unlikely to be a highly specific effect.

In retinal capillary pericytes, DNA synthesis can be directly stimulated by addition of I(1,4,5)P$_3$ and DAG to permeabilized cells (Carney et al., 1985). An exciting development by Sylvia, Curtin, Norman, Stec, and Busbee (1988) offers a biochemical mechanism that links the phosphoinositide signaling system to proliferation. These authors showed that a low-activity form of DNA polymerase α, immunoaffinity purified from adult-derived human fibroblasts, was activated by interaction with PIP, whereas a high-affinity form of the enzyme did not interact with PIP or its derivatives. PIP was apparently hydrolyzed in the presence of a highly purified, low-activity form of DNA polymerase α, effecting the release of DAG and the retention of IP$_2$ by the enzyme complex. The resulting IP$_2$/protein complex exhibited both increased affinity of binding to DNA template/primer and increased deoxynucleotidyltransferase activity. These data indicate that IP_2 may function as an effector molecule in the activation of a low-activity form of human DNA polyermase α and suggest that it may function as a second messenger during the initiation of mitosis in stimulated cells. This interesting study awaits confirmation and further development.

Phosphoinositide Turnover in Transformed Cells

An association between changes in PI turnover and cellular transformation was first suggested by Diringer and Friis (1977). They observed that under conditions that limited normal but not transformed cell growth, such as serum omission or high cell density, there was a decline in the PI turnover rate in normal quail cells but not in the Rous sarcoma-transformed cells. These studies suggested that PI turnover in transformed cells might be constitutively activated. More recently, similar observations linking phosphoinositide turnover to transformation have been made in other cell types (Macara et al., 1985; Whitman et al., 1985).

Relationship between Growth Factors and Oncogenes

In recent years, a great deal of attention has been focused on the identification and characterization of oncogenes and their products in order to identify the biochemical steps at which normal, positive-growth regulatory signals are mimicked or bypassed in transformed cells. Many such genes, postulated to have arisen by viral transduction and mutational activation of normal cellular genes (proto-oncogenes), have been identified in transforming viruses and tumor cells. A detailed discussion of the characteristics of these oncogenes is beyond the scope of this review. The interested reader should consult one or more of several reviews on this subject (Bishop, 1985; Macara et al., 1985; Whitman et al., 1986; Berridge, 1987a). The basic theme of current oncogene research is that oncogene products in transformed cells free them from proliferative restraints of their untransformed counterparts. Characterization of oncogenes and their products has revealed five basic mechanisms by which oncogene products appear to generate positive-growth regulatory signals. These are (1) stimulation of growth-factor production and secretion by viral oncogenes (e.g. *abl*, *src*, and *mos*); (2) coding for a protein highly homologous to the β subunit of PDGF (e.g., by *sis*); (3) constitutive activation of a growth factor membrane receptor by viral oncogenes *erb* B, *fms*, and *neu*, which code for products that share structural homology with known growth factor receptors; (4) constitutive activation of an intermediate signal transducer, such as G proteins, by *ras*; and (5) constitutive expression or activation of nuclear targets for growth-factor signals (e.g., *myc* and *fos*). It has been felt that the evidence for phosphoinositide involvement is particularly strong for *ras*, because this gene product (a G protein) serves to couple growth-factor receptors to $I(1,4,5)P_3$ production (Wakelam et al., 1986). However, a study by Downward, deGunzburg, Riehl, and Weinberg (1988) does not support this view. They showed that transfection of Rat-1 fibroblasts with the p21 *ras* gene increases bradykinin receptor numbers rather than directly coupling p21 *ras* between the receptor and phospholipase C.

Possible Targets for Phosphoinositide Messengers

Despite the fact that most mitogens stimulate phosphoinositide turnover, the mechanisms by which this turnover eventually alters nuclear transcription or triggers mitosis are not adequately understood. One might anticipate that Ca^{2+} and DAG would act synergistically to activate C-kinase, but the critical targets for this enzyme remain unknown. There is some evidence to suggest that C-kinase phosphorylates RNA polymerase II *in vitro*, enhancing its binding to substrate (Chuang et al., 1987). Many proteins appear to be phosphorylated by this enzyme (see C-kinase), but it is not known which ones are important for proliferation.

It appears likely that one important distal target for both Ca^{2+} and DAG signals may be the nuclear oncogenes *myc* and *fos*. A link between these oncogenes and the normal growth factor–signaling pathway appears to be supported by observations that the transcription of the cellular homologues of these oncogenes was stimulated when quiescent cells were treated with mitogens that activate phosphoinositide turnover.

Relationship Between Tyrosine Kinases and Phosphoinositide Kinases

A biochemical link between oncogenes and controlling elements of the phosphoinositide signal transduction pathway was first suspected when it was reported that two protein tyrosine kinase oncogene products, pp60*src* and pp68*ros*, also possessed phosphoinositide kinase activity (Macara et al., 1984; Sugimoto et al., 1984). These reports were soon followed by the demonstration of an association of phosphoinositide kinase activities with two more transforming protein tyrosine kinases, middle t/pp60*src* and p120*abl* (Fry et al., 1985). However, studies of inactivation profiles and immunoprecipitation of tyrosine kinase and phosphoinositide kinase activities in normal or transformed cells and studies of cloned tyrosine kinases expressed in bacterial vectors indicated that tyrosine kinase-associated phosphoinositide kinase activity is unlikely to contribute significantly to the total phosphoinositide kinase activity in cells (MacDonald et al., 1985). Nevertheless, oncogene-transformed tyrosine kinases do appear to consistently possess a small fraction of the total PI-kinase activity. The significance of this association between the two activities remains to be established.

Studies indicate that there are two types of PI kinases and that the tyrosine kinase-associated PI kinase specifically phosphorylates the D-3 position of the inositol in PI to generate phosphatidylinositol 3-phosphate [PI(3)P] (Whitman et al., 1988). The detection of PI kinase activity with specificity for the D-3 position in immunoprecipitates of the protein product of polyoma middle T/pp60*src* competent for transformation, or of the ligand-activated PDGF receptor, indicates that PI(3)P itself could be a critical mediator of mitogenic signals.

Besides the plasma membrane, the nuclear membrane also appears to contain PI kinase, and, as discussed earlier, it has been suggested that PIP may be a potent activator of DNA polymerase. Polyphosphoinositides also appear to be associated with chromatin and undergo marked increases at the time of differentiation. Finally, it should be noted that despite the extensive correlations between phosphoinositide metabolism and oncogenesis, a true cause- and-effect relationship between these events is yet to be firmly established.

Glycosyl-Phosphatidylinositol in Membranes

Within recent years, a novel structural feature by which certain proteins are anchored to membranes has been elucidated. This structure involves a covalent linkage between the protein and an ethanolamine-containing oligosaccharide, which in turn is glycosidically linked to PI. Glycosyl-PIs have now been detected in a number of eukaryotic cells (for review, see Low and Saltiel, 1988). The covalent nature of the linkage to PI is indicated by the fact that the proteins so anchored are released into a water-soluble form when intact cells are treated with the glycosyl-PI-specific phospholipase C (GPI-PLC). The proteins so released retain full activity but are unable to reassociate with the membrane. There is a growing list of proteins that are covalently attached to glycosyl-PI, based on their susceptibility to GPI-PLC–induced release from membranes. So far, over 30 such proteins have been identified. Included among these are several hydrolytic enzymes, mammalian and protozoan antigens, coat proteins, cell adhesion proteins, and others. Undoubtedly, more proteins will be found to share this anchoring mechanism. The existence of the lipid anchor and the specific enzymes that are capable of releasing the anchored proteins indicates that the release process may have physiological significance, especially in those diseases where the circulating levels of hydrolytic enzymes anchored to glycosyl-PIs appear to be altered.

The proteins are linked to an ethanolamine residue that, via a phosphodiesteric linkage, is attached to a glycan consisting of several mannose, galactose, and glucosamine residues. The terminal glucosamine of the glycan is glycosidically linked at its C-1 position to the 6-OH of the inositol ring of the PI molecule. Although the monosaccharide sequence of the glycan exhibits a certain degree of heterogeneity, some basic features of the structure appear to be shared by proteins anchored via PI.

In addition to a phospholipase C with specificity toward glycosyl-PI, a specific phospholipase D has also been identified (Davitz et al., 1987; Low and Prasad, 1988). Hydrolysis with phospholipase C or phospholipase D leaves in the membrane a DAG or a PA moiety, respectively.

Mammalian cells have also been shown to contain free glycosyl-PI, which does not contain ethanolamine or an attached protein. Preliminary studies in BC3H1 cells and liver microsomes support the notion that biosynthesis of these free lipids temporally follows that of PI, possibly via glycosylation of a specific pool of PI (Farese et al., 1987). These lipids contain 1,2-DAG or 1,2-alkylacylglycerol as part of a hydrophobic domain, and they appear to play a role in signal transduction. This seems likely, at least in the case of insulin action, where these lipids appear to act as a source of enzyme-modulating second messengers (Saltiel and Sorbora-Cazan, 1987). The mechanism by which insulin regulates the cleavage of glycosyl-PI is unclear, but, on the basis of preliminary studies in BC3H1 cells, it appears to involve activation of a glycosylated-PI–specific phospholipase C, which leads to the formation of DAG and inositol-glycan (Saltiel et al., 1987). Studies with toxins and G protein-specific antibodies indicate that the insulin-induced activation of this specific phospholipase C may be mediated by a guanine nucleotide binding protein (Korn et al., 1987). The product of phospholipase C action, inositol-glycan, has been shown to regulate the activities of several insulin-sensitive enzymes measured in cell-free assays, including cAMP-phosphodiesterase (Saltiel et al., 1986), adenylate cyclase and pyruvate dehydrogenase, and phospholipid methyltransferase (Saltiel and Cuatrecasas, 1987). Furthermore, inositol-glycans in intact adipocytes appear to mimic the action of insulin in promoting glucose utilization and lipolysis (Kelly et al., 1987; Saltiel and Sorbora-Cazan, 1987) but not glucose transport.

The glycosyl-PIs involved in protein anchoring or insulin action appear to have fatty acid compositions that are different from each other and also from free PI. This raises the possibility that distinct DAG species may result from phospholipase C-induced hydrolysis of glycosylated-PI, and this may account for the limited activation of C-kinase by this mechanism. Most of the work on glycosylated-PIs is recent in origin, and further investigations will certainly be forthcoming (i.e., isolation and characterization of various glycosylated-PI species, biosynthetic pathways, their second-messenger functions, and regulatory aspects of specific phospholipases involved in their metabolism).

Pharmacological Action of Li^+

Lithium has been successfully used in the treatment of both the manic and depressive phases of bipolar affective disorders. Although lithium has been in clinical use for over 30 years, the biochemical basis of its pharmacological action is still unclear. Advances in deciphering the role of inositol lipids in signal transduction and the observation that lithium perturbs inositol lipid metabolism suggested to some that there is a causal link between the effects of Li^+ on phosphoinositide metabolism and its pharmacological action. Inhibition of inositol phosphatase by lithium was reported by Allison and Stewart (1971) several years ago. This action of lithium appears to interfere with the inositol-lipid cycle in a selective manner, leading to depletion of free inositol and elevation of specific inositol phosphates in the brain of animals given lithium (Sherman et al., 1986).

The selectivity of this depletion in the CNS cells arises from the fact that neural cells have limited access to plasma inositol (Berridge, 1984) and must depend upon the *de novo* synthesis from glucose via D-inositol-3-phosphate, which is also sensitive to lithium due to the inhibition of D-inositol-3-phosphatase. Thus, it is postulated that in lithium-treated patients production of inositol, either by *de novo* synthesis or by stimulated turnover of inositol lipids, will be suppressed. The entrapment of cellular inositol as inositol monophosphates would be particularly rapid in neurons that might manifest a chronically active stimulation of the phosphoinositide cycle (Berridge, 1984). The depletion of inositol in these cells would tend to eventually limit the amount of PIP_2 and result in inhibition of the transduction of signals originating from pathogenic neurons. Some support for this theory is indicated by the demonstration that prolonged stimulation of GH3 cells in the presence of lithium does appear to reduce inositol lipid levels to the extent that both agonist-induced phosphoinositide hydrolysis and intracellular Ca^{2+} mobilization are impaired (Drummond et al., 1987).

Although this hypothesis is attractive, it is probably an oversimplification and remains unproven. For example, if inositol depletion were the sole mechanism underlying the pharmacological action of lithium, then extracellular inositol should protect against the effects of lithium. However, this protection either does not occur or only occurs at unphysiologically high concentrations of inositol in many cell systems, including neural cells (Nahorski et al., 1986). Also, the study by Downes and Stone (1986) in parotid glands indicates that lithium, in combination with carbachol, greatly reduces the labeling of PI and PA, but the synthesis of PIP_2 is not appreciably affected. Furthermore, more recent developments have indicated that in many types of cells, treatment with lithium not only leads to accumulation of inositol monophosphates but also D-I(1,3,4)P_3 and D-I(1,4)P_2 as well (for review, see Drummond, 1987) due to inhibition of phosphatases that act on these metabolites. These actions of lithium may also contribute to the pharmacological action of lithium.

Another possible site for the action of lithium may involve its interaction with G proteins. Lithium, at therapeutically efficacious concentrations, has been shown to inhibit both adrenergic and cholinergic agonist-induced increases in the binding of GTP to membranes from rat cerebral cortex (Avissar et al., 1988). Thus, lithium may exert its pharmacological effects by altering G protein receptor coupling.

Recently, Lee et al. (1992) reported that if suitably fortified with inositol, guinea pig, rat, and mouse cerebral cortex slices show rises in I(1,4,5)P_3 and I(1,3,4,5)P_4 in the presence of lithium. This may have therapeutic implications.

Fertilization

Ca^{2+} ions play a central role in fertilization. The possibility that a transient increase in intracellular free Ca^{2+} is a pivotal event during fertilization was suggested as early as 1974, when treatment of sea urchin eggs with the Ca^{2+}

ionophore A23187 was shown to initiate changes in protein and DNA synthesis, mimicking the initiating effects of sperm penetration (Steinhardt and Epel, 1974; Whitaker and Steinhardt, 1982). Depending on the type of egg, the Ca^{2+} source can be either intracellular or extracellular. Jaffe (1983) has generalized that the eggs of protostomes (e.g., worms, insects, crustacea, molluscs, and others) rely on an immediate source of extracellular Ca^{2+}, whereas the eggs of deuterostomes (sea urchins, starfish, vertebrates) are activated at fertilization by Ca^{2+} released from an intracellular store.

Evidence implicating the phosphoinositide messenger system in fertilization has been obtained by studying (1) the effect of intracellular injection of $I(1,4,5)P_3$; (2) the effect of pretreatment of eggs with phorbol esters or synthetic cell-permeable DAGs; and (3) the effect of non-hydrolyzable analogues of GTP. Most of the experiments using these approaches have been carried out with sea urchin eggs and *Xenopus* oocytes. We briefly summarize these studies here (for detailed reviews, see Whitaker and Steinhardt, 1985; Whitaker, 1989).

Evidence for a Role of I(1,4,5)P3

In unfertilized sea urchin eggs (deuterosomes), microinjection of $I(1,4,5)P_3$ produces many of the changes normally associated with fertilization (Whitaker and Irvine, 1984). The transient increase in $[Ca^{2+}]_i$ elicited by $I(1,4,5)P_3$ follows a time course similar to the Ca^{2+} transient following fertilization and is not affected by removing external Ca^{2+}. Furthermore, the eggs contain substantial amounts of phosphoinositides that begin to turn over rapidly at fertilization, and there are increases in the levels of $I(1,4,5)P_3$ immediately after insemination (Crossley et al., 1988). $I(1,4,5)P_3$ activates voltage-dependent Ca^{2+} influx in permeabilized *Xenopus* oocytes (Parker and Miledi, 1987), and a rise in intracellular free Ca^{2+} has been correlated with nuclear envelope breakdown in sea urchin embryos and stimulation of chromatin condensation immediately after fertilization (Steinhardt and Alderton, 1988). Thus, it appears highly likely that $I(1,4,5)P_3$ is the crucial second messenger for fertilization.

An intracellular wave of Ca^{2+} occurs at fertilization. The $[Ca^{2+}]_i$ first increases at the point of sperm entry and spreads throughout the cytoplasm (Jaffe, 1983). This Ca^{2+} wave in sea urchin eggs can be initiated by a very localized microinjection of $I(1,4,5)P_3$ (Swann and Whitaker, 1986). Somewhat similar conclusions have also been reached from studies with *Xenopus* oocytes (Busa et al., 1985; see also Intracellular Ca^{2+} Oscillations).

Evidence for a Role of DAG

DAG, the other second messenger produced by hydrolysis of PIP_2, is also involved in the transduction of a fertilization signal, although the mechanism for this is not as well defined as with $I(1,4,5)P_3$. DAG is produced, along with $I(1,4,5)P_3$, after fertilization of sea urchin eggs (Ciapa and Whitaker, 1986). Also, treatment of sea urchin eggs with dioctanoylglycerol induces an increase in cytoplasmic pH (Shen and Burgart, 1986), which is likely to be a cytoplasmic signal responsible for stimulating protein and DNA synthesis (Rosengurt, 1985).

Treatment of unfertilized mouse eggs with PMA causes oscillations in $[Ca^{2+}]_i$ similar to those that occur at fertilization (Cuthbertson and Cobbold, 1985).

Evidence for a Role of G Proteins

The existence of G proteins in the plasma membrane of eggs is indicated by the presence of cholera toxin and pertussis toxin substrates (Turner et al., 1987) and a protein that reacts with a monoclonal antibody raised against the amino-terminal region of mammalian p21-H *ras*. Further support is indicated by the demonstration that microinjection of GTP-γ-S causes a transient increase in $[Ca^{2+}]_i$ very similar to that seen at fertilization (Swann et al., 1987). Also, microinjection of GTP-γ-S causes cortical granule exocytosis in sea urchin eggs, and GDP-β-S prevents envelope elevation on fertilization (Turner et al., 1986).

Intracellular Ca^{2+} Oscillations

Numerous publications have documented that the intracellular concentration of Ca^{2+} often oscillates when cells are exposed to agonist concentrations close to threshold. Such oscillations in Ca^{2+} can be measured directly (Cuthbertson and Cobbold, 1985) or inferred indirectly from the measurement of the membrane potential (Igusa and Miyazaki, 1986; Woods et al., 1987). $I(1,4,5)P_3$ may play a role in inducing these oscillations because several agonists that are known to stimulate the hydrolysis of inositol lipids also initiate oscillatory activity in appropriate cells (Sauve et al., 1987). Furthermore, direct injection of $I(1,4,5)P_3$ into *Xenopus* oocytes (Oron et al., 1985; Parker and Miledi, 1986) and guinea pig hepatocytes (Capiod et al., 1987) leads to oscillations of membrane current. The action of $I(1,4,5)P_3$ appears to be mediated via intracellular release of Ca^{2+} because it can be inhibited by injection of EGTA into target cells prior to injection of $I(1,4,5)P_3$ or by exposure to an appropriate agonist. It is not affected by removal of extracellular Ca^{2+} or by addition of Ca^{2+} channel antagonists.

In *Xenopus* oocytes, Berridge, Cobbold, and Cuthbertson (1988) found that iontophoretic injection of $I(1,4,5)P_3$ led to an immediate single peak of depolarization (dependent on intracellular Ca^{2+}), which was followed after a latency of about 60 sec by a burst of oscillatory activity. The frequency, amplitude, and duration of the oscillatory burst were dependent on the amount of $I(1,4,5)P_3$ injected (i.e., the duration of iontophoresis). Attempts to induce the oscillatory response by varying the frequency rather than the duration of the iontophoresis were unsuccessful.

However, the frequency of $I(1,4,5)P_3$ application had a remarkable effect on the initial peak of depolarization. This peak of depolarization was markedly attenuated when $I(1,4,5)P_3$ injections were repeated at shorter intervals. There was no attenuation if the injections were repeated every 2 min. The authors' interpretation of these results is that the initial depolarization results from the release of Ca^{2+} from an $I(1,4,5)P_3$-sensitive compartment, which is rapidly desensitized to $I(1,4,5)P_3$. The subsequent oscillatory activity is suggested to reflect the periodic uptake and release of Ca^{2+} by an $I(1,4,5)P_3$-insensitive compartment. They propose that the $I(1,4,5)P_3$-insensitive compartment, when overfilled, releases Ca^{2+}, possibly as a result of Ca^{2+}-dependent Ca^{2+} release. The lag between the initial $I(1,4,5)P_3$-induced depolarization and subsequent oscillatory activity represents the time required for the overfilling of the $I(1,4,5)P_3$-insensitive compartment.

It should be noted that oscillatory patterns, although showing considerable variations within a population of cells, exhibit remarkable similarities in the same cell upon repetitive stimulation with the same agonist. This phenomenon has been interpreted to indicate that each cell has its own specific "calcium fingerprint" (Prentki et al., 1988). However, the physiological significance of these "calcium fingerprints" remains to be elucidated.

ACKNOWLEDGMENTS

The author wishes to thank Rajendra Rana and Ellen Bergstrom for assistance in preparation of the figures and Karen Wipperfurth for her dedication and skill in preparation of the manuscript. This review was written during the tenure of NIH Grants HL16318, GM33850, and DA03699.

REFERENCES

Abdel-Latif, A.A. (1986). Calcium-mobilizing receptors, polyphosphoinositides, and the generation of second messengers. Pharmacol. Rev. 38, 227–272.

Abdel-Latif, A.A., Akhtar, R.A., & Hawthorne, J.N. (1977). Acetylcholine increases the breakdown of trisphosphoinositide in rabbit iris muscle prelabelled with [^{32}P]phosphate. Biochem. J. 162, 61–73.

Ackermann, K.E., Gish, B.G., Honchar, M.P., & Sherman, W.R. (1987). Evidence that inositol 1-phosphate in brain of lithium-treated rats results mainly from phosphatidylinositol metabolism. Biochem. J. 242, 517–524.

Akhtar, R.A., & Abdel-Latif, A.A. (1980). Requirement for calcium ions in acetylcholine-stimulated phosphodiesteratic cleavage of phosphatidylmyo-inositol 4,5-bisphosphate in rabbit iris smooth muscle. Biochem. J. 142, 599–604.

Akhtar, R.A., & Abdel-Latif, A.A. (1984). Carbachol causes rapid phosphodiesteratic cleavage of phosphatidylinositol 4,5-bisphosphate and accumulation of inositol phosphates in rabbit iris smooth muscle; prazasin inhibits noradrenaline and ionophore A23187-stimulated accumulation of inositol phosphates. Biochem. J. 224, 291–300.

Allison, J.H., & Stewart, M.A. (1971). Reduced brain inositol in lithium-treated rats. Nature 233, 267–268.

Andersson, T., Schlegel, W., Monod, A., Krause, K.H., Stendahl, O., & Lew, D.P. (1986). Leukotriene

B₄ stimulation of phagocytes results in the formation of inositol 1,4,5-trisphosphate. A second messenger for Ca^{2+} mobilization. Biochem. J. 240, 333–340.

Ashendel, C.L., Staller, J.M., & Boutwell, R.K. (1983). Protein kinase activity associated with a phorbol ester receptor purified from rat brain. Cancer Res. 43, 4333–4337.

Avissar, S., Schreiber, G., Danon, A., & Belmaker, R.H. (1988). Lithium inhibits adrenergic and cholinergic increases in GTP binding in rat cortex. Nature 331, 440–442.

Baker, R., & Thompson, W. (1972). Potential distribution and turnover of fatty acids in phosphatidic acid, phosphoinositides, phosphatidylcholine and phosphatidylethanolamine in rat brain *in vivo*. Biochim. Biophys. Acta 270, 489–503.

Bansal, V.S., Inhorn, R.C., & Majerus, P.W. (1987). The metabolism of inositol 1,3,4-trisphosphate to inositol 1,3-bisphosphate. J. Biol. Chem. 262, 9444–9447.

Banschbach, M.W., Geison, R.L., & Hokin-Neaverson, M. (1974). Acetylcholine increases the level of diglyceride in mouse pancreas. Biochem. Biophys. Res. Commun. 58, 714–718.

Banschbach, M.W., Geison, R.L., & Hokin-Neaverson, M. (1981). Effects of cholinergic stimulation on levels and fatty acid composition of diacylglycerols in mouse pancreas. Biochim. Biophys. Acta 663, 34–45.

Banschbach, M.W., & Hokin-Neaverson, M. (1980). Acetylcholine promotes the synthesis of prostaglandin E in mouse pancreas. FEBS Lett. 117, 131–132.

Batty, I.R., Nahorski, S.R., & Irvine, R.F. (1985). Rapid formation of inositol 1,3,4,5-tetrakisphosphate following muscarinic receptor stimulation of rat cerebral cortical slices. Biochem. J. 232, 211–215.

Bauduin, H., Galand, N., & Boeynaems, J.M. (1981). *In vitro* stimulation of prostaglandin synthesis in the rat pancreas by carbamylcholine, caerulein, and secretin. Prostaglandins 22, 35–51.

Baukal, A.J., Guillemette, G., Rubin, R., Spat, A., & Catt, K.J. (1985). Binding sites for inositol trisphosphate in the bovine adrenal cortex. Biochem. Biophys. Res. Commun. 133, 532–538.

Bell, J.D., Buxton, I.L.O., & Brunton, L.L. (1985). Enhancement of adenylate cyclase activity in S49 lymphoma cells by phorbol esters. Putative effect of C-kinase on α-S-GTP catalytic subunit interaction. J. Biol. Chem. 260, 2625–2628.

Berridge, M.J. (1984). Inositol trisphosphate and diacylglycerol as second messengers. Biochem. J. 220, 345–360.

Berridge, M.J. (1986). Inositol phosphates as second messengers. In: Receptor Biochemistry and Methodology (Venter, J.C., & Harrison, L.C., eds.), Vol. 7, Phosphoinositides and Receptor Mechanisms (Putney, J.W. Jr., ed.), pp. 25–45. Alan R. Liss, New York.

Berridge, M.J. (1987a). Inositol lipids and cell proliferation. Biochim. Biophys. Acta 907, 33–45.

Berridge, M.J. (1987b). Inositol trisphosphate and diacylglycerol: Two interacting second messengers. Annu. Rev. Biochem. 56, 159–193.

Berridge, M.J., Cobbold, P.H., & Cuthbertson, K.S.R. (1988). Spatial and temporal aspects of cell signalling. Phil. Trans. R. Soc. Lond. B320, 325–343.

Berridge, M.J., Heslop, J.P., Irvine, R.F., & Brown, K.D. (1984). Inositol trisphosphate formation and calcium mobilization in Swiss 3T3 cells in response to platelet-derived growth factor. Biochem. J. 222, 195–201.

Berridge, M.J., & Irvine R.F. (1984). Inositol trisphosphate, a novel second messenger in cellular signal transduction. Nature 312, 314–321.

Biden, T.J., Vallar, L., & Wollheim, C.B. (1988). Regulation of inositol 1,4,5-trisphosphate metabolism in RINm5F cells. Biochem. J. 251, 435–440.

Biden, T.J., & Wollheim, C.B. (1986). Ca^{2+} regulates the inositol tris/tetrakisphosphate pathway in intact and broken preparations of insulin-secreting RINm5F cells. J. Biol. Chem. 261, 11931–11934.

Biden, T.J., Wollheim, C.B., & Schlegel, W. (1986). Inositol 1,4,5-trisphosphate and intracellular Ca^{2+} homeostasis in clonal pituitary cells (GH3). Translocation of Ca^{2+} into mitochondria from a functionally discrete portion of the nonmitochondrial store. J. Biol. Chem. 261, 7223–7229.

Billah, M.M., Lapetina, E.G., & Cuatrecasas, P. (1980). Phospholipase A₂ and phospholipase C activities of platelets. Differential substrate specificity, Ca^{2+} requirement, pH dependence, and cellular localization. J. Biol. Chem. 255, 10227–10231.

Bills, T.K., Smith, J.B., & Silver, M.J. (1977). Selective release of arachidonic acid from the phospholipids of human platelets in response to thrombin. J. Clin. Invest. 60, 1–6.

Bishop, J.M. (1985). Viral oncogenes. Cell 42, 23–38.

Bocckino, S.B., Blackmore, P.F., & Exton, J.H. (1985). Stimulation of 1,2-diacylglycerol accumulation in hepatocytes by vasopressin, epinephrine, and angiotensin II. J. Biol. Chem. 260, 14201–14207.

Bond, M., Kitazawa, T., Somlyo, A.P., & Somlyo, A.V. (1984). Release and recycling of calcium by the sarcoplasmic reticulum in guinea-pig portal vein smooth muscle. J. Physiol. 355, 677–695.

Bradford, P.G., & Rubin, R.P. (1986). Guanine nucleotide regulation of phospholipase C activity in permeabilized rabbit neutrophils. Inhibition by pertussis toxin and sensitization to submicromolar calcium concentrations. Biochem. J. 239, 97–102.

Brockerhoff, H., & Ballou, C.E. (1962). Phosphate incorporation in brain phosphoinositides. J. Biol. Chem. 237, 49–52.

Burch, R.M., Luini, A., & Axelrod, J. (1986a). Phospholipase A_2 and phospholipase C are activated by distinct GTP-binding proteins in response to $\alpha 1$-adrenergic stimulation in FRTL5 thyroid cells. Proc. Natl. Acad. Sci. USA 83, 7201–7205.

Burch, R.M., Luini, A., Mais, D.E., Corda, D., Vanderhoek, J.Y., Kohn, L.D., & Axelrod, J. (1986b). $\alpha 1$-Adrenergic stimulation of arachidonic acid release and metabolism in a rat thyroid cell line. Mediation of cell replication by prostaglandin E_2. J. Biol. Chem. 261, 11236–11241.

Burgess, G.M., McKinney, J.S., Irvine, R.F., & Putney, J.W. Jr. (1985). Inositol 1,4,5-trisphosphate and inositol 1,3,4-trisphosphate formation in Ca^{2+}-mobilizing-hormone-activated cells. Biochem. J. 232, 237–243.

Busa, W.B., Ferguson, J.E., Joseph, S.K., Williamson, J.R., & Nuccitelli, R. (1985). Activation of frog (Xenopus laevis) eggs by inositol trisphosphate. I. Characterization of Ca^{2+} release from intracellular stores. J. Cell Biol. 101, 677–682.

Capiod, T., Field, A.C., Ogden, D.C., & Sandford, C.A. (1987). Internal perfusion of guinea–pig hepatocytes with buffered Ca^{2+} or inositol 1,4,5-trisphosphate mimics noradrenaline activation of K^+ and Cl conductances. FEBS Lett. 217, 247–252.

Carney, D.H., Scott, D.L., Gordon, E.A., & Labelle, E.F. (1985). Phosphoinositides in mitogenesis: Neomycin inhibits thrombin-stimulated phosphoinositide turnover and initiation of cell proliferation. Cell 42, 488–497.

Castanga, M., Takai, Y., Kaibuchi, K., Sano, K., Kikkawa, U., & Nishizuka, Y. (1982). Direct activation of calcium-activated, phospholipid-dependent protein kinase by tumor-promoting phorbol esters. J. Biol. Chem. 257, 7847–7851.

Chan, C.C., Duhamel, L., & Ford-Hutchinson, A. (1985). Leukotriene B_4 and 12-hydroxyeicosatetraenoic acid stimulate epidermal proliferation in vivo in the guinea pig. J. Invest. Dermatol. 85, 333–334.

Chan, K.M., & Turk, J. (1987). Mechanism of arachidonic acid-induced Ca^{2+} mobilization from rat liver microsomes. Biochim. Biophys. Acta 928, 186–193.

Cheah, A.M. (1981). Effect of long-chain unsaturated fatty acids on the calcium transport of sarcoplasmic reticulum. Biochim. Biophys. Acta 648, 113–119.

Chuang, L.F., Cooper, R.H., Yau, P., Bradbury, E.M., & Chuang, R.Y. (1987). Protein kinase C phosphorylates leukemia RNA polymerase II. Biochem. Biophys. Res. Commun. 145, 1376–1383.

Ciapa, B., & Whitaker, M. (1986). Two phases of inositol polyphosphate and diacylglycerol production at fertilization. FEBS Lett. 195, 347–351.

Cockcroft, S., & Allan, D. (1984). The fatty acid composition of phosphatidylinositol, phosphatide and 1,2-diacylglycerol in stimulated human neutrophils. Biochem. J. 222, 557–559.

Colard, O., Breton, M., & Bereziat, G. (1986). Arachidonate mobilization in diacyl, alkylacyl and alkenylacyl phospholipids on stimulation of rat platelets by thrombin and the Ca^{2+} ionophore A23187. Biochem. J. 233, 691–695.

Colodzin, M., & Kennedy, E.P. (1965). Biosynthesis of diphosphoinositide in brain. J. Biol. Chem. 240, 3771–3780.

Connolly, T.M., Bansal, V.S., Bross, T.E., Irvine, R.F., & Majerus, P.W. (1987). The metabolism of tris- and tetraphosphates of inositol by 5-phosphomonoesterase and 3-kinase enzymes. J. Biol. Chem. 262, 2146–2149.

Connolly, T.M., Bross, T.E., & Majerus, P.W. (1985). Isolation of a phosphomonoesterase from human platelets that specifically hydrolyses the 5-phosphate of inositol 1,4,5-trisphosphate. J. Biol. Chem. 260, 7868–7874.

Connolly, T.M., Lawing, W.J. Jr., & Majerus, P.W. (1986a). Protein kinase C phosphorylates human platelet inositol trisphosphate 5′-phosphomonoesterase, increasing the phosphatase activity. Cell 46, 951–958.

Connolly, T.M., Wilson, D.B., Bross, T.E., & Majerus, P.W. (1986b). Isolation and characterization of the inositol cyclic phosphate products of phosphoinositide cleavage by phospholipase C. Metabolism in cell-free extracts. J. Biol. Chem. 261, 122–126.

Cosgrove, D.J. (1969). Ion-exchange chromatography of inositol polyphosphates. Ann. N.Y. Acad. Sci. 165, 677–686.

Coussens, L., Parker, P.J., Rhee, L., Yang-Feng, T.L., Waterfield, E., Chen, M.D., Francke, U., & Ullrich, A. (1986). Multiple, distinct forms of bovine and human protein kinase C suggest diversity in cellular signaling pathways. Science 233, 859–866.

Creba, J.A., Downes, C.P., Hawkins, P.T., Brewster, G., Michell, R.H., & Kirk, C.J. (1983). Rapid breakdown of phosphatidylinositol 4-phosphate and phosphatidylinositol 4,5-bisphosphate in rat hepatocytes stimulated by vasopressin and other Ca^{2+}-mobilizing hormones. Biochem. J. 212, 733–747.

Crossley, I., Swann, K., Chambers, E., & Whitaker, M. (1988). Activation of sea urchin eggs by inositol phosphates is independent of external calcium. Biochem. J. 252, 257–262.

Cuthbertson, K.S.R., & Cobbold, P.H. (1985). Phorbol ester and sperm activate mouse oocytes by inducing sustained oscillations in cell Ca^{2+}. Nature 316, 541–542.

Davitz, M.A., Herald, D., Shek, S., Krakow, J., Englund, P.T., & Nussenzweig, V. (1987). A glycan-phosphatidylinositol-specific phospholipase D in human serum. Science 238, 81–84.

Dawson, R.M.C. (1959). Studies on the enzymic hydrolysis of monophosphoinositide by phospholipase preparations from *P. notatum* and ox pancreas. Biochim. Biophys. Acta 33, 68–77.

Deckmyn, H., Tu, S.-M., & Majerus, P.W. (1986). Guanine nucleotides stimulate soluble phosphoinositide-specific phospholipase C in the absence of membranes. J. Biol. Chem. 261, 16553–16558.

Diringer, H., & Friis, R.R. (1977). Changes in phosphatidylinositol metabolism correlated with growth state of normal and Rous sarcoma virus-transformed Japanese quail cells. Cancer Res. 37, 2978–2984.

Dixon, J.F., & Hokin, L.E. (1984). Secretagogue-stimulated phosphatidylinositol breakdown in the exocrine pancreas liberates arachidonic acid, stearic acid, and glycerol by sequential actions of phospholipase C and diacylglycerol lipase. J. Biol. Chem. 259, 14418–14425.

Dixon, J.F., & Hokin, L.E. (1985). The formation of inositol 1,2-cyclic phosphate on agonist stimulation of phosphoinositide breakdown in mouse pancreatic minilobules. J. Biol. Chem. 260, 16068–16071.

Dixon, J.F., & Hokin, L.E. (1987a). Inositol 1,2-cyclic 4,5-trisphosphate is formed in the rat parotid gland on muscarinic stimulation. Biochem. Biophys. Res. Commun. 149, 1208–1213.

Dixon, J.F., & Hokin, L.E. (1987b). Inositol 1,2-cyclic 4,5-trisphosphate: Concentration relative to inositol 1,4,5-trisphosphate in pancreatic minilobules on stimulation with carbamylcholine in the absence of lithium. Possible role as a second messenger in long- but not short-term responses. J. Biol. Chem. 262, 13892–13895.

Dixon, J.F., & Hokin, L.E. (1989). Kinetic analysis of the formation of inositol 1:2-cyclic phosphate in carbachol-stimulated pancreatic minilobules. Half is formed by direct phosphodiesteratic cleavage of phosphatidylinositol. J. Biol. Chem. 264, 11721–11724.

Dougherty, R.W., & Niedel, J.E. (1986). Cytosolic calcium regulates phorbol diester binding affinity in intact phagocytes. J. Biol. Chem. 261, 4097–4100.

Douglas, W.W. (1974). Involvement of calcium in exocytosis and the exocytosis-vesiculation sequence. Biochem. Soc. Symp. 39, 1–28.

Downes, C.P., & Michell, R.H. (1985). Inositol phospholipid breakdown as a receptor-controlled generator of second messengers. In: Molecular Mechanisms of Transmembrane Signalling (Cohen, P., & Houslay, M.D., eds.), pp. 4–56. Elsevier Science Publishing Co., Amsterdam.

Downes, C.P., Mussat, M.C., & Michell, R.H. (1982). The inositol trisphosphate phosphomonoesterase of the human erythrocyte membrane. Biochem. J. 203, 169–177.

Downes, C.P., & Stone, M.A. (1986). Lithium-induced reduction in intracellular inositol supply in cholinergically stimulated parotid gland. Biochem. J. 234, 199–204.

Downward, J., deGunzburg, J., Riehl, R., & Weinberg, R.A. (1988). p21 ras-induced responsiveness of phosphatidylinositol turnover to bradykinin is a receptor number effect. Proc. Natl. Acad. Sci. USA 85, 5774–5778.

Drummond, A.H. (1985). Bidirectional control of cytosolic free calcium by thyrotropin-releasing hormone in pituitary cells. Nature 315, 752–755.

Drummond, A.H. (1987). Lithium and inositol lipid-linked signalling mechanisms. Trends Pharmacol. Sci. 8, 129–133.

Drummond, A.H., Joels, L.A., & Hughes, P.J. (1987). The interaction of lithium ions with inositol lipid signalling systems. Biochem. Soc. Trans. 15, 32–35.

Durell, J., Garland, J.T., & Friedel, R.O. (1969). Acetylcholine action: Biochemical aspects. Science 165, 862–866.

Enyedi, P., & Williams, G.H. (1988). Heterogenous inositol tetrakisphosphate binding sites in the adrenal cortex. J. Biol. Chem. 263, 7940–7942.

Fain, J.N. (1987). Activation of phosphoinositide specific phospholipase C by ligands in the presence of guanine nucleotides. In: Mechanisms of Signal Transduction by Hormones and Growth Factors (Cabot, M.C., & McKeehan, W.L., eds.), pp. 133–147. Alan R. Liss, New York.

Fain, J.N., & Berridge, M.J. (1979a). Relationship between hormonal activation of phosphatidylinositol hydrolysis, fluid secretion and calcium flux in the blowfly salivary gland. Biochem. J. 178, 45–58.

Fain, J.N., & Berridge, M.J. (1979b). Relationship between phosphatidylinositol synthesis and recovery of 5-hydroxytryptamine-responsive Ca^{2+} flux in blowfly salivary glands. Biochem. J. 180, 655–661.

Farese, R.V., Konda, T.S., Davis, J.S., Standaert, M.L., Pollet, R.J., & Cooper, R. (1987). Insulin rapidly increases diacylglycerol by activating de novo phosphatidic acid synthesis. Science 236, 586–589.

Fein, A., Payne, R., Corson, D.W., Berridge, M.J., & Irvine, R.F. (1984). Photoreceptor excitation and adaptation by inositol 1,4,5-trisphosphate. Nature 311, 157–160.

Fisher, D.B., & Mueller, G.C. (1971). Studies of the mechanism by which PHA stimulates phospholipid metabolism of human lymphocytes. Biochim. Biophys. Acta 248, 434–448.

Fry, M.J., Gebhardt, A., Parker, P.J., & Foulkes, G. (1985). Phosphatidylinositol turnover and transformation of cells by Abelson murine lukemia virus. EMBO J. 4, 3173–3178.

Garland, L.G., Bonser, R.W., & Thompson, N.T. (1987). Protein kinase C inhibitors are not selective. Trends Pharmacol. Sci. 8, 162–164.

Gilman, A.G. (1987). G proteins: Transducers of receptor-generated signals. Annu. Rev. Biochem. 56, 615–649.

Goldberg, N.D., & Haddox, M. (1977). Cyclic GMP metabolism and involvement in biological regulation. Annu. Rev. Biochem. 46, 823–896.

Goldman, D.W., Gilford, L.A., Olson, D.M., & Goetzl, E.J. (1985). Tranduction by leukotriene B4 receptors of increases in cytosolic calcium in human polymorphonuclear leukocytes. J. Immunol. 135, 525–530.

Gomperts, B.D. (1983). Involvement of guanine nucleotide-binding protein in the gating of Ca^{2+} by receptors. Nature 306, 64–66.

Graham, R.A., Meyer, R.A., Szwergold, B.S., & Brown, T.R. (1987). Observation of myo-inositol 1,2-(cyclic) phosphate in a Morris hepatoma by ^{31}P NMR. J. Biol. Chem. 262, 35–37.

Habenicht, A.J.R., Glomset, J.A., Goerig, M., Gronwald, R., Grulich, J., Loth, U., & Schettler, G. (1985). Cell cycle-dependent changes in arachidonic acid and glycerol metabolism in Swiss 3T3 cells stimulated by platelet-derived growth factor. J. Biol. Chem. 260, 1370–1373.

Habenicht, A.J.R., Glomset, J.A., King, W.C., Nist, C., Mitchell, C.D., & Ross, R. (1981). Early changes in phosphatidylinositol and arachidonic acid metabolism in quiescent Swiss 3T3 cells stimulated to divide by platelet-derived growth factor. J. Biol. Chem. 256, 12329–12335.

Hansen, C.A., Mah, S., & Williamson, J.R. (1986). Formation and metabolism of inositol 1,3,4,5-tetrakisphosphate in liver. J. Biol. Chem. 261, 8100–8103.

Hansson, A., Serhan, C.N., Haeggstrom, J., Ingelman-Sundberg, M., & Samuelsson, B. (1986). Activation of protein kinase C by lipoxin A and other eicosanoids. Intracellular action of oxygenation products of arachidonic acid. Biochem. Biophys. Res. Commun. 134, 1215–1222.

Harwood, J.L., & Hawthorne, J.N. (1969). The properties and subcellular distribution of phosphatidylinositol kinase in mammalian tissues. Biochim. Biophys. Acta 171, 75–88.

Haslam, R.J., & Davidson, M.M.L. (1984). Guanine nucleotides decrease the free Ca^{2+} required for secretion of serotonin from permeabilized platelets. Evidence of a role of GTP-binding protein in platelet activation. FEBS Lett. 174, 90–95.

Hawkins, P.T., Stephens, L., & Downes, C.P. (1986). Rapid formation of inositol 1,3,4,5-tetrakisphosphate and inositol 1,3,4-trisphosphate in rat parotid glands may both result indirectly from receptor-stimulated release of inositol 1,4,5-trisphosphate from phosphatidylinositol 4,5-bisphosphate. Biochem. J. 238, 507–516.

Heslop, J.P., Irvine, R.F., Tashjian, A.H. Jr., & Berridge, M.J. (1985). Inositol tetrakis- and pentakis-phosphates in GH$_4$ cells. J. Exp. Biol. 119, 395–401.

Hill, T.D., Dean, N.M., & Boyenton, A.L. (1988). Inositol 1,3,4,5-tetrakisphosphate induces Ca^{2+} sequestration in rat liver cells. Science 242, 1176–1178.

Hirata, M., Sasaguri, T., Hamachi, T., Hashimoto, T., Kukita, M., & Koga, T. (1985). Irreversible inhibition of Ca^{2+} release in saponin-treated macrophages by the photoaffinity derivative of inositol 1,4,5-trisphosphate. Nature 317, 723–725.

Hokin, L.E. (1966). Effects of calcium omission on acetylcholine-stimulated amylase secretion and phospholipid synthesis in pigeon pancreas slices. Biochim. Biophys. Acta 115, 219–221.

Hokin, L.E. (1968). Dynamic aspects of phospholipids during protein secretion. Int. Rev. Cytol. 23, 187–208.

Hokin, L.E. (1985). Receptors and phosphoinositide-generated second messengers. Annu. Rev. Biochem. 54, 205–235.

Hokin, L.E. (1987). The road to the phosphoinositide-generated second messengers. Trends Pharmacol. Sci. 8, 53–56.

Hokin, L.E., & Hokin, M.R. (1955). Effects of acetylcholine on the turnover of phosphoryl units in individual phospholipids of pancreas slices and brain cortex slices. Biochim. Biophys. Acta 18, 102–110.

Hokin, L.E., & Hokin, M.R. (1958a). Phosphoinositides and protein secretion in pancreas slices. J. Biol. Chem. 233, 805–810.

Hokin, L.E., & Hokin, M.R. (1958b). Acetylcholine and the exchange of inositol and phosphate in brain phosphoinositide. J. Biol. Chem. 233, 818–821.

Hokin, L.E., & Hokin, M.R. (1964a). The incorporation of ^{32}P from [γ-^{32}P]adenosine triphosphate into polyphosphoinositides and phosphatidic acid in erythrocyte membranes. Biochim. Biophys. Acta 84, 563–575.

Hokin, M.R., & Hokin, L.E. (1953). Enzyme secretion and the incorporation of ^{32}P into the phospholipids of pancreas slices. J. Biol. Chem. 203, 967–977.

Hokin, M.R., & Hokin, L.E. (1954). Effects of acetylcholine on phospholipids in the pancreas. J. Biol. Chem. 209, 549–558.

Hokin, M.R., & Hokin, L.E. (1964b). Interconversions of phosphatidylinositol and phosphatidic acid involved in the response to acetylcholine in the salt gland. In: Metabolism and Physiological Significance of Lipids (Dawson, R.M.C., & Rhodes, D.N., eds.), pp. 423–434. John Wiley and Sons, New York.

Hokin-Neaverson, M. (1974). Acetylcholine causes a net decrease in phosphatidylinositol and a net increase in phosphatidic acid in mouse pancreas. Biochem. Biophys. Res. Commun. 58, 763–768.

Hokin-Neaverson, M. (1977). Metabolism and role of phosphatidyl-inositol in acetylcholine-stimulated membrane function. Adv. Exp. Biol. Med. 83, 429–446.

Hughes, A.R., Takemura, H., & Putney, J.W. Jr. (1988). Kinetics of inositol 1,4,5-trisphosphate and inositol cyclic 1:2,4,5-trisphosphate metabolism in intact rat parotid acinar cells. Relationship to calcium signalling. J. Biol. Chem. 263, 10314–10319.

Ian, M.G. (1985). Oncogenes, ions and phospholipids. Amer. J. Physiol. 248, C3–C11.

Igusa, Y., & Miyazaki, S.I. (1986). Periodic increase of cytoplasmic free calcium in fertilized hamster eggs measured with calcium–sensitive electrodes. J. Physiol. 377, 193–205.

Imai, A., & Gershengorn, M.C. (1986). Phosphatidylinositol 4,5-bisphosphate turnover is transient while phosphatidylinositol turnover is persistent in thyrotropin-releasing hormone-stimulated rat pituitary cells. Proc. Natl. Acad. Sci. USA 83, 8540–8544.

Imboden, J.B., Shoback, D.M., Pattison, G., & Stobo, J.D. (1986). Cholera toxin inhibits the T-cell antigen receptor-mediated increases in inositol trisphosphate and cytoplasmic free calcium. Proc. Natl. Acad. Sci. USA 83, 5673–5677.

Inhorn, R.C., Bansal, V.S., & Majerus, P.W. (1987). Pathway for inositol 1,3,4-trisphosphate and 1,4-bisphosphate metabolism. Proc. Natl. Acad. Sci. USA 84, 2170–2174.

Irvine, R.F. (1982). How is the level of free arachidonate controlled in mammalian cells? Biochem. J. 204, 3–16.

Irvine, R.F., Letcher, A.J., Heslop, J.P., & Berridge, M.J. (1986a). The inositol tris/tetrakisphosphate pathway-demonstration of Ins(1,4,5)P3 3-kinase activity in animal tissues. Nature 320, 631–634.

Irvine, R.F., Letcher, A.J., Lander, D.J., & Berridge, M.J. (1986b). Specificity of inositol phosphate-stimulated Ca^{2+} mobilization from Swiss-mouse 3T3 cells. Biochem. J. 240, 301–304.

Irvine, R.F., Letcher, A.J., Lander, D.J., & Downes, C.P. (1984). Inositol trisphosphates in carbachol-stimulated rat parotid glands. Biochem. J. 223, 237–243.

Irvine, R.F., & Moor, R.M. (1986). Micro-injection of inositol 1,3,4,5-tetrakisphosphate activates sea urchin eggs by a mechanism dependent on external Ca^{2+}. Biochem. J. 240, 917–920.

Ishii, H., Connolly, T.M., Bross, T.E., & Majerus, P.W. (1986). Inositol cyclic trisphosphate [inositol l,2-(cyclic)-4,5-trisphosphate] is formed upon thrombin stimulation of human platelets. Proc. Natl. Acad. Sci. USA 83, 6397–6401.

Jackson, T.R., Hallam, T.J., Downes, C.P., & Hanley, M.R. (1987). Receptor coupled events in bradykinin action: Rapid production of inositol phosphates and regulation of cytosolic free Ca^{2+} in a neural cell line. EMBO J. 6, 49–54.

Jaffe, L.F. (1983). Sources of calcium in egg activation: A review and a hypothesis. Dev. Biol. 99, 256–276.

Jones, L.M., & Michell, R.H. (1974). Breakdown of phosphatidylinositol provoked by muscarinic cholinergic stimulation in rat parotid-gland fragments. Biochem. J. 142, 583–590.

Joseph, S.K., Hansen, C.A., & Williamson, J.R. (1987). Inositol 1,3,4,5-tetrakisphosphate increases the duration of the inositol 1,4,5-trisphosphate-mediated Ca^{2+} transient. FEBS Lett. 219, 125–129.

Joseph, S.K., & Williamson, J.R. (1986). Characteristics of inositol trisphosphate-mediated Ca^{2+} release from permeabilized hepatocytes. J. Biol. Chem. 261, 14658–14664.

Jy, W., & Haynes, D.H. (1987). Thrombin-induced calcium movements in platelet activation. Biochim. Biophys. Acta 929, 88–102.

Katada, T., Gilman, A.G., Watanabe, Y., Bauer, S., & Jakobs, K.H. (1985). Protein kinase C phosphorylates the inhibitory guanine-nucleotide-binding regulatory component and apparently suppresses its function in hormonal inhibition of adenylate cyclase. Eur. J. Biochem. 151, 431–437.

Katz, B. (1969). The Release of Neural Transmitter Substances, pp. 1–60. Charles C. Thomas, Springfield.

Keenan, R.W., & Hokin, L.E. (1964). The positional distribution of the fatty acids in phosphatidyl inositol. Biochim. Biophys. Acta 84, 458–460.

Kelly, K.L., Mato, J.M., Merida, I., & Jarett, L. (1987). Glucose transport and antilipolysis are differentially regulated by the polar head group of an insulin-sensitive glycophospholipid. Proc. Natl. Acad. Sci. USA 84, 6404–6407.

Kemp, P., Hubscher, G., & Hawthorne, J.N. (1961a). A liver phospholipase hydrolysing phosphoinositides. Biochim. Biophys. Acta 31, 585–586.

Kemp, P., Hubscher, G., & Hawthorne, J.N. (1961b). Phosphoinositides. 3. Enzymic hydrolysis of inositol-containing phospholipids. Biochem. J. 79, 193–200.

Kennerly, D.A. (1987). Diacylglycerol metabolism in mast cells. J. Biol. Chem. 262, 16305–16313.

Kikkawa, U., Takai, Y., Tanaka, Y., Miyaka, R., & Nishizuka, Y. (1983). Protein kinase C as a possible receptor protein of tumor-promoting phorbol esters. J. Biol. Chem. 258, 11442–11445.

Kim, D., & Clapham, D.E. (1989). Potassium channels in cardiac cells activated by arachidonic acid and phospholipids. Science 244, 1174–1176.

King, W.G., & Rittenhouse, S.E. (1989). Inhibition of protein kinase C by staurosporine promotes elevated accumulations of inositol trisphosphates and tetrakisphosphate in human platelets exposed to thrombin. J. Biol. Chem. 264, 6070–6074.

Kitano, T., Hashimoto, T., Kikkawa, U., Ase, K., Saito, N., Tanaka, C., Ichimori, Y., Tsukamoto, K., & Nishizuka, Y. (1987). Monoclonal antibodies against rat brain protein kinase C and their application to immunochemistry in nervous tissues. J. Neurochem. 7, 1520–1525.

Knopf, J.L., Lee, M.H., Sultzman, L.A., Kriz, R.W., Loomis, C.R., Hewick, R.M., & Bell, R.M. (1986). Cloning and expression of multiple protein kinase C cDNAs. Cell 46, 491–502.

Kojima, I., Kojima, K., & Rasmussen, H. (1985). Role of calcium fluxes in the sustained phase of angiotensin II-mediated aldosterone secretion from adrenal glomerulosa cells. J. Biol. Chem. 260, 9177–9184.

Kolesnick, R.N., Musacchio, I., Thaw, C., & Gershengorn, M.C. (1984). Arachidonic acid mobilizes calcium and stimulates prolactin secretion from GH_3 cells. Amer. J. Physiol. 246, E458–E462.

Kolesnick, R.N., & Paley, A.E. (1987). 1,2-Diacylglycerols and phorbol esters stimulate phosphatidylcholine metabolism in GH_3 pituitary cells. J. Biol. Chem. 262, 9204–9210.

Korn, L.J., Siebel, C.W., McCormick, F., & Roth, R.A. (1987). Ras p21 as a potential mediator of insulin action in *Xenopus* oocytes. Science 236, 840–843.

Kraft, A.S., & Anderson, W.B. (1983). Phorbol esters increase the amount of Ca^{2+}, phospholipid-dependent protein kinase associated with plasma membrane. Nature 301, 621–623.

Kuo, J.F., Schatzman, R.C., Turner, R.S. & Mazzei, G.J. (1984). Phospholipid-sensitive Ca^{2+}-dependent protein kinase: A major protein phosphorylation system. Mol. Cell Endocrinol. 35, 65–73.

Lands, W.E.M., & Samuelsson, B. (1968). Phospholipid precursors of prostaglandins. Biochim. Biophys. Acta 164, 426–429.

Lapetina, E.G., Billah, M.M., & Cuatrecasas, P. (1980). Rapid acylation and deacylation of arachidonic acid into phosphatidic acid of horse neutrophils. J. Biol. Chem. 255, 966–970.

Lapetina, E.G., Reep, R., Ganong, B.R., & Bell, R.M. (1985). Exogenous sn-1,2-diacylglycerols containing saturated fatty acids function as bioregulators of protein kinase C in human platelets. J. Biol. Chem. 260, 1358–1361.

Laychock, S.G., & Putney, J.W. Jr. (1982). Roles of phospholipid metabolism in secretory cells. In: Cellular Regulation of Secretion and Release (Conn, P.M., ed.), pp. 53–105. Academic Press, New York.

Lee, C.H., Dixon, J.F., Reichman, M., Moummi, C., Los, G., & Hokin, L.E. Li^{+} increases accumulation of inositol 1,4,5-trisphosphate and inositol 1,3,4,5-tetrakisphosphate in cholinergically stimulated brain cortex slices in guinea pig, mouse and rat. Biochem. J. In press.

Lee, C.H., & Hokin, L.E. (1989). Inositol 1,2-cyclic 4,5-trisphosphate is an order of magnitude less potent than inositol 1,4,5-trisphosphate in mobilizing intracellular stores of calcium in mouse pancreatic acinar cells. Biochem. Biophys. Res. Commun. 159, 561–565.

Litosch, I., & Fain, J.N. (1986). Minireview. Regulation of phosphoinositide breakdown by guanine nucleotides. Life Sci. 39, 187–194.

Low, M.G., & Prasad, A.R.S. (1988). A phospholipase D specific for the phosphatidylinositol anchor of cell-surface proteins is abundant in plasma. Proc. Natl. Acad. Sci. USA 85, 980–984.

Low, M.G., & Saltiel, A.R. (1988). Structural and functional roles of glycosyl-phosphatidylinositol in membranes. Science 239, 268–275.

Macara, I.G., Marinetti, G.V., & Balduzzi, P.C. (1984). Transforming protein of avian sarcoma virus UR2 is associated with phosphatidylinositol kinase activity: Possible role in tumorigenesis. Proc. Natl. Acad. Sci. USA 81, 2728–2732.

Macara, I.G., Marinetti, G.V., Livingston, J.N., & Balduzzi, P.C. (1985). Lipid phosphorylating activities and tyrosine kinases: A possible role for phosphatidylinositol turnover in transformation. In: Growth Factors and Transformation (Feramisco, F., ed.), pp. 365–368. Cold Spring Harbor Laboratory, New York.

MacDonald, M.L., Kuenzel, E.A., Glomset, J.A., & Krebs, E.G. (1985). Evidence from two transformed cell lines that the phosphorylations of peptide and phosphatidylinositol are catalyzed by different proteins. Proc. Natl. Acad. Sci. USA, 82, 3993–3997.

Majerus, P.W., Connolly, T.M., Bansal, V.S., Inhorn, R.C., Ross, T.S., & Lips, D.L. (1988). Inositol phosphates: Synthesis and degradation. J. Biol. Chem. 263, 3051–3054.

Majerus, P.W., Connolly, T.M., Deckmyn, H., Ross, T.S., Bross, T.E., Ishii, H., Bansal, V.S., & Wilson, D.B. (1986). The metabolism of phosphoinositide-derived messenger molecules. Science 234, 1519–1526.

Majerus, P.W., Neufeld, E.J., & Wilson, D.B. (1984). Production of phosphoinositide-derived messengers. Cell 37, 701–703.

Matuoka, K., Fukami, K., Nakanishi, O., Kawai, S., & Takenawa, T. (1988). Mitogenesis in response to PDGF and bombesin abolished by microinjection of antibody to PIP_2. Science 239, 640–643.

McNeil, P.L., McKenna, M.P., & Taylor, D.L. (1985). A transient rise in cytosolic calcium follows stimulation of quiescent cells with growth factors and is inhibitable with phorbol myristate acetate. J. Cell Biol. 101, 372–379.

Meade, C.J., Turner, G.A., & Bateman, P.E. (1986). The role of polyphosphoinositides and their breakdown products in A23187-induced release of arachidonic acid from rabbit polymorphonuclear leucocytes. Biochem. J. 238, 425–431.

Messing, R.O., Carpenter, C.L., & Greenberg, D.A. (1986). Inhibition of calcium flux and calcium channel antagonist binding in the PC12 neural cell line by phorbol esters and protein kinase C. Biochem. Biophys. Res. Commun. 136, 1049–1056.

Meyer, T., Holowka, D., & Stryer, L. (1988). Highly cooperative opening of calcium channels by inositol 1,4,5-trisphosphate. Science 240, 653 656.

Michell, R.H. (1975). Inositol phospholipids and cell surface receptor function. Biochim. Biophys. Acta 415, 81–147.

Michell, R.H., & Hawthorne, J.N. (1965). The site of diphosphoinositide synthesis in rat liver. Biochem. Biophys. Res. Commun. 21, 333–338.

Morris, A.P., Gallacher, D.V., Irvine, R.F., & Petersen, O.H. (1987). Synergism of inositol trisphosphate and tetrakisphosphate in activating Ca^{2+}-dependent K^+ channels. Nature 330, 653–655.

Muallem, S., Schoeffield, M., Pandol, S., & Sachs, G. (1985). Inositol trisphosphate modification of ion transport in rough endoplasmic reticulum. Proc. Natl. Acad. Sci. USA 82, 4433–4437.

Nahorski, S.R., Kendall, D.A., & Batty, I. (1986). Receptors and phosphoinositide metabolism in the central nervous system. Biochem. Pharmacol. 35, 2447–2453.

Nakamura, T., & Ui, M. (1983). Suppression of passive cutaneous anaphylaxis by pertussis toxin, an islet-activating protein, as a result of inhibition of histamine release from mast cells. Biochem. Pharmacol. 32, 3435–3441.

Nakamura, T., & Ui, M. (1985). Simultaneous inhibitions of inositol phospholipid breakdown, arachidonic acid release, and histamine secretion in mast cells by islet-activating protein, pertussis toxin. J. Biol. Chem. 260, 3584–3593.

Nakashima, S., Tohmatsu, T., Hattori, H., Suganuma, A., & Nozawa, Y. (1987a). Guanine nucleotides stimulate arachidonic acid release by phospholipase A2 in saponin-permeabilized human platelets. J. Biochem. 101, 1055–1058.

Nakashima, S., Tohmatsu, T., Shirato, L., Takenaka, A., & Nozawa, Y. (1987b). Neomycin is a potent agent for arachidonic acid release in human platelets. Biochem. Biophys. Res. Comm. 146, 820–826.

Neufeld, E.J., & Majerus, P.W. (1983). Arachidonate release and phosphatidic acid turnover in stimulated platelets. J. Biol. Chem. 258, 2461–2467.

Niedel, J.E., & Blackshear, P.J. (1986). Protein kinase C. In: Receptor Biochemistry and Methodology (Venter, J.C., & Harrison, L.C., eds.), Vol. 7, Phosphoinositides and Receptor Mechanisms (Putney, J.W. Jr., ed.), pp. 47–88. Alan R. Liss, New York.

Niedel, J.E., Kuhn, L., & Vandenbark, G.R. (1983). Phorbol diester receptor copurifies with protein kinase C. Proc. Natl. Acad. Sci. USA 80, 36–40.

Nishizuka, Y. (1984). Turnover of inositol lipids and signal transduction. Science 225, 1365–1370.

Nishizuka, Y. (1986). Studies and perspectives of protein kinase C. Science 233, 305–312.

Nishizuka, Y. (1988). The heterogeneity and differential expression of multiple species of the protein kinase C family. Biofactors 1, 17–20.

Nishizuka, Y., Takai, Y., Kishimoto, A., Kikkawa, U., & Kaibuchi, K. (1984). Phospholipid turnover in hormone action. Recent Prog. Horm. Res. 40, 301–341.

Okano, Y., Ishizuka, Y., Nakashima, S., Tohmatsu, T., Takagi, H., & Nozawa, Y. (1985). Arachidonic acid release in rat peritoneal mast cells stimulated with antigen, ionophore A23187, and compound 48/80. Biochem. Biophys. Res. Commun. 127, 726–732.

Ono, Y., Kikkawa, U., Ogita, K., Fujii, T., Kurokawa, T., Asaoka, Y., Sekiguchi, K., Ase, K., Igarashi, K., & Nishizuka, Y. (1987). Expression and properties of two types of protein kinase C: Alternative splicing from a single gene. Science 236, 1116–1120.

Ordway, R.W., Walsh, J.V. Jr., & Singer, J.J. (1989). Arachidonic acid and other fatty acids directly activate potassium channels in smooth muscle cells. Science 244, 1176–1179.

Orellana, S.A., Solski, P.A., & Brown, J.H. (1985). Phorbol ester inhibits phosphoinositide hydrolysis and calcium mobilization in cultured astrocytoma cells. J. Biol. Chem. 260, 5236–5239.

Oron, Y., Dascal, N., Nadler, E., & Lupu, M. (1985). Inositol 1,4,5-trisphosphate mimics muscarinic response in *Xenopus* oocytes. Nature 313, 141–143.

Parker, I., & Miledi, R. (1986). Changes in intracellular calcium and in membrane currents evoked by injection of inositol trisphosphate into *Xenopus* oocytes. Proc. R. Soc. Lond. 228, 307–315.

Parker, I., & Miledi, R. (1987). Inositol trisphosphate activates a voltage-dependent calcium influx in *Xenopus* oocytes. Proc. R. Soc. Lond. 231, 27–36.

Parker, J., Daniel, L.W., & Waite, M. (1987). Evidence of protein kinase C involvement in phorbol diester-stimulated arachidonic acid release and prostaglandin synthesis. J. Biol. Chem. 262, 5385–5393.

Pasti, G., Lacal, J.-C., Warren, B.S., Aaronson, S.A., & Blumberg, P.M. (1986). Loss of mouse fibroblast cell response to phorbol esters restored by microinjected protein kinase C. Nature 324, 375–377.

Preiss, J., Bell, R.M., & Niedel, J.E. (1987). Diacylglycerol mass measurements in stimulated HL-60 phagocytes. J. Immunol. 138, 1542–1545.

Prentki, M., Glennon, M.C., Thomas, A.P., Morris, R.L., Matschinsky, F.M., & Corkey, B. (1988). Cell-specific patterns of oscillating free Ca^{2+} in carbamylcholine-stimulated insulinoma cells. J. Biol. Chem. 263, 11044–11047.

Rana, R.S., & Hokin, L.E. (1990). Role of phosphoinositides in transmembrane signalling. Physiol. Rev. 70, 115–164.

Rasmussen, H., Apfeldorf, W., Barrett, P., Takuwa, N., Zawalich, W., Kreutter, D., Park, S., & Takuwa, Y. (1986). Inositol lipids: Integration of cellular signalling systems. In: Receptor Biochemistry and Methodology (Venter, J.C., & Harrison, L.C., eds.), Vol. 7, Phosphoinositides and Receptor Mechanisms (Putney, J.W. Jr., ed.), pp. 109–147. Alan R. Liss, New York.

Rasmussen, H., & Barrett, P.Q. (1984). Calcium messenger system: An integrated view. Physiol. Rev. 64, 938–984.

Renard, D., Poggioli, J., Berthon, B., & Claret, M. (1987). How far does phospholipase C activity depend on the cell calcium concentration? A study of intact cells. Biochem. J. 243, 391–398.

Reynolds, E.E., & Dubyak, G.R. (1985). Activation of calcium mobilization and calcium influx by $\alpha 1$-adrenergic receptors in a smooth muscle cell line. Biochem. Biophys. Res. Commun. 130, 627–632.

Ringer, S. (1883). A further contribution regarding the influence of different constituents of the blood on the contraction of the heart. J. Physiol. 4, 29–42.

Rittenhouse, S.E. (1982). Inositol lipid metabolism in the responses of stimulated platelets. Cell Calcium 3, 311–322.

Rittenhouse, S.E., & Sasson, J.P. (1985). Mass changes in myoinositol trisphosphate in human platelets stimulated by thrombin. Inhibitory effects of phorbol ester. J. Biol. Chem. 60, 8657–8660.

Rittenhouse-Simmons, S. (1979). Production of diglyceride from phosphatidylinositol in activated human platelets. J. Clin. Invest. 63, 580–587.

Roman, I., Gmaj, P., Noweicka, C., & Angielski, S. (1979). Regulation of Ca^{2+}-efflux from kidney and liver mitochondria by unsaturated fatty acids and Na^{+} ions. Eur. J. Biochem. 102, 615–623.

Rosengurt, E. (1985). The mitogenic response of cultured 3T3 cells: Integration of early signals and synergistic effects in a unified framework. In: Molecular Aspects of Cellular Regulation (Cohen, P., & Houslay, M.D., eds.) Vol. 4, Molecular Mechanisms of Transmembrane Signalling, pp. 429–452. Elsevier Science Publishing Co., Amsterdam.

Ross, T.S., & Majerus, P.W. (1986). Isolation of D-myo-inositol 1:2-cyclic phosphate 2-inositol phosphohydrolase from human placenta. J. Biol. Chem. 261, 11119–11123.

Safran, A., Sagi-Eisenberg, R., Neumann, D., & Fuchs, S. (1987). Phosphorylation of the acetylcholine receptor by protein kinase C and identification of the phosphorylation site within the receptor δ subunit. J. Biol. Chem. 262, 10506–10510.

Saltiel, A.R., & Cuatrecasas, P. (1987). Insulin stimulates the generation from hepatic plasma membranes of modulators derived from an inositol glycolipid. Proc. Natl. Acad. Sci. USA 83, 5793–5797.

Saltiel, A.R., Fox, J.A., Sherline, P., & Cuatrecasas, P. (1986). Insulin-stimulated hydrolysis of a novel glycolipid generates modulators of cAMP phosphodiesterase. Science 233, 967–972.

Saltiel, A.R., Sherline, P., & Fox, J.A. (1987). Insulin-stimulated diacylglycerol production results from the hydrolysis of a novel phosphatidylinositol glycan. J. Biol. Chem. 262, 1116–1121.

Saltiel, A.R., & Sorbora-Cazan, L.R. (1987). Inositol glycan mimics the action of insulin on glucose utilization in rat adipocytes. Biochem. Biophys. Res. Commun. 149, 1084–1092.

Sano, K., Takai, Y., Yamanishi, J., & Nishizuka, Y. (1983). A role of calcium-activated phospholipid-dependent protein kinase in human platelet activation. Comparison of thrombin and collagen actions. J. Biol. Chem. 258, 2010–2013.

Santiago-Calvo, E., Mule, S., Redman, C.M., Hokin, M.R., & Hokin, L.E. (1964). The chromatographic separation of polyphosphoinositides and studies on their turnover in various tissues. Biochim. Biophys. Acta 84, 550–562.

Sautebin, L., Caruso, D., Galli, G., & Paoletti, R. (1983). Preferential utilization of endogenous arachidonate by cyclo-oxygenase in incubations of human platelets. FEBS Lett. 157, 173–178.

Sauve, R., Simoneau, C., Parent, L., Monette, R., & Roy, G. (1987). Oscillatory activation of calcium-dependent channels in HeLa cells induced by histamine H1 receptor stimulation: A single channel study. J. Membr. Biol. 96, 199–208.

Schulz, I., & Stolze, H.H. (1980). The exocrine pancreas: The role of secretagogues, cyclic nucleotides, and calcium in enzyme secretion. Annu. Rev. Physiol. 42, 127–156.

Sekar, M.C., Dixon, J.F., & Hokin, L.E. (1987). The formation of inositol 1,2-cyclic 4,5-trisphosphate and inositol 1,2-cyclic 4-bisphosphate on stimulation of mouse pancreatic minilobules with carbamylcholine. J. Biol. Chem. 262, 340–344.

Sekar, M.C., & Hokin, L.E. (1986). The role of phosphoinositides in signal transduction. J. Membr. Biol. 89, 193–210.

Sekar, M.C., & Hokin, L.E. (1987). Inhibitors of diacylglycerol lipase and diacylglycerol kinase inhibit carbamylcholine-stimulated responses in guinea pig pancreatic minilobules. Arch. Biochem. Biophys. 256, 509–514.

Shah, J., & Pant, H.C. (1988). Potassium-channel blockers inhibit inositol trisphosphate-induced calcium release in the microsomal fractions isolated from the rat brain. Biochem. J. 250, 617–620.

Shears, B. (1989). Metabolism of the inositol phosphates produced upon receptor activation. Biochem. J. 260, 313–324.

Shen, S.S., & Burgart, L.J. (1986). 1,2-Diacylglycerols mimic phorbol 12-myristate 13-acetate activation of the sea urchin egg. J. Cell. Physiol. 127, 330–340.

Sherman, W.R., Gish, B.G., Honchar, M.P., & Munsell, L.Y. (1986). Effects of lithium on phosphoinositide metabolism *in vivo*. Fed. Proc. 45, 2639–2646.

Siess, W. (1985). Evidence for the formation of inositol-4-monophosphate in stimulated human platelets. FEBS Lett. 185, 151–156.

Siess, W., Siegel, F.L., & Lapetina, E.G. (1983). Arachidonic acid stimulates the formation of 1,2-diacylglycerol and phosphatidic acid in human platelets. Degree of phospholipase C activation correlates with protein phosphorylation, platelet shape change, serotonin release, and aggregation. J. Biol. Chem. 253, 11236–11242.

Snyder, P.M., Krause, K.H., & Welsh, M.J. (1988). Inositol trisphosphate isomers, but not inositol 1,3,4,5-tetrakisphosphate, induce calcium influx in *Xenopus* oocytes. J. Biol. Chem. 263, 11048–11051.

Solanki, V., & Slaga, T.J. (1984). Phorbol ester tumor promoter receptors and their down-modulation. In: Mechanisms of Tumor Promotion (Slaga, T.J., ed.), pp. 97–111. CRC Press, Boca Raton, FL.

Spat, A., Bradford, P.G., McKinney, J.S., Rubin, R.P., & Putney, J.W. Jr. (1986). A saturable receptor for ^{32}P-inositol-1,4,5-trisphosphate. Nature 319, 514–516.

Steinhardt, R.A., & Alderton, J. (1988). Intracellular free calcium rise triggers nuclear envelope breakdown in the sea urchin embryo. Nature 332, 364–366.

Steinhardt, R.A., & Epel, D. (1974). Activation of sea urchin eggs by a calcium ionophore. Proc. Natl. Acad. Sci. USA 71, 1915–1919.

Stephens, L.R., Hawkins, P.T., Morris, A.J., & Downes, P.C. (1988). L-myo-Inositol 1,4,5,6-tetrakis-phosphate (3-hydroxy)kinase. Biochem. J. 249, 283–292.

Storey, D.J., Shears, S.B., Kirk, C.J., & Michell, R.H. (1984). Stepwise enzymatic dephosphorylation of inositol 1,4,5-trisphosphate to inositol in liver. Nature 312, 374–376.

Streb, H., Bayerdorffer, E., Haase, W., Irvine, R.F., & Schulz, I. (1984). Effect of inositol-1,4,5-trisphosphate on isolated subcellular fractions of rat pancreas. J. Membr. Biol. 81, 241–253.

Streb, H., Irvine, R.F., Berridge, M.J., & Schulz, I. (1983). Release of Ca^{2+} from a nonmitochondrial intracellular store in pancreatic acinar cells by inositol 1,4,5-trisphosphate. Nature 306, 67–69.

Sugimoto, Y., Whitman, M., Cantley, L.C., & Erikson, R.L. (1984). Evidence that the Rous sarcoma transformed gene product phosphorylates phosphatidylinositol and diacylglycerol. Proc. Natl. Acad. Sci. USA 81, 2117–2121.

Supattapone, S., Worley, P.F., Baraban, J.M., & Snyder, S.H. (1988). Solubilization, purification, and characterization of an inositol trisphosphate receptor. J. Biol. Chem. 263, 1530–1534.

Sutherland, C.A., & Amin, D. (1982). Relative activities of rat and dog platelet phospholipase A$_2$ and diglyceride lipase. Selective inhibition of diglyceride lipase by RHC 80267. J. Biol. Chem. 257, 14006–14010.

Swann, K., Ciapa, B., & Whitaker, M.J. (1987). Cellular messengers and sea urchin egg activation. In: The Molecular Biology of Invertebrate Development (O'Connor, D., ed.), pp. 45–69. Alan R. Liss, New York.

Swann, K., & Whitaker, M.J. (1986). The part played by inositol trisphosphate and calcium in the propagation of the fertilization wave in sea urchin eggs. J. Cell Biol. 103, 2333–2342.

Sylvia, V., Curtin, G., Norman, J., Stec, J., & Busbee, D. (1988). Activation of a low specific activity form of DNA polymerase α by inositol-1,4-bisphosphate. Cell 54, 651–658.

Takai, Y., Kishimoto, A., Kikkawa, U., Mori, T., & Nishizuka, Y. (1979). Unsaturated diacylglycerol as a possible messenger for the activation of calcium-activated, phospholipid-dependent protein kinase system. Biochem. Biophys. Res. Commun. 91, 1218–1224.

Tashjian, A.H. Jr., Heslop, J.P., & Berridge, M.J. (1987). Subsecond and second changes in inositol polyphosphates in GH4C1 cells induced by thyrotropin-releasing hormone. Biochem. J. 243, 305–308.

Thomas, A.P., Marks, J.S., Coll, K.E., & Williamson, J.R. (1983). Quantitation and early kinetics of inositol lipid changes induced by vasopressin in isolated and cultured hepatocytes. J. Biol. Chem. 258, 5716–5725.

Thompson, W., & Dawson, R.M.C. (1964). The triphosphoinositide phosphodiesterase of brain tissue. Biochem. J. 91, 237–243.

Tilly, B.C., van Paridon, P.A., Verlaan, I., Wirtz, K.W.A., de Laat, S.W., & Moolenaar, W.H. (1987). Inositol phosphate metabolism in bradykinin-stimulated human A431 carcinoma cells. Biochem. J. 244, 129–135.

Turner, P.R., Jaffe, L.A., & Fein, A. (1986). Regulation of cortical vesicle exocytosis in sea urchin eggs by inositol 1,4,5-trisphosphate and GTP-binding protein. J. Cell Biol. 102, 70–76.

Turner, P.R., Jaffe, L.A., & Primakoff, P. (1987). A cholera toxin sensitive G-protein stimulates exocytosis in sea urchin eggs. Dev. Biol. 120, 577–583.

Ui, M. (1986). Pertussis toxin as a probe of receptor coupling to inositol lipid metabolism. In: Receptor Biochemistry and Methodology (Venter J.C., & Harrison, L.C., eds.), Vol. 7, Phosphoinositides and Receptor Mechanisms (Putney, J.W. Jr., ed.), pp. 163–195. Alan R. Liss, New York.

Vallejo, M., Jackson, T., Lightman, S., & Hanley, M.R. (1987). Occurrence and extracellular actions of inositol pentakis- and hexakisphosphate in mammalian brain. Nature 330, 656–658.

Waite, M. (1989). Handbook of Lipid Research (Hanahan, D.J., series ed.), Vol. 5, The Phospholipases. Plenum Publishing Corporation, New York.

Wakelam, M.J.O., Houslay, M.D., Davies, S.A., Marshall, C.J., & Hall, A. (1986). The role of N-ras p21 in the coupling of growth factor receptors to inositol phospholipid turnover. Biochem. Soc. Trans. 15, 45–47.

Walsh, C.E., Waite, B.M., Thomas, M.J., & Dechatelet, L.R. (1981). Release and metabolism of arachidonic acid in human neutrophils. J. Biol. Chem. 256, 7228–7234.

Watson, S.P., & Lapetina, E.G. (1985). 1,2-Diacylglycerol and phorbol ester inhibit agonist-induced formation of inositol phosphates in human platelets: Possible implications for negative feedback regulation of inositol phospholipid hydrolysis. Proc. Natl. Acad. Sci. USA 82, 2623–2626.

Weinstein, I.B. (1987). Growth factors, oncogenes, and multistage carcinogenesis. J. Cell Biochem. 33, 213–224.

Weiss, S.J., McKinney, J.S., & Putney, J.W. (1982). Receptor-mediated net breakdown of phosphatidylinositol 4,5-bisphosphate in parotid acinar cells. Biochem. J. 206, 555–560.

Whitaker, M. (1989). Phosphoinositide second messengers in eggs and oocytes. In: Inositol Lipids in Cellular Signalling (Michell, R.H., Downes, C.P., & Drummond A., eds.), pp. 459–479. Academic, New York.

Whitaker, M.J., & Irvine, R.F. (1984). Inositol 1,4,5-trisphosphate microinjection activates sea urchin eggs. Nature 312, 636–639.

Whitaker, M.J., & Steinhardt, R.A. (1982). Ionic regulation of egg activation. Q. Rev Biophys. 15, 593–666.

Whitaker, M.J., & Steinhardt, R.A. (1985). Ionic signaling in the sea urchin egg at fertilization. In: Biology of Fertilization (Metz, C.B., & Monroy, A., eds.), Vol. 3, pp. 167–222. Academic Press, New York.

Whitman, M., Downes, C.P., Keeler, M., Keller, T., & Cantley, L. (1988). Type I phosphatidylinositol kinase makes a novel inositol phospholipid, phosphatidylinositol-3-phosphate. Nature 332, 644–646.

Whitman, M., Fleischman, L., Chahwala, S.B., Cantley, L., & Rosoff, P. (1986). Phosphoinositides, mitogenesis, and oncogenesis. In: Receptor Biochemistry and Methodology (Venter, J.C., & Harrison, L.C., eds.), Vol. 7, Phosphoinositides and Receptor Mechanisms (Putney, J.W. Jr., ed.), pp. 197–217. Alan R. Liss, New York.

Whitman, M., Kaplan, D.R., Schaffhausen, B., Cantley, L., & Roberts, T.M. (1985). Association of phosphatidylinositol kinase activity with polyoma middle-T competent for transformation. Nature 315, 239–242.

Willcocks, A.L., Strupish, J., Irvine, R.F., & Nahorski, S.R. (1989). Inositol 1:2-cyclic,4,5-trisphosphate is only a weak agonist at inositol 1,4,5-trisphosphate receptor. Biochem. J. 257, 297–300.

Wilson, D.B., Bross, T.E., Sherman, W.R., Berger, R.A., & Majerus, P.W. (1985a). Inositol cyclic phosphates are produced by cleavage of phosphatidylphosphoinositols (polyphosphoinositides) with purified sheep seminal vesicle phospholipase C enzymes. Proc. Natl. Acad. Sci. USA 84, 4013–4017.

Wilson, D.B., Connolly, T.M., Bross, T.E., Majerus, P.W., Sherman, W.R., Tyler, A.N., Rubin, L.J., & Brown, J.E. (1985b). Isolation and characterization of the inositol cyclic phosphate products of polyphosphoinositide cleavage by phospholipase C. Physiological effects in permeabilized platelets and Limulus photoreceptor cells. J. Biol. Chem. 260, 13496–13501.

Witters, L.A., & Blackshear, P.J. (1987). Protein kinase C–mediated phosphorylation in intact cells. In: Methods in Enzymology (Means, A.R., & Conn, P.M., eds.), pp. 412–424. Academic Press, New York.

Wolf, B.A., Turk, J., Sherman, W.R., & McDaniel, M.L. (1986). Intracellular Ca^{2+} mobilization by arachidonic acid. Comparison with myo-inositol 1,4,5-trisphosphate in isolated pancreatic islets. J. Biol. Chem. 261, 3501–3511.

Wollheim, C.B., & Biden, T.J. (1986). Second messenger function of inositol 1,4,5-trisphosphate. Early changes in inositol phosphates, cytosolic Ca^{2+}, and insulin release in carbamylcholine-stimulated RIN5F cells. J. Biol. Chem. 261, 8314–8319.

Woods, N.M., Cuthbertson, K.S.R., & Cobbold, P.H. (1987). Phorbol-ester-induced alterations of free calcium ion transients in single rat hepatocytes. Biochem. J. 246, 619–623.

Yamada, K., Okano, Y., Miura, K., & Nozawa, Y. (1987). A major role for phospholipase A_2 in antigen-induced arachidonic acid release in rat mast cells. Biochem. J. 247, 95–99.

Yoshida, Y., Huang, F.L., Nakabayashi, H., & Huang, K.P. (1988). Tissue distribution and developmental expression of protein kinase C isozymes. J. Biol. Chem. 263, 9868–9873.

Zeitler, P., & Handwerger, S. (1985). Arachidonic acid stimulates phosphoinositide hydrolysis and human placental lactogen release in an enriched fraction of placental cells. Mol. Pharmacol. 28, 549–554.

Author Index

Subject Index